CONSERVATION OF GREAT PLAINS ECOSYSTEMS:
CURRENT SCIENCE, FUTURE OPTIONS

Ecology, Economy & Environment

VOLUME 5

The titles published in this series are listed at the end of this volume.

Conservation of Great Plains Ecosystems: Current Science, Future Options

Edited by

S. R. Johnson

Center for Agricultural and Rural Development,
Iowa State University, Ames, U.S.A.

and

Aziz Bouzaher

Formerly with the Center for Agricultural and Rural Development,
Iowa State University, Ames, U.S.A.;
now at the World Bank, Washington, DC, U.S.A.

KLUWER ACADEMIC PUBLISHERS

DORDRECHT / BOSTON / LONDON

Library of Congress Cataloging-in-Publication Data

Conservation of Great Plains ecosystems : current science, future
options / edited by S.R. Johnson and Aziz Bouzaher.
p. cm. -- (Ecology, economy & environment ; v. 5)
Includes index.
ISBN 0-7923-3747-6 (hb : acid free paper)
1. Ecosystem management--Great Plains--Congresses.
2. Conservation of natural resources--Great Plains--Congresses.
3. Great Plains--Congresses. I. Johnson, Stephen R., 1943-
II. Bouzaher, Aziz. III. Series.
QH76.5.G75C65 1995
333.7'2'0978--dc20 95-38230

ISBN 0-7923-3747-6

Published by Kluwer Academic Publishers,
P.O. Box 17, 3300 AA Dordrecht, The Netherlands.

Kluwer Academic Publishers incorporates
the publishing programmes of
D. Reidel, Martinus Nijhoff, Dr W. Junk and MTP Press.

Sold and distributed in the U.S.A. and Canada
by Kluwer Academic Publishers,
101 Philip Drive, Norwell, MA 02061, U.S.A.

In all other countries, sold and distributed
by Kluwer Academic Publishers Group,
P.O. Box 322, 3300 AH Dordrecht, The Netherlands.

Printed on acid-free paper

Printed in the Netherlands

Contents

Climate and Biological Resources

Land Resources

Water Resources

Energy and Mineral Resources

Agriculture in the Great Plains

Environmental Management Initiatives

Acknowledgments

Planning, organizing, and conducting the symposium, "Conservation of Great Plains Ecosystems: Current Science, Future Options," April 7–9, 1993, Kansas City, Missouri, required the cooperation of a number of organizations and individuals who gave unselfishly of their time and resources. The symposium was sponsored by the U.S. Environmental Protection Agency (USEPA) Region VII, the Western Governors' Association, the U.S. Fish and Wildlife Service, and the Center for Agricultural and Rural Development (CARD) at Iowa State University. CARD's participation was supported by a cooperative agreement with USEPA Region VII, under contract No. X 007822-01-1. Those working through the agreement to develop the symposium were Aziz Bouzaher and S. R. Johnson, CARD; and Susan Gordon, Kerry Herndon, and Cathy Tortorici, USEPA Region VII.

We also acknowledge the important contributions of the members of key committees supporting the symposium. The Science Committee identified topics and speakers, and assisted with developing the major themes of the symposium. The committee members were Aziz Bouzahcr, CARD; Peter Buesseler, Minnesota Department of Natural Resources; Jo Clark, Western Governors' Association; Dick Gersib, Nebraska Games and Parks Commission; Kerry Herndon, USEPA Region VII; S. R. Johnson, CARD; Fritz Knopf, U.S. Fish and Wildlife Service; Hal Salwasser, University of Montana; Al Steuter, The Nature Conservancy; Margaret Thomas, Midwest Research Institute; Cathy Tortorici, USEPA Region VII; and John Wilson, USEPA/OPA. The Media Committee assisted with local arrangements for media representatives and with disseminating results from the symposium. Members of the committee were Brian Meyer, College of Agriculture Information Service, Iowa State University; and Scott Pendelton, *Christian Science Monitor*.

The USEPA Region VII office supplied recorders and facilitators who played an essential role in conducting the symposium and in synthesizing ideas and concepts that emerged. They were: Craig Bernstein, Mark Bogina, Shellie Brodie, Terry Dean, Roxanne DiLaura, Marla Downing, Billy Fairless, Bob Fenemore, Kathyleen Fenton, Walt Foster, Brenda Groskinsky, Gene Gunn, John Helvig, Charles Hensley, Charlie Hensley, Diane Herschberger, Marian Hess, Chris Hess, Gayle Hoopes, DeWayne Knott, Lynn Kring, Greg McCabe, Ruben McCullers, Rowena Michaels, Diana Newman, Delores Platt, Bill Rice, Ron Ritter, Donna Sefton, Larry Shepard, Gerry Shimek, Michele Stewart, Dick Sumpter, Dave Wagoner, Alan Wehmeyer, Gary Welker, and Steve Wurtz.

At CARD, the staff coordinators for the symposium were Judith Gildner Pim, Jan Holtkamp, Betty Hempe, and Kathy Gardner. They were responsible for an array of activities including preparing and distributing the symposium brochure and summary, and they handled the invitations for participants and local arrangements at the Westin Crown Center in Kansas City. Production of this book was managed by Jan Nordlund and Kathy Gardner. We and the authors of the individual chapters worked closely with Jan and Kathy in preparing this manuscript for publication. The book benefited significantly from their contribution in improving the organization and presentation of the material. Donna Fincham and Tom Hiett from Iowa State's Media Graphics unit prepared camera-ready pages for publication.

—*Aziz Bouzaher*
S. R. Johnson

Introduction

Introduction

Aziz Bouzaher and S. R. Johnson
Iowa State University

The Great Plains region includes all or portions of 13 states, more than 60 tribal reservations and Indian trust land, numerous rural communities experiencing secular decline, and growing urban complexes. The differences within the Great Plains, from the viewpoint of political and economic organizations, are matched by its biological diversity. The Great Plains includes prairie grasslands, wetlands, riparian areas, forests, savannas, and a wildlife community that is correspondingly diverse. On first observation, these features would seem to discourage integrated management approaches that concurrently conserve natural resources and frail unique ecosystems, stimulate economic growth, and contribute to a lifestyle more satisfying to the inhabitants and the many visitors that enjoy the landscape and the cultural and biological texture of the region.

The symposium, "Conservation of Great Plains Ecosystems: Current Science, Future Options," forms the basis for this book. This symposium was organized to help define the new Great Plains regional environmental initiative. This integrated management initiative is being sponsored by a growing number of partners including state and federal resource agencies, nongovernmental organizations, private sector trade and industry groups, and federal agencies in the United States, Canada, and Mexico. The Western Governors' Association coordinates these institutions and agencies and is serving as liaison to encourage broadened participation by state and local governments.

The objective of the symposium was to synthesize available scientific and descriptive information and to provide a more solid foundation for launching the Great Plains regional environmental initiative. Representatives of the Western Governors' Association, the regional offices of the U.S. Environmental Protection Agency (USEPA) that represent the Great Plains states, the U.S. Fish and Wildlife Service of the Department of the Interior, and scientists from academic and research institutions in the Great Plains recommended that the symposium also assess evolving concepts of integrated environmental management and the major resource complexes of the Great Plains.

The resource complexes addressed in the symposium include "Community and Economic Resources" (Chapters 4, 5, 6, 7, and 8); "Climate and Biological Resources" (Chapters 9, 10, 11, and 12); "Land Resources" (Chapters 13, 14, and 15); "Water Resources" (Chapter 16, 17, and 18); and "Energy and Mineral Resources" (Chapter 19).

The chapters in the section on Community and Economic Resources chronicle the population dynamics of the Great Plains as they relate to economic development and to emerging agricultural technologies that support large-scale, monoculture land use. Populations are becoming more concentrated in urban areas. Rural areas are declining in both population and economic activity. Major issues identified were differences in infrastructure for the rural and urban communities, and the economic disparity that is emerging among the associated populations. Using natural resources to enhance the landscape and to create opportunities for alternative forms of economic activity that support rural cultures and communities were identified as important to the success of the integrated environmental management initiative in the Great Plains. Proceeding along the current development path, especially in rural areas, may increase risks for the related ecosystems and resources and, at the same time, decrease the capacity of the economies to support current communities and population.

The section on Climate and Biological Resources emphasizes the quality and diversity of the major natural systems in the Great Plains. From a distance, the Great Plains appears highly uniform. Upon closer observation, however, a vast texture of different ecosystems, species, and cultures emerges. This diversity and texture are perhaps more frail than in other resource and climatic areas. The Great Plains depends heavily on water quality and quantity. Agriculture is a major user of water and has had a significant impact on this resource, and perhaps even on the climate. Climate change, if it in fact materializes, will have far-reaching consequences for the Great Plains. Alternatively, at the micro-climate level, resource patterns have significant implications for biological diversity and other resource quality in the Great Plains. The issue of declining resource quality and increasing environmental risk emerges as a likely consequence of continued use and development patterns.

Land resources and the landscape of the Great Plains have been significantly affected by agricultural use, forestry, and economic development patterns. The chapters in the Land Resources section provide a view that emphasizes the landscape of the Great Plains. Often, land use has been narrowly defined by farm, watershed, or ecosystem. From the perspective of integrated management, the aggregate is more than the sum of its parts. The large, comparatively uniform land resource areas of the Great Plains are linked by climatic and other features that define a landscape. The landscape and land use patterns have major implications for the sustainability of large area ecosystems. Rangeland use, agriculture, and urbanization patterns have altered this resource as well as the capacities of the natural systems.

Water is the limiting resource in the Great Plains for economic activity and for sustaining many ecosystems. The chapters in the Water Resources section point to the fact that the Great Plains, as presently developed, is a net water consumer. The recharge of the natural water supply system is not sufficient to sustain use at current levels. This imbalance has long-term implications for the natural ecosystems and the communities in the Great Plains. The imbalance also is linking the economy of the Great

Plains. Agriculture is a particularly heavy water user. Future uses of the Great Plains resource base will be governed more and more by water availability and quality.

Energy and mineral resources are less abundant in the Great Plains than in other regions of the United States. Still, these resources have been overexploited in the Great Plains. Oil and natural gas reserves have been depleted. Other mineral resources are important to the Great Plains economy, but also have been depleted. Policies guiding the use of these mineral resources in ways that more adequately reflect their true opportunity costs are suggested. Highly exploitive approaches to the development of natural resources have left a legacy that will require attention in the future.

The first section—"Defining and Valuing the Ecosystems in the Great Plains" (Chapters 1, 2, and 3)—and the final two parts— "Agriculture in the Great Plains" (Chapters 20, 21, and 22) and "Environmental Management Initiatives" (Chapters 23 and 24)—form the "bookends" for the rest of the book. These sections discuss the concepts and perspectives needed to assess the condition and trends in resource quality as well as describe the important features of new, more comprehensive management approaches.

Chapter 1 discusses the history of the Great Plains and provides a useful context for current concerns about the sustainability of development in the region, along with an appreciation for the precursors of its population's cultures. Other chapters on valuation and approaches to defining *ecosystem* provide concepts useful as foundations for the discussion in the subsequent sections and for more comprehensive approaches to management or policy. Integrated environmental management approaches incorporate explicitly and implicitly concepts of valuation and ecosystem performance and integrity. The importance of factoring human populations, economies, and landscapes into workable ecosystem definitions, resource conservation and enhancement, environmental protection, and community development are emphasized. Valuing the features of ecosystems and schemes for using markets as a basis for economic valuation were also addressed. As management efforts and policies aimed at conservation and development become more integrated, issues of valuation emerge as critical to choices of short- and long-term strategies.

Agriculture is clearly the predominant "user" of the Great Plains natural resources, land, water and, indirectly, various animal and plant species. Perhaps because of the uniformity of Great Plains agriculture, it is characterized by a high degree of crop and animal specialization. The current agriculture's lack of diversity has had a strong interaction with the natural diversity of the local ecosystems and the quality of resources in the Great Plains. A recurrent theme of the chapters in this section is the importance of recognizing the implications of monoculture cultivation in the Great Plains and how this system affects the sustainability of agriculture. Opportunities provided by emerging agricultural technologies for more specialized and more sensitive use of Great Plains resources were also assessed. There are important possibilities for evolving to a technology for Great Plains agriculture that will better accommodate resource conservation and ecosystem protection, two of the building blocks of sustainability.

The last section addresses the new, more comprehensive and incentive-based environmental management approaches. Ideas on successful regulatory and nonregulatory approaches to ecosystem conservation are undergoing fundamental change. Increasingly, regulatory and nonregulatory management approaches are for systems rather than the quality of particular media—water, land, or air. The integrated management of ecosystems, while more demanding and complex, is becoming recognized as more efficient. These integrated management systems are inclusive and participatory, in the sense that for most applications, the human population and the communities are integral components. Thus, ecosystem management on a landscape scale must involve local organizations, the social and economic activities of the residents, the culture and heritage of these populations, as well as the biological and natural resources. The costs of dealing more fully with these complexities in integrated management initiatives are balanced by the benefits of achieving the multiple objectives of the residents and other constituents more efficiently.

Conservation, resource enhancement, and economic development do not necessarily compete as objectives in more integrated ecosystem or environmental management contexts. Managing resources to ensure the sustainability of economic and community development is becoming more broadly recognized as consistent with conserving resources and preserving the integrity of ecosystems. Striking a sustainable balance between sensitive ecosystem equilibria and economic and community development requires the use of market and other incentives as well as regulation in associated management. Managing the landscape to conserve resources and to sustain and encourage a higher quality of life for residents and other users must also involve active participation by local citizens and communities. The opportunity presented by modern integrated management concepts is one of achieving more efficiently harmony among environmental and development interests in the Great Plains.

Of course, interest in Great Plains resources and the use of this natural endowment for human activity is not unique to this century. Great Plains history includes the rise and fall of numerous human civilizations and animal species. Much can be learned from these episodes as we work to develop more sustainable resource use and integrate our environmental management approaches. Perhaps it is the uniformity of the Great Plains and the feasibility of organizing its resources for narrowly defined uses that is both its strength and its weakness. Themes repeated during the symposium were the risk of narrowing our use of resources, underestimating the frailty of these resources once organized for specialized use, and the problems of sustaining economic activities and resource quality.

Viewed in the long term, the settlement of the Great Plains during the past 150 years, and the use of its resources for agriculture, has occupied a relatively brief period. The challenge to current residents and users of Great Plains resources is to learn both from the past and from modern scientific developments of the importance of integrating resource conservation with economic and development activity. The future of the Great Plains and the populations that will use and appreciate the resources are tied to current use patterns and development. Informed and cooperative public and private sector participation in developing community-based integrated management and ecosystem approaches is the key to a future with more opportunity for those who will live in and use the Great Plains.

Defining and Valuing Ecosystems of the Great Plains

1. History, Environment, and the Future of the Great Plains

Dan Flores
University of Montana

Historical Background

On the coasts Americans refer to the Great Plains, with more than a hint of contempt, as *flyover country.* Looking out the window of a jet or an air-conditioned auto, it seems at first view to be vast, flat, and mostly empty. Although it can be green with the rains preceding summer solstice, more often the operative color of the place is yellow. This neutral tone combined with the two dimensions of linear horizons and bowl-shaped skies gives the Plains one of its ancient names: the Horizontal Yellow. In fact, many names have been given to this enormous expanse of country stretching from Texas to Saskatchewan. *Llanos,* a Spanish word, has been in common use for 400 years and was eventually adopted and translated as "plains" by Northern Europeans, who had no suitable word for dry, grassy steppes. Upon first approaching the region, English speakers had appropriated a French term, *Gran Prairie,* to name a place as alien to Scots and Englishmen as the British Highlands would be to a Bedouin.

Llanos and Gran Prairie both persist today, and in juxtaposition capture the essence of this unique country in a name used since the 1820s: the Great Plains. Nonetheless, the implications of the long multiplicity of terms for the Great Plains ought to be noted. Despite the snide references to flyover country, and the common plaint by outsiders that "there's nothing there," for humans the Great Plains is an old country, evidently the earliest extensively occupied part of what is now the United States. People have been eking out a living on the Horizontal Yellow for more than 11,000 years. Empty though it may appear, minimalist as its available resources may seem, troubled as its future may strike us at this moment in time, the long history of the Great Plains does provide one lesson of crystalline clarity: People have been living, loving, and transforming nature in flyover country for a very long time and are not likely to cease doing so for a long time to come.

Many Americans may not know that over this long span of time, human history has more than once intersected explosively with the environment of the Great Plains. While these patterns of Great Plains history are reasonably clear, and certainly suggestive enough to be sobering, the future of the present society on the Plains is

Stanley R. Johnson and Aziz Bouzaher (ed.), *Conservation of Great Plains Ecosystems,* 3–9.

more problematic. History shares with the sciences some ability to analyze cause and effect, but unfortunately, unlike the hard sciences, history is not a predictive discipline. Who can say what new matrix of circumstances, knowledge, values, and technology will affect the decisions influencing the future?

Since the time of the early nineteenth-century Great Plains explorers—Lewis and Clark, Zebulon Pike, and the Stephen Long Expedition of 1819–20—the Great Plains has been regarded as an environmental anomaly in North America. Far more than Spaniards like Cabeza de Vaca or Coronado, who had previous experience with the Spanish Meseta or the deserts of Mexico, or the Comanches with origins on the edge of the Great Basin, Northern Europeans saw the Great Plains as a country of extremes and deficiencies.

Half a century ago, historians Walter Prescott Webb (1931) and James Malin (1956) summarized these so-called deficiencies: Because of the rainshadow effect of the Rockies, the Great Plains is deficient in moisture; because of this semi-aridity, the area is a grassland, deficient in trees. Its rivers are shallow, seasonal, and unsuitable for navigation. For modern-day Americans, whose landscape aesthetic has been shaped by pictorial representations of mountains and other topography of high relief, who tend to regard wilderness as synonymous with forests, and whose nineteenth-century sense of cultural worth often depended on the presence of monumental scenery such as the Rockies, Yosemite, or the Grand Canyon, the Great Plains is regarded as deficient in aesthetics. This is despite the fact that the Great Plains actually contains dozens of different grassland complexes; at least three mountain ranges; rivers, potholes, and playa lakes rich in wildlife; or that the Badlands of the Northern Plains and the Llano Estacado Canyonlands of the south are among the most visually striking places on the continent. As geographer Neil Evernden (1983) commented, our aesthetic sense has not so much been filled by the Great Plains as the Great Plains' aesthetic has been denied. For that and other reasons, the Great Plains today has fewer parks and preserved ecosystems than any other part of the West.

Environmental History

The story of sequential human cultures interacting with the Great Plains environment demonstrates some interesting patterns, and patterns are probably the best to hope for in trying to see into the future. One pattern that lends weight to the need for extreme care on the part of late twentieth-century Plains inhabitants is simply this: Because the Plains environment is made fragile by drought, the hold of human societies on the Plains has often been not only disruptive, but tenuous. No part of the continent invites such easy human environmental alteration yet can collapse so quickly and completely under that transformation as the Great Plains.

Compared with the Eastern woodlands or the Rockies, the Great Plains is ecologically a simple system, with few of the safeguards that are built into more diverse systems. Historically, this has made the Plains deceptively fragile. The result of human interaction has been a series of ecological crashes and simplifications, many with profound consequences for human societies that were unwilling or unable to adapt to accelerating change. Conversely, human habitation has transformed the Great Plains, so that for 11,000 years it has been an environment much shaped by human actions.

The most far-reaching Great Plains ecological collapse since humans arrived on the continent was unquestionably the first, the Pleistocene extinctions, which saw thirty-two genera of mammals disappear, due in large part to the intrusion of highly skilled hunters from Siberia into an ecological setting that had never experienced human hunting pressure before. These extinctions, which peaked between 9,000 and 10,000 years ago, not only eliminated most of the American-evolved megafauna, they also eliminated the Great Plains' first human cultures. The Paleolithic cultures who occupied the Plains for almost 3,000 years, which is 200 times longer than we have, collapsed when the giant animals they hunted were gone. This ancient phase of Plains history ended a millennium later with another major ecological disruption: the great 2,000-year drought of the Altithermal. This drought occurred when the climate warmed and dried subtly, yet enough in a place so delicately balanced between precipitation and evaporation that it reduced plant diversity by as much as 50 percent. Human adaptation evidently could not keep pace with this climatic change; the Plains were all but abandoned during the Altithermal. A sensible strategy of both animals and people on the Great Plains, abandonment was also the response of the Plains Villager and Plains Woodland cultures around 1,000 years ago. These cultures had pushed their riverside farm villages far out onto the Plains; the villages then fell back to the prairie perimeter when drought struck in 1299 A.D.

A more recent lesson of fragility and marginality is the ranching collapse of the 1880s. It demonstrates that even in a place that evolved with native grazers, a policy of land privatization coupled with commercial grazing was capable of bringing on short-term cultural ruin and long-term ecological alteration. Although traditional ranching historians argue that a variety of human adaptive responses emerged from the Big Die-up, modern environmentalists have questioned the grazing adaptation in light of a much transformed rangeland ecology on the Plains over the last century.

Another large pattern in the long-term environmental history of the Great Plains has been the tendency for Plains cultures to respond to the few but apparently limitless resources of the great grasslands with narrow economic specializations. The Paleolithic big-game hunters inhabited the richest, most diverse Great Plains environment that people have been privileged to see, a North American version of today's Masai-Mara of East Africa. Yet the archeological record indicates that the response to that relatively wetter, lusher, far more diverse environment of 100 centuries ago was a narrow specialization on big game. Similarly, when the Pleistocene extinctions and a stressed climate regime produced the dwarfed modern bison, a "weed" animal that converted solar energy on the Plains into a huge biomass of a single species, the situation existed for Plains cultures to specialize in hunting bison. Bison specialization thus became a feature of the Plains when the European arrival made new technology available in the form of the horse (an original American native that reoccupied its old niche in an instant of ecological time) and imposed new systems in the form of the market economy. At that point in history, many cultures along the periphery of the Plains abandoned horticulture or gathering and became bison specialists. This pressure contributed to yet another great environmental crash and the demise of still another way of life a century ago.

However, the historical record provides a contrasting example: the legacy of the Archaic peoples, who moved onto the Plains after the Altithermal ended and resided there for 4,500 years thereafter. Although their story remains one of the least understood on the Plains, the evidence so far is that the Archaic peoples occupied the grasslands longer and more successfully than any other people before or since. Their secret is instructive, but truly there was nothing magical about it. Consciously, as an act of policy, they seem to have kept their numbers small. They were generalist gatherers and hunters whose economies were diverse, so that their effect was spread across a wide range of resources. Although the Archaic peoples certainly did alter their environment, using broadcast fire that suppressed shrubs and brush and enlarged the grassland area, their real genius seems to have been not so much alteration of local bioregions as adaptation to them. If cultural longevity and integration with intact natural ecosystems make up the essential environmental criteria for human success, then the Archaic lifestyle probably provides the most successful model on the American plains.

That all these patterns continue into our own time folds us like layers into the fabric of history. Narrow economic specializations, such as the monocrop dry farming of wheat or the dripping of irrigation water onto skinned prairies that have been revegetated with exotic crops like sorghum and cotton, are no more diversified an approach than was buffalo hunting. And according to many monitors of Great Plains health, mounting internal pressures on the key resources of soil and water, coupled with external pressures from the market and from global climate change, keep these modern economic specializations on the familiar, slippery slope of Plains history, with periodic ruin always a threat.

That the Plains is still a fragile and uneasy place to live has been seared into the modern consciousness by the Dust Bowl of the 1930s, which was triggered by an ordinary drought but magnified by five decades of the most massive human ecological transformation of the Plains since the Pleistocene Epoch. By the early twentieth century the Plains culture had driven out the buffalo and wolves. Now much of the grass is likewise gone. The ecological collapse and epic human abandonment of the Dust Bowl have become the great historical experience of the modern Great Plains. Yet it is also an historical experience that many traditionalists have used to argue against doomsday projections for the Plains, using it as an example of how hard times have acted as a natural selection process to fix more appropriate institutions in place.

Nonetheless, today many people believe that the Great Plains is facing yet another watershed. There are serious doubts about the success of modern adaptations to the Plains. From Texas to Kansas and Colorado, half a century of irrigated agriculture has increased the human population of the Plains twenty times beyond what it was under Comanche hegemony. Yet that far larger population has drained the resource on which this society rests—the fossil Ogallala Aquifer—so much that in Texas alone by the 1990s the wells have stopped pumping on 20 percent of the acreage irrigated just a decade before. On the Southern Plains the aquifer's expected economic life is estimated at anywhere from twenty to an optimistic fifty years, but no more. Compared with the 4,500-year life of the Archaics, deep-well irrigated agriculture's brief century on the Plains may not be long enough for future archeologists even to bother with. Some researchers fear that the ordinary cycle of droughts in combination with the global climate changing at a rate far faster than humans have ever experienced before will call for waves of adjustment on the Plains in the next century.

There are even more direct portents of trouble. Demographic patterns of rural out-migration and an aging population on the Great Plains stand in sharp contrast to most other parts of the country, especially the rest of the West. In the 1980s large parts of rural west Texas, Montana, and Kansas suffered Dust Bowl-sized (up to 16 percent) population losses. Thirty-eight of North Dakota's forty-one counties suffered 10 percent declines, as did fifty of Nebraska's fifty-two counties, and twenty-two of Oklahoma's twenty-three western counties.

Finally, the modern Plains has become an imperceptible echo of what it once was. The tallgrass prairie has suffered the most complete devastation of any major ecosystem on the continent, while the mixed prairie and High Plains grasslands have endured enormous losses—approaching 75 percent of their original extent in Nebraska and North Dakota—to agriculture. As a result, in addition to the collapse of its core ecology centered on bison, wolves, and prairie dogs, 55 species from the Great Plains are currently listed as threatened or endangered, and a whopping 728 are potential additions. Fred Samson and Fritz Knopf (1994) have declared that in "the larger context of conserving biological diversity in . . . natural ecosystems in North America, prairies are a priority, perhaps the highest priority."

Future Directions

As serious reappraisal of the burden of modern Plains history has begun, voices like those of Donald Worster (1979, 1992, 1993), Bret Wallach (1985), and Frank and Deborah Popper (1987, 1988) have called modern Great Plains society a tragic mistake. In the much-quoted phrase of the Poppers, it is "the largest, longest-running agricultural and environmental miscalculation in American history."

In a 1991 article in *Great Plains Research,* geographer William Riebsame explored the use of evolutionary adaptation models to explain how modern society might respond to future environmental change on the Great Plains. Half a century ago, Walter Prescott Webb's *The Great Plains* (1931) argued that technological solutions to Plains life such as windmills, barbed wire, and Colt revolvers were examples of adaptive genius. Considering long-term sustainability, Riebsame also pursued the theme of adaptation, or the formulation of a set of cultural traits closely fitted to the environmental conditions of a given place. But he contrasted true adaptation with resiliency, or a society's tendency to cling or rebound stubbornly to its institutions even following natural disturbances that cast their suitability into doubt. Along with historian Donald Worster, Riebsame thinks that many of the institutions that have characterized modern society's response to the Plains have been resilient rather than adaptive, technological refinements of the status quo, enabling modern Plains society to avoid or forestall making true adaptations based on the limits the Great Plains imposes.

Viewing the future in the context of aquifer drawdown, rapid climate change, mounting losses in ecological diversity, and a flowering biocentric environmentalism, it seems clear to some—though not to others—that the Great Plains is on the threshold of a paradigm shift. Whether this shift is brought on by an eventual disillusionment with the continuing pattern of cyclical ruin, or by more planned, truly adaptive strategies built around the natural systems and cycles of the grasslands, future society on the Horizontal Yellow is going to look very different.

What might the Plains look like half a century from now? There are some models, of course. Phil Burgess, of Denver's Center for the New West, is conducting a study called "A New Vision of the Heartland: The Great Plains in Transition" (Matthews 1992), which posits what Burgess calls an "urban-archipelago" Plains society, a high-tech oasis country with population islands supported by a service economy. As for the rural lands in between, perhaps Wes Jackson's Kansas Land Institute has an answer for that in Jackson's search for a native grass (maybe a genetically manipulated grama grass) that could replace wheat as a commercial crop. The Poppers' Buffalo Commons and Bob Scott's Big Open, both of which envision reintroducing the native Plains fauna (including predators) in enormous wildlife refuges, are romantic visions but hold out a promise of natural sustainability and even ecotourism.

Others (Riebsame 1991; Flores 1990) believe that the future of the Great Plains will be based on patterns of historical continuity observable since the 1930s. While there might be some rural water importation and agricultural cropping in a few select locations, and while private ranching (probably increasingly with buffalo) will continue on a reduced scale, it seems likely that the Great Plains will follow the land use model of the Rocky Mountain West. This reversal of privatization was a process initiated by the Dust Bowl, with the federal reacquisition of 11 million acres that became the nucleus of the national grasslands system, and a start in the creation of large park preserves. Rather than a private lands empire, the future Plains will feature far more publicly owned and managed lands, some used principally for commercial grazing, while others will be used for restoring the natural ecological diversity that has proved capable of weathering natural change on the Plains so well. The great national preserves now existing on the Northern Plains—Saskatchewan's Prairie National Park and Theodore Roosevelt and Badlands National Parks—would be augmented eventually by similar parks, or more likely national preserves, on the Central and Southern Plains and in the tallgrass prairie. Far fewer people will live on the Great Plains a century from now—but they will occupy a much more diverse and interesting environment than today.

As a consequence of its inherent environmental conditions, for more than a hundred centuries the Great Plains has appeared as an alluring landscape for human societies, but one where sustained habitation has proved challenging. The Great Plains seems to invite human manipulation, but over the long span of history these North American steppes with their sweeping grasslands and great skies have expelled human cultures again and again. Twentieth-century Great Plains society certainly represents a significant overshoot of what this simple and severe landscape can sustain. By being more flexible, more sensitive, and ready to embrace some considerable downsizing, humans have within their grasp a modern adaptation to living well on the Great Plains. The real question is whether they will embrace those adaptations willingly, or resist until the Plains treats them roughly and forces the issue.

References

Evernden, Neil. 1983. Beauty and nothingness: Prairie as failed resource. *Landscape* 27:3-8.

Flores, Dan. 1990. *Caprock canyonlands: Journeys into the heart of the Southern Plains.* Austin: University of Texas Press.

Malin, James. 1956. *The grassland of North America: Prolegomena to its history.* Lawrence, Kansas.

Matthews, Ann. 1992. *Where the buffalo roam.* New York: Grove Weidenfeld.

Popper, Frank, and Deborah Popper. 1987. A daring proposal for dealing with an inevitable disaster. *Planning* (Dec.):12-18.

———. 1988. The fate of the plains. *High Country News* (Sept. 26):15-19.

Riebsame, William. 1991. Sustainability of the Great Plains in an uncertain environment. *Great Plains Research* 1:133-151.

Samson, Fred, and Fritz Knopf. 1994. Prairie conservation in North America. *BioScience* 44(June): 418-421.

Wallach, Bret. 1985. The return of the prairie. *Landscape* 28:1-5.

Webb, Walter Prescott. 1931. *The Great Plains: A study in institutions and environment.* Boston: Ginn and Company.

Worster, Donald. 1979. *Dust Bowl: The Southern Plains in the 1930s.* New York: Oxford University Press.

———. 1992. *Under western skies: Nature and history in the American West.* New York: Oxford University Press.

———. 1993. *The wealth of nature: Environmental history and the ecological imagination.* New York: Oxford University Press.

2. Ecosystems of the Great Plains: Scales, Kinds, and Distributions

W. A. Reiners
University of Wyoming

Perceptions about the Great Plains of the United States and Canada have varied widely through the historical development of the region. Emerging from the forests of Appalachia, early settlers initially noted the absence of trees in the Prairie Peninsula, where grassland extended into Wisconsin, Missouri, Illinois, Indiana, and Ohio. Because they viewed this as evidence of infertility, the settlers avoided the open prairies in favor of the scattered oak openings and floodplain forests of that area. In fact, some of the former glacial lake basins of the Prairie Peninsula were not drained and plowed until the early twentieth century, and draining of pothole lakes in the Northern Plains continues even now (Van der Valk 1989). As early as 1820, explorers of the western Great Plains perceived the region west of the 100th meridian as *The Great American Desert* and did not suppose that it would ever support much of a civilization (Frazier 1989; Lewis 1979). But, around mid-century, the Great Plains west of the Mississippi was settled rapidly, especially after the extension of railroads. As part of the commercial development by the railroads, the Great Plains was extolled as an agricultural paradise (Frazier 1989). But in the 1930s, the great drought and resulting dust storms recast the Great Plains as a dubious environment for agriculture after all—perhaps a kind of desert lurking between seductively good precipitation years.

These contrasting views about the nature and exploitability of the environment and ecosystems of the Great Plains persist today (Riebsame 1990, 1991, 1993). On one hand, the Great Plains is perceived as an area of enormous productivity—one of the world's great breadbaskets as well as a wholesome place to live. Others have a more tragic view as they consider the 200-year destruction of native cultures and natural resources (Frazier 1989).

In fact, both viewpoints are, or have been, correct for particular times and places. This vast inland empire is a global breadbasket that has suffered a long history of environmental damage. But to what extent are environmentally destructive events reversible? To what extent can the remaining native biota be conserved? And can we now sustainably manage this uniquely productive region for the enjoyment and welfare of future generations?

Stanley R. Johnson and Aziz Bouzaher (ed.), *Conservation of Great Plains Ecosystems*, 11–32.

The Conservation of the Great Plains Ecosystems Symposium addressed the goal of conserving Great Plains ecosystems: an attempt to deal with some of these questions. This chapter explores the ecosystem concept, describes generally measured properties of ecosystems, shows how some of these properties are mapped as independent geographical variables over the extent of the Great Plains and, of most importance, defines at a particular scale the ecosystems constituting the Great Plains.

The Ecosystem Concept

The ecosystem concept is a difficult scientific abstraction that has been absorbed into environmental applications by nonscientists. Meanings are highly variable at both the explicit and subtle, semantic level. To promote a common usage, a definition of ecosystem and remarks characterizing ecological thought today are provided.

A simple definition of *ecosystem* is an assemblage of abiotic and biotic components making up an interdependent system (Tansley 1935). Abstractions of nature—such as the ecosystem concept—can be used in both a general way and a local, specific way. For example, one can discuss the tallgrass prairie ecosystem as an abstract representation of an ecosystem type that once covered much of the former Prairie Peninsula. But one can also use the ecosystem concept to describe, analyze, and predict behavior of a specific representation of a general ecosystem type, such as a prairie relict consisting of only a few acres along a railroad in Iowa.

One of the convenient and, at the same time, troublesome properties of the ecosystem concept is that it is dimensionless. An ecosystem can be a rotting acorn or the biosphere—it is any system fulfilling the minimal components of the definition. Thus, an ecosystem can be defined across a broad range of scales as long as the system fulfills the minimal requirements of this definition. In working across scales, however, there is an obligation to be conscientiously explicit about defining the scale of the defined ecosystem and about using appropriately scaled criteria in analyzing such an ecosystem. If, for example, ecosystems are characterized at the scale of a woodlot, or a watershed, or the entire Great Plains (Figure 2.1), each must be described and analyzed in appropriately scaled physiographic and edaphic terms such as local landscape position for the woodlot, regional river basin for the watershed, or the subcontinental geological context for the entire Great Plains.

Since ecosystems exist at multiple scales, they can and must be dealt with at more than one scale. A system of nested ecosystems at different scales is not only possible but extremely workable. Later in this chapter, a classification of Great Plains ecosystems at a particular scale will be suggested.

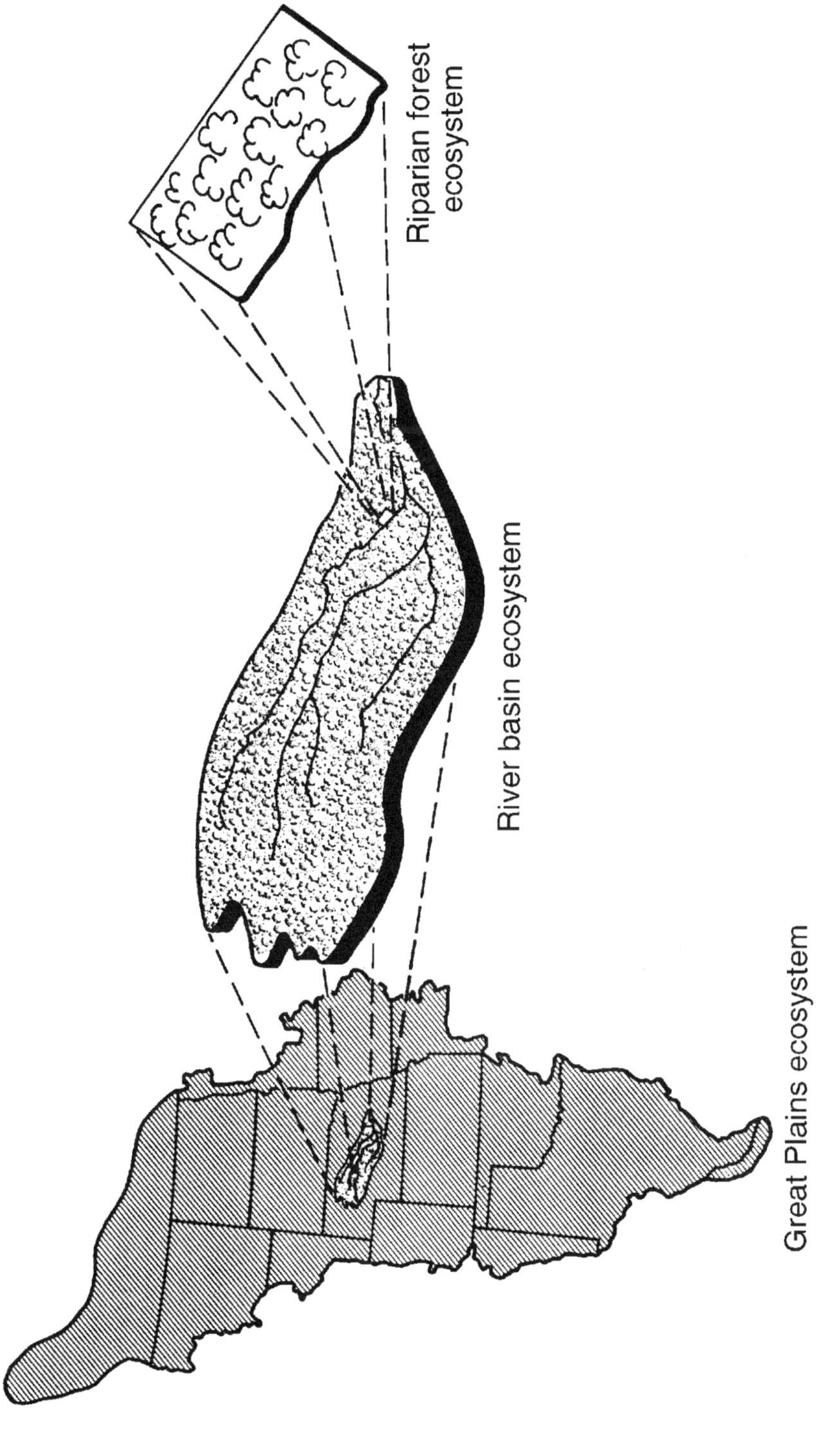

Figure 2.1. Nested ecosystems of different scales

So how does one recognize ecosystems and on what grounds can they be defined? Practical definitions of ecosystems are almost always determined by the phenomenon of greatest interest, and different phenomena often do not map congruently across space. The phenomenon may be some aspect of structural homogeneity such as uniformity of a crop field; or it may be a unit of landscape having hydrologic integrity, such as a watershed of a first-order stream; or it may represent an area affected by a particular episodic event such as a fire, windstorm, or flood. Figure 2.2 illustrates how ecosystems defined by different criteria may be mapped as noncongruent, overlapping areas.

While there can be flexibility in the criteria used to define ecosystems, the spatial definition for one phenomenon will not necessarily be realistic for another. However, for many purposes, land cover (vegetation plus surface soils) serves as a convenient basis of identification and definition of terrestrial ecosystems. Together, the soils and vegetation component largely controls the microclimate, water regime, entry of solar energy through primary production to the ecosystem, and much of the biogeochemical properties of the local ecosystem. Land cover also has the advantage of being visually obvious and readily characterized by ground or remote-sensing methods. Finally, land cover is roughly synonymous with upland habitat, which closely links to the occurrence and well-being of native biota (Scott et al. 1993).

Ecosystem Properties

When analyzing and modeling environmental systems at the ecosystem level, it is useful to focus on certain structural and functional properties. A list of very general ecosystem properties, each of which can be subdivided for more detail and definition of particular ecosystems, follows:

- Climate at appropriate space and time scales
- Soils
- Water regime (hydrology, chemistry)
- Species composition
- Physical structure (dimensions, distribution of live and dead biomass, photosynthetic area)
- System energetics (physical energy budgets, biological energy flows)
- Nutrient budgets and cycles
- Regulatory processes
- Topological relationships (influences from neighboring ecosystems)

These functional properties may be generalized across all ecosystems, although they may differ according to the situations, such as in terrestrial versus aquatic ecosystems.

Climate is a fundamental variable that drives and constrains ecosystem processes. At small scales, climate can be modified by the ecosystem itself and interpreted as the microclimate, and at mesoscales, changes in regional land cover can alter the climate

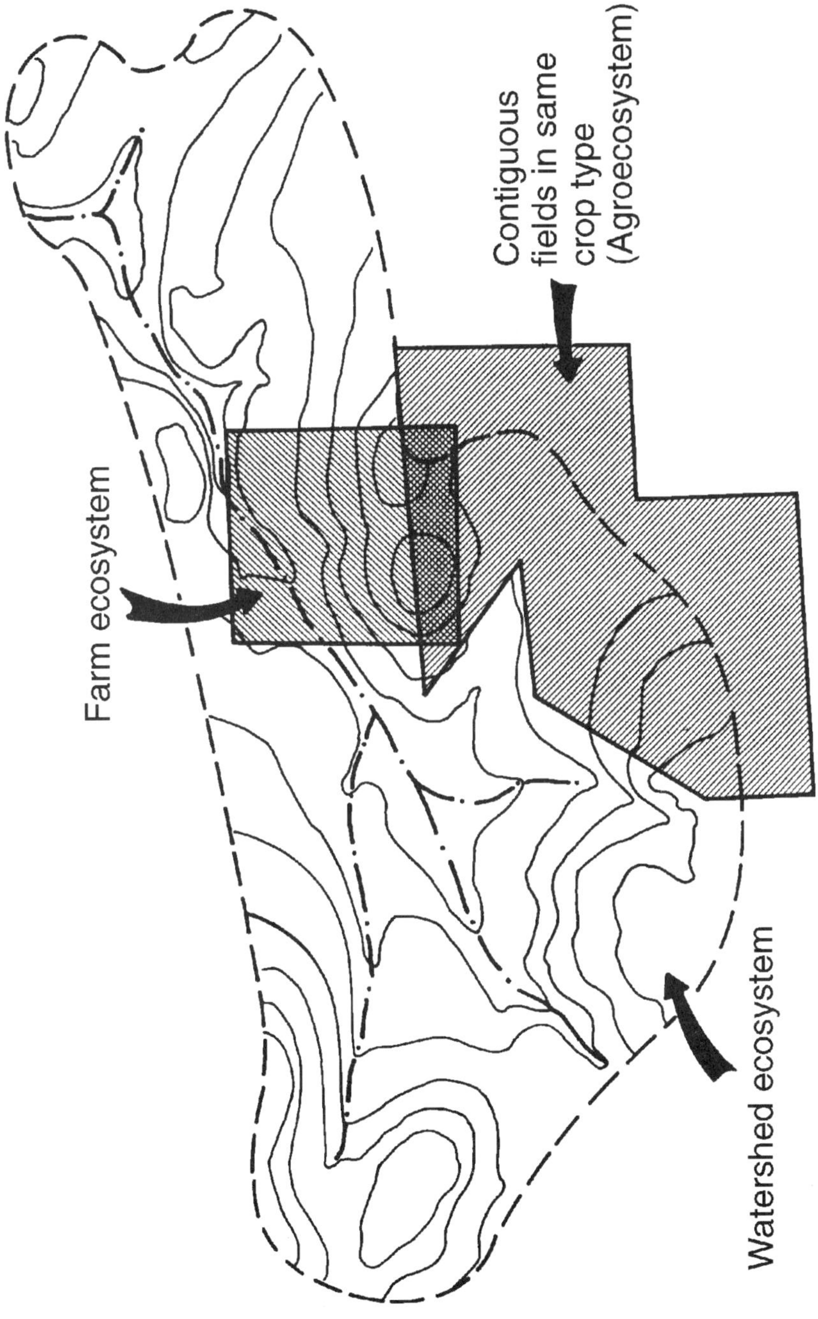

Figure 2.2. Ecosystems of the same scale mapped independently according to different criteria

to an unknown degree (Pielke et al. 1993). At the level of this presentation, however, macroclimate can be viewed as being driven by processes so large as to be independent of the local ecosystem.

Soils are a product of climate, parent material, vegetation, topography, and management. At large scale they are principally dominated by climate, but these other driving factors, particularly parent material, become increasingly important at smaller scales (Jenny 1980).

Water regime refers to the spatial and temporal distribution of soil moisture storage, surface and subsurface flows, and the chemistry of these soil solutions. For aquatic and wetland systems, water takes on additional properties, such as the seasonal patterns of flow through alluvial systems, or the rise and fall of water levels in lakes and ponds. As with terrestrial ecosystems, variability in chemical concentrations is critical to the nature of aquatic and wetland ecosystems.

A central biological property of ecosystems is species composition—the total list of species that live within the system, all or part of the time. These species may have multiple functions, and they usually have widely varying degrees of importance in the productivity and self-regulation (homeostasis) of the ecosystem. These species essentially drive the biological functions of the ecosystem. Some of these may be rare and endangered species.

Collectively, the soil, biota, and dead organic matter compose much of the physical structure of terrestrial ecosystems. Physical structure includes canopy height, leaf area distribution with depth, below-ground distribution of roots, total biomass, and distribution of organic matter on and in the soil. Physical structure largely dictates the distribution of sunlight in space and time, turbulent transfer of materials into and out of the canopy, disposition of precipitation throughout the canopy and soil profile, and temperature profiles. For aquatic ecosystems, different structural variables such as stream channel, lake, or pond form and water flows and depths are relevant.

Ecosystems are always open systems as far as energy is concerned, and usually for matter as well. Ecosystem energetics actually have two components: the physical energy budget consisting of the exchange of short- and long-wave radiation, and energy exchange through sensible heat, latent heat, and conduction to soil or water. The biological energy budget involves the conversion of solar energy to photosynthate, and the fate of that photosynthate as it flows through plants, herbivores, predators, and decomposers.

In addition to energy, ecosystems require critical, limiting material resources, most often water, carbon dioxide, nitrogen, phosphorus, and other nutrients. The acquisition, recycling, and loss of these materials (excluding water that flows through ecosystems) is analyzed and modeled in the biogeochemical descriptions of ecosystems, often referred to as *nutrient cycling* or *nutrient budgets* of locally defined ecosystems.

The combination of ecosystem energetics, biogeochemistry, and various mechanisms of information transfer make up the regulatory processes that order population sizes, ecosystem process rates, physical structure, and the ability to recover from disturbances that lead to metastable or quasi-stable states in ecosystems. Although stability is very difficult to define or measure, it is clear that ecosystems do persist over time, representing at least a minimal notion of stability (Botkin and Sobel 1975).

Applied ecologists, such as foresters and range and wildlife managers, have long understood the importance of spatial relations among ecosystems on the landscape. Applying more formal means to evaluating the topological, or spatial, relationships of neighboring ecosystems on one another is maturing rapidly and receiving attention from basic ecologists in a format sometimes termed *landscape ecology* (Turner and Gardner 1991). Such spatially specific influences of one ecosystem on another, such as fire spread, surface water flow, or animal usage of different resources (including forage, cover, and water), undoubtedly will be incorporated in future ecosystem analyses.

From this discussion, it should be clear that the term *ecosystem* is not restricted to assemblages operating in the absence of human influence, either historically or presently. The Great Plains is not composed of just remnant plots of tallgrass prairie ecosystems, or isolated riparian woodland ecosystems, or residual prairie pothole ecosystems. Among its many ecosystems, it also includes fields of wheat, sorghum, and alfalfa; giant reservoirs; engineered farm ponds and channelized rivers; rangelands planted in exotic grasses; solid waste dumps; and cities.

No broadly accepted measure exists for assessing how well ecosystems function. Therefore, claims of ecosystem dysfunctionality because of the loss of a species or some human intervention require specific definition and justification. Such justification may be no more than a value judgment. If an ecosystem is entirely dysfunctional, it is not an ecosystem at all—by definition.

Geographic Variables of the Great Plains

Ecosystem properties can be quantified and characterized as properties of locally defined ecosystems, such as a wheat field or prairie pothole. But some of these properties can also be treated as independent variables with independent geographic patterns of variation. By mapping these variables, we gain a picture of the complex pattern of overlapping conditions that makes up the Great Plains. Maps of some germane geographic variables are presented to provide the framework for ecosystem definition and location in the Great Plains.

Unfortunately, maps covering both the Canadian and U.S. portions of the Great Plains are not always available, so in some cases compromises were made in choosing maps to cover the Canadian section. Where this was not readily practicable, the Canadian section was sacrificed. Another limitation is that many excellent regional maps use a different eastern boundary than that approximately defined for this symposium.

Physiography, defined here as regionally dominating land forms, is a principal variable for climate, soil, and vegetation features. In general, the Great Plains slopes gently from the Rockies to the Mississippi drainage and the Gulf of Mexico and is etched by past and present fluvial erosion patterns producing isolated uplands and alluvial plains (Figure 2.3). A northern portion of the Great Plains has been glaciated, obscuring fluvial erosion patterns and pocking the landscape with lakes and ponds (see glacial boundary in Figure 2.3). Limited relief and gentle topography are unifying features of the Great Plains, but isolated areas of steep relief exist along valley walls, in badlands, and within the Black Hills. Relief is slight to moderate in the undulating uplands and virtually nonexistent on alluvial flood plains and lacustrine basins. Generally, the less extensive the local relief, the more important microrelief is to ecosystem structure and function. Prairie potholes and playa lakes are not deep and are not located in areas of extensive relief, but differences of a few meters in areas where they are found are critical to their existence. Unfortunately, such relief cannot be appreciated on a map of the scale shown in Figure 2.3.

Geology underlies the physiography of the Great Plains at large and small scales. The limited relief, as illustrated in Figure 2.3, results from the relatively low tectonic activity in the Central Lowlands and Great Plains that make up this area of study. This region has been characterized by subsidence and shallow marine sedimentation followed by gentle uplift above sea level. In this kind of geological regime, surface geology mainly consists of spatially extensive, horizontally disposed, Paleozoic and Mesozoic sedimentary rocks (Figure 2.4). This bedrock geology is veneered in many places by extensive, tertiary alluvial fans extending eastward from the Rockies, by quaternary glacial till in the north, and by recent alluvium in the many floodplains etching the region. This geological variability, while subtle compared with mountainous regions, is extremely important in determining local relief and soil properties.

The position of the Great Plains with respect to latitude, the Rocky Mountains, and the Gulf of Mexico gives it a characteristically continental climate (Rosenberg 1987; Harrington and Harman 1991). Winter temperatures range from frigid in the northern half of the region to mild, but variable, in the southern half. Summer temperatures are quite warm over the entire region. Of greatest interest from the point of view of ecosystem distribution is that the growing season is roughly defined by the frost-free period, decreasing with latitude and proximity to the mountains (Figure 2.5). However, precipitation is the preeminent climatic consideration for the Great Plains. This varies longitudinally (Figure 2.6) and, in combination with temperature, sets the water budgets available for natural and agricultural ecosystems of the Great Plains. Perhaps the most fortunate natural phenomenon in the Great Plains is that precipitation is concentrated in the summer during the period of vegetative growth. Thus, although total precipitation is relatively low, the concentration of precipitation during the summer permits highly successful, precipitation-based agriculture, at least up to the 100th meridian, which

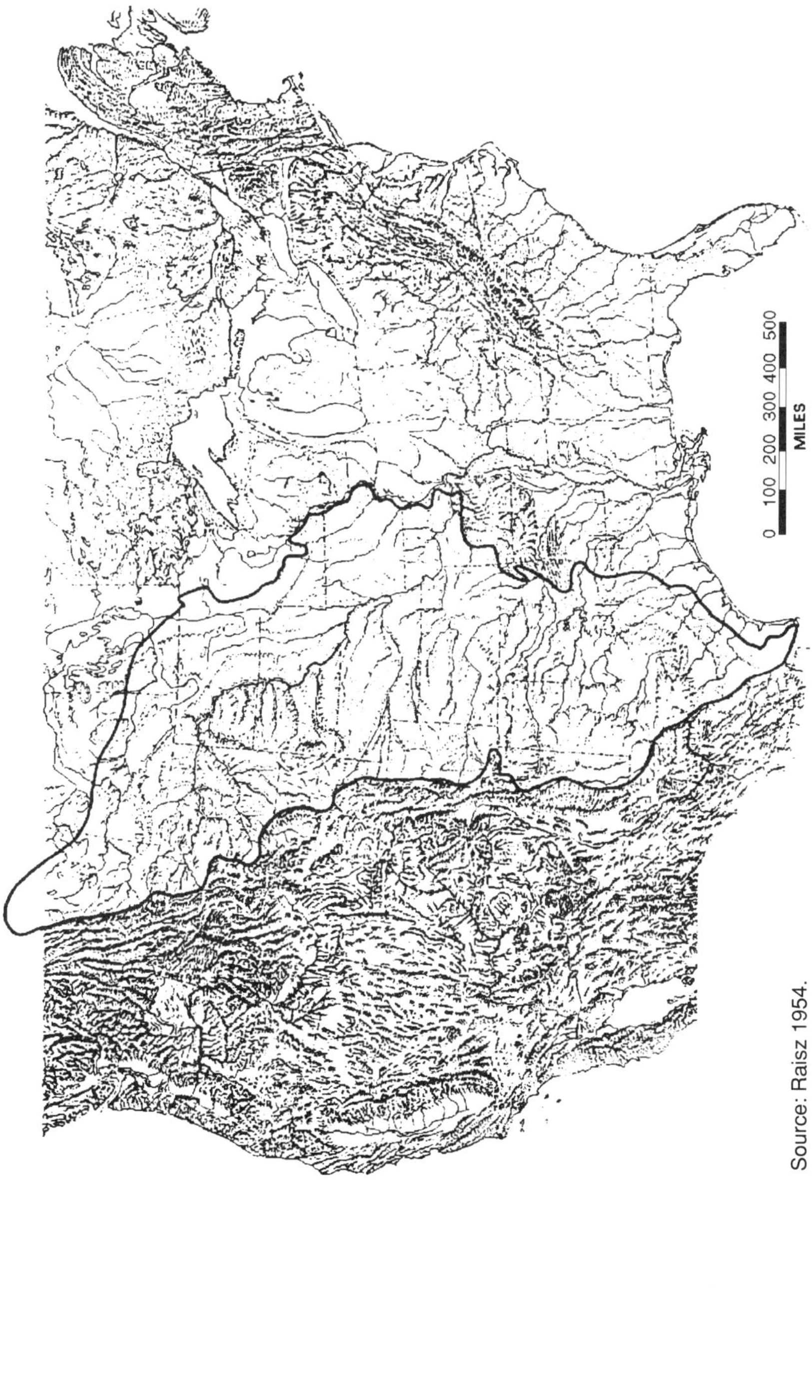

Source: Raisz 1954.

Figure 2.3. Physiographic map of the United States

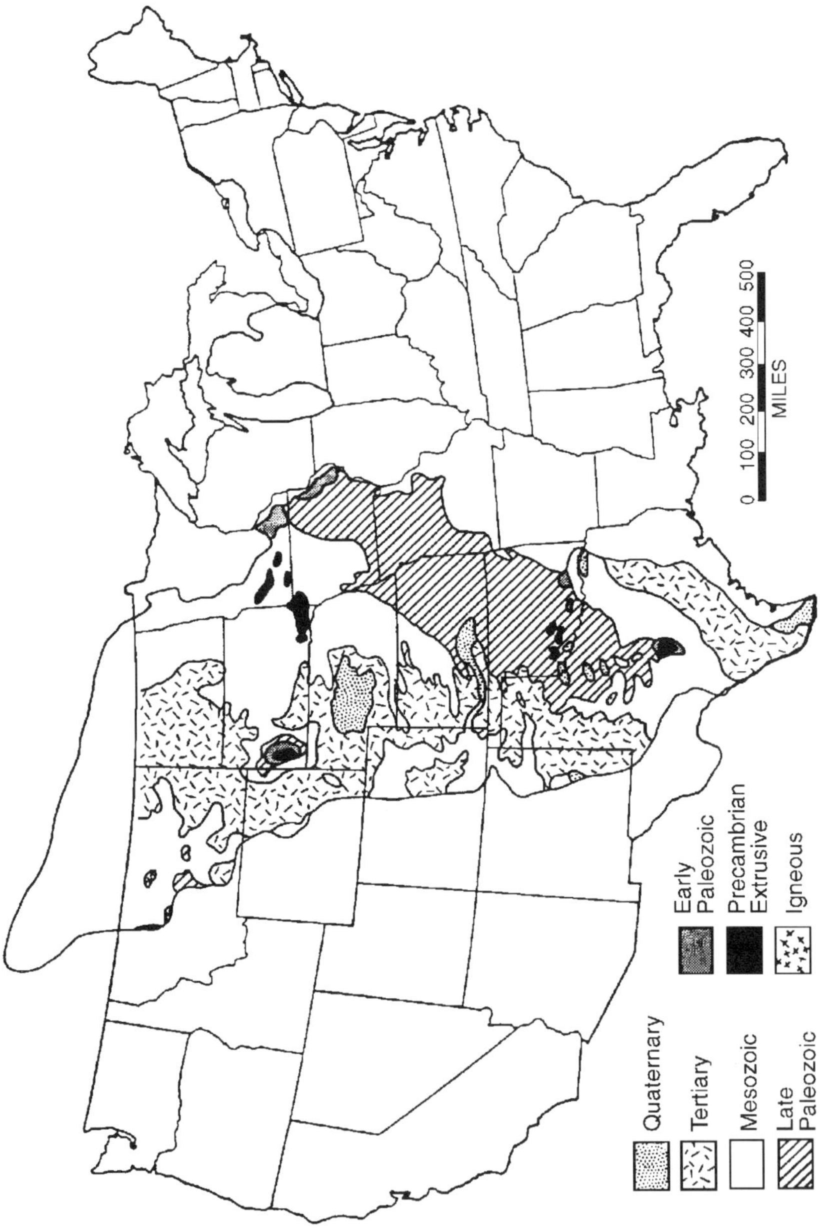

Source: Stokes 1966.

Figure 2.4. Geologic map of the United States

basically is the 20-inch precipitation isoline. Annual variability of precipitation is perhaps the largest problem with agriculture in the nonirrigated portions of the Great Plains.

An important resource of the Great Plains is the excellent soil complex dominating most of the region. These soils are the product of parent material, climate, and to a large degree, the grassland-dominated vegetation that formerly covered the region. The low resolution map of soils shown in Figure 2.7 illustrates some patterns of soil order and group distribution in the Great Plains. In general, soil depth and amount of organic matter diminish from east to west in parallel with effective precipitation. These variables, plus textural variability superimposed on this climatic pattern by geological substrates, provide much of the basis for soil differentiation (Akin 1991). With the onset of cultivation in the nineteenth century, these soils have undergone extensive changes as described in Cole et al. (1988, 1989).

Originally, the Great Plains was mostly grassland interspersed with riparian forests extending all the way from Illinois and Indiana in the east to the Rocky Mountains in the west. Grasslands varied from tallgrass prairie in the east to mid-grass prairies in the central Plains, to the short-grass prairies in the lee of the Rockies. The original vegetation of the eastern portion of the region is mainly of historical interest, except for the heritage imposed by vegetation on soil properties (Akin 1991). In the western portion, much of the original vegetation still survives, although it has been altered to unknown degrees by changes in grazing and fire regimes. If humankind were eliminated from the Great Plains, it is not known how the vegetation would readjust given changes in ambient carbon dioxide levels in today's atmosphere and the existence of introduced Eurasian plants such as smooth brome or Russian thistle.

The Great Plains has been characterized by large-scale conversion to agriculture within the last 150 years so that ecosystems of the regime are, to a large extent, agro-ecosystems. Virtually all of the region east of the 100th meridian has been converted to agriculture except for areas too steep to plow, too wet to cultivate, or otherwise unsuitable for agriculture. The pattern of land use conversion becomes irregular west of the 100th meridian, depending on soil conditions and irrigation water supply. The western-most portion of the region is largely covered with original vegetation but with dryland farming, alluvial floodplain farming, and some center-pivot irrigation farming dispersed within the matrix of grazing land. Distribution of major crops is represented in the maps of Figure 2.8. These distributions are dictated by climate and soil conditions just as were the native ecosystems of the past.

Other alterations of ecosystems in the Great Plains have occurred with a large-scale redistribution of water. Very large reservoirs now punctuate the courses of major rivers in the region and, in some areas such as central Texas, small catchbasin ponds dot the landscape with a high density. Irrigation by groundwater in the central portion (Figure 2.9a) and by surface water in the western portion (Figure 2.9b) has extended the range of agriculture well west of the 100th meridian. In contrast, considerable areas of natural lakes, ponds, and wetlands of the glaciated northeastern portions of the region have

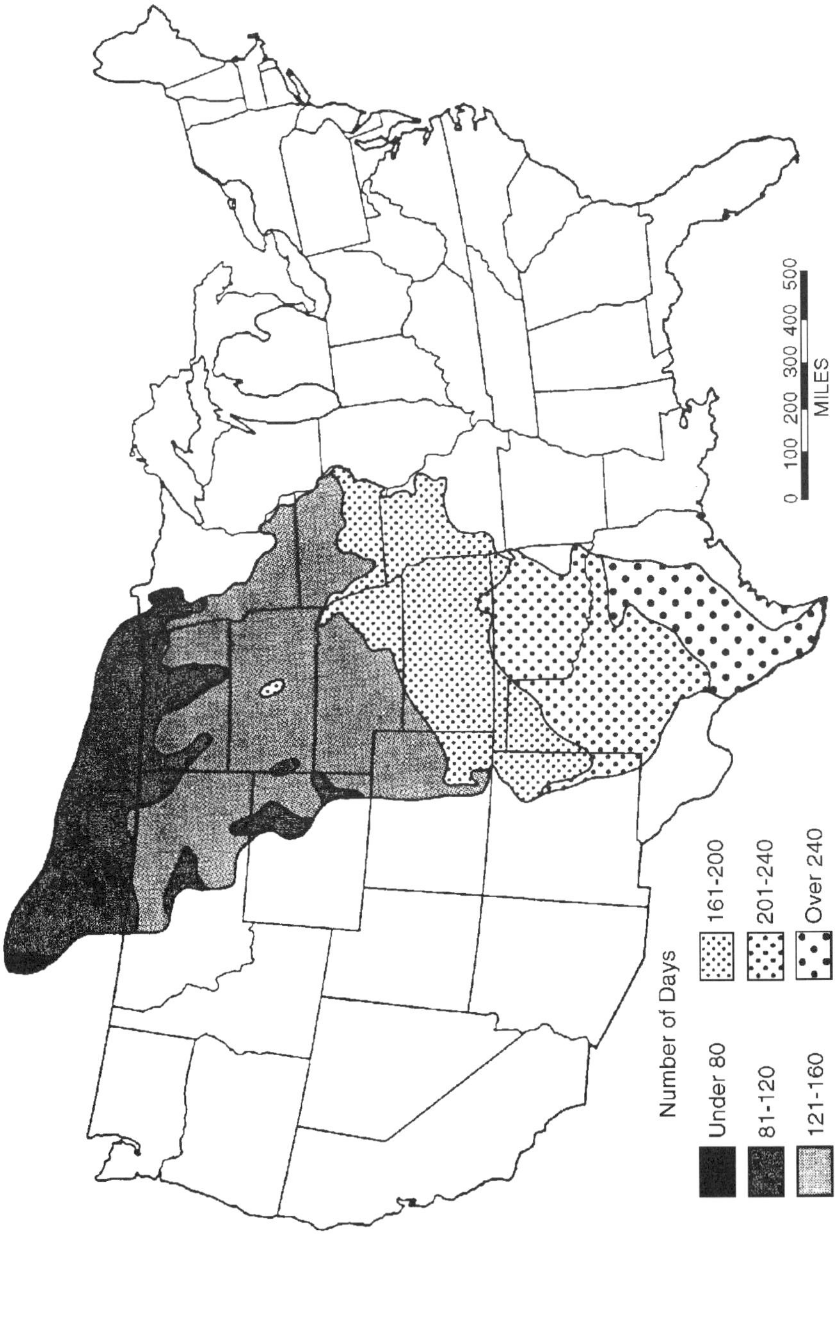

Source: Espenshade 1986.

Figure 2.5. Frost-free season in the Great Plains

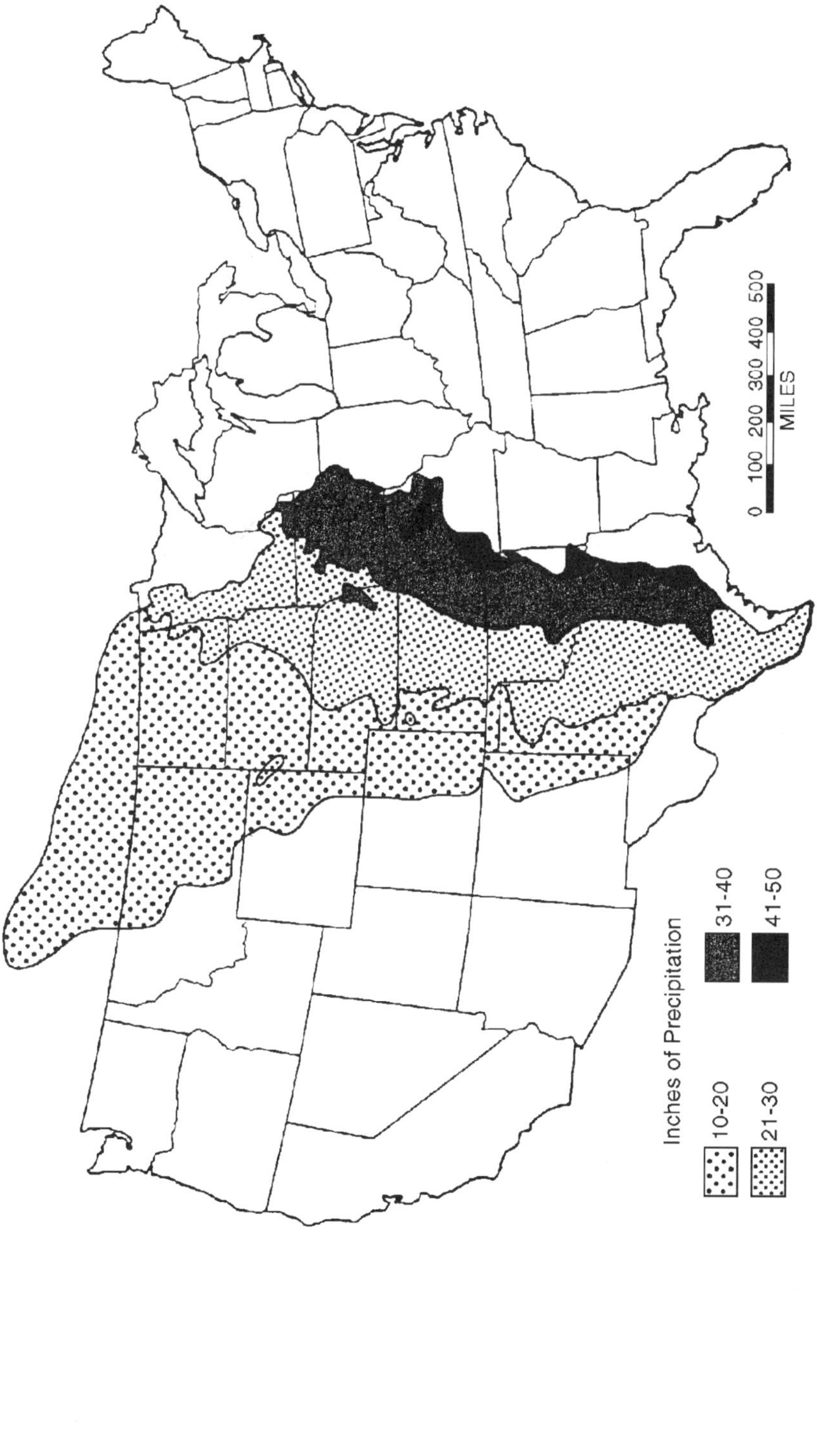

Source: Espenshade 1986.

Figure 2.6. Average annual precipitation in the Great Plains

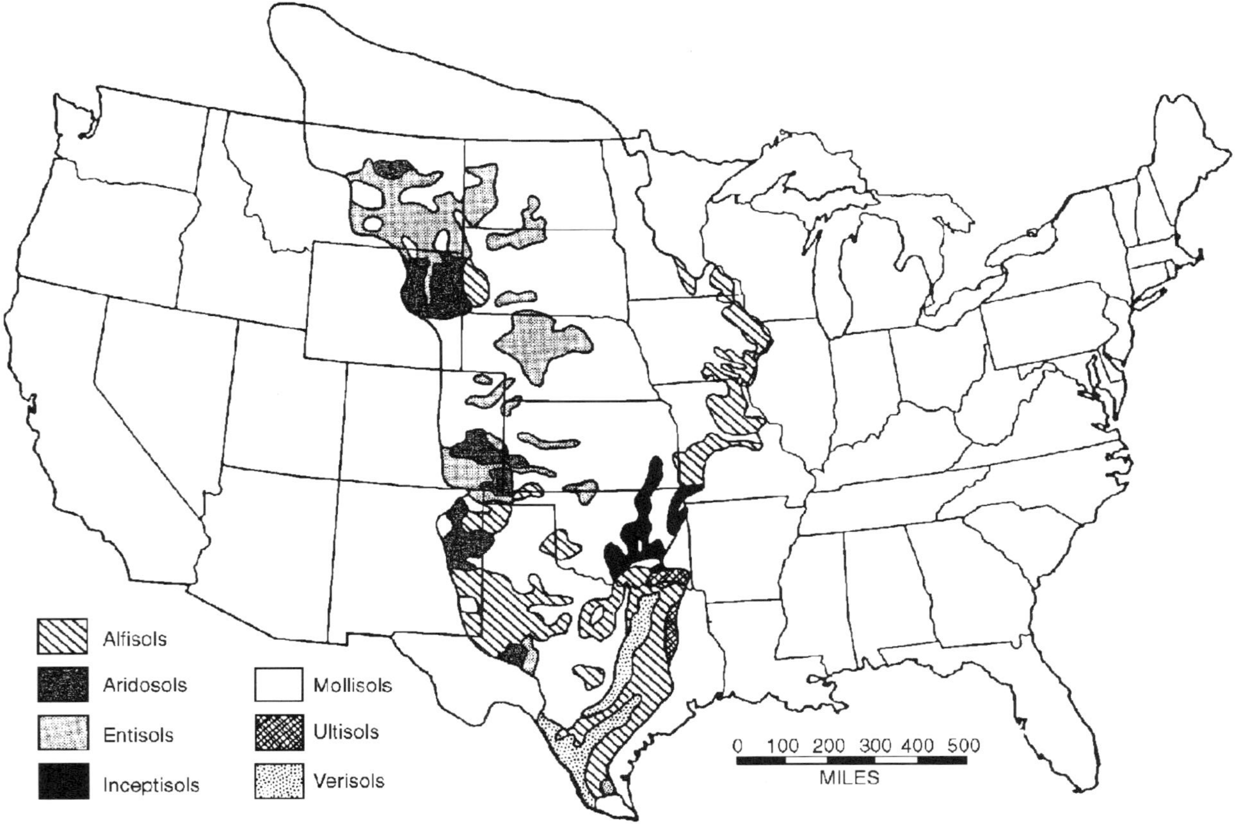

Source: USDA 1967, revised 1985.

Figure 2.7. Principal soils in the Great Plains

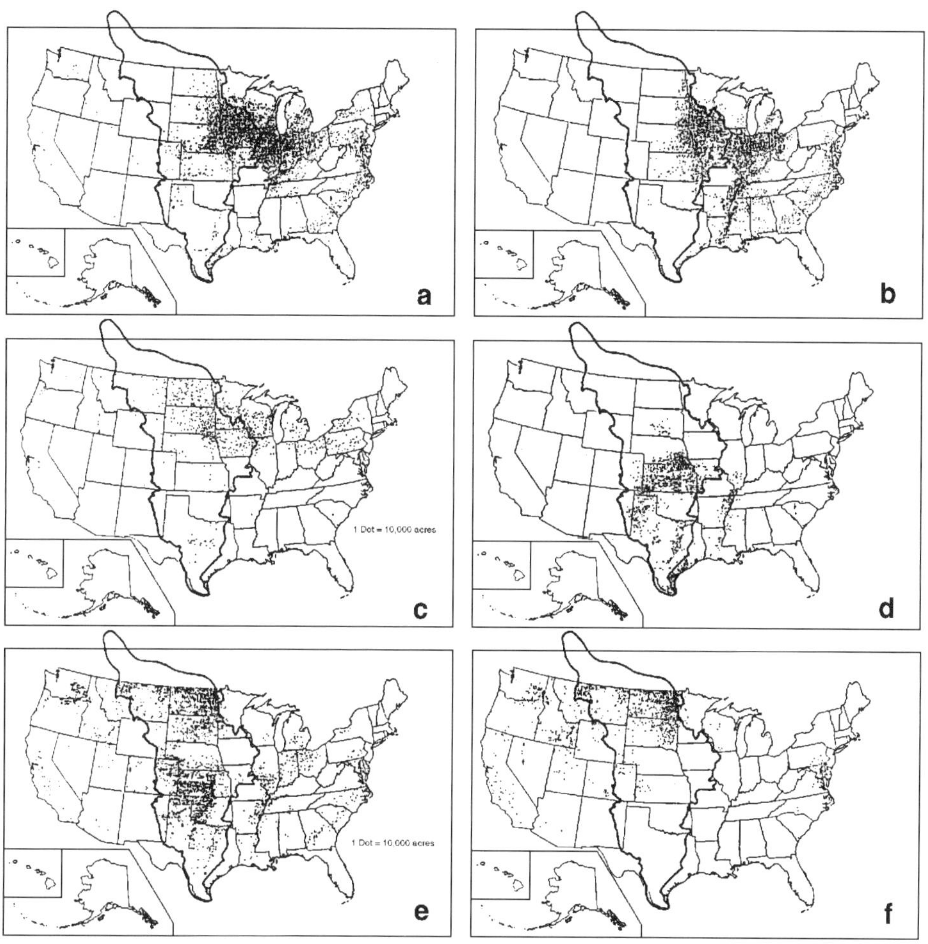

Source: U.S. Bureau of the Census 1990.

Figure 2.8a-f. Composited maps of some major crop distributions in the Great Plains

a. Corn for grain or seed, 1987 (each dot equals 10,000 acres)

b. Soybeans for beans, 1987 (each dot equals 10,000 acres)

c. Oats, 1987 (each dot equals 5,000 acres)

d. Sorghum for grain or seed, 1987 (each dot equals 5,000 acres)

e. Wheat for grain, 1987 (each dot equals 10,000 acres)

f. Barley for grain, 1987 (each dot equals 5,000 acres)

been drained to extend tillage onto muck soils (Van der Valk 1989) (Figure 2.9c). Playa lakes of western Texas are also being depleted by water manipulations (Bolen et al. 1979). All of these hydrological alterations have enormous impact on the ecosystems of their respective areas (Great Plains Agricultural Council 1979).

Structure of the Great Plains Ecosystems

The intersection of geographic variables described in the preceding section leads to a spatial pattern of contemporary ecosystems. *Contemporary* ecosystems refer to those that cover Great Plains landscapes today, not the historical or relict ecosystems that are believed to have covered the landscapes before extensive intervention by European culture. Terrestrial ecosystems account for the vast majority of area constituting the Great Plains. But riparian, wetland, and aquatic systems are extremely important in the study of Great Plains ecosystems. Therefore, a two-scale approach is necessary for defining contemporary ecosystems for this huge region (nearly 2.2 million square kilometers for the U.S. section alone). For upland ecosystems, representatives of an ecosystem type must collectively comprise more than 0.75 percent of the region's total area. For aquatic, wetland, and riparian ecosystems, the general classes and regional locations are merely listed. These will have to be located geographically on a subregional basis.

To identify and locate upland ecosystems of the Great Plains as defined here, I have, with the cooperation of Dr. Thomas R. Loveland of the U.S. Geological Survey's EROS Data Center in Sioux Falls, South Dakota, reprocessed the land cover type classified imagery prepared and published by Loveland et al. (1991). I delineated the outline of the Great Plains region as defined for this book on the EROS image, then aggregated the 113 land cover types occurring within the region into 20 more generalized types as shown in Figure 2.10a-f.

Table 2.1 lists the land covers for the Great Plains along with areas and the percentages of the total area occupied by each. Types of land cover identified by this process can be viewed as terrestrial ecosystems resolvable at pixel sizes of 1 kilometer. With images as shown at the scale of Figure 2.10, the subregional patterns of land cover types can be readily distinguished by the merger of like pixels into apparent polygons of regional ecosystem types. It should be emphasized that the images in Figure 2.10 can readily be decomposed to 1.1 kilometer pixels to better resolve spatial types and the spatial distribution of types within local areas. Even at this level of resolution, however, several ecosystem types are mixtures of natural to seminatural vegetation with cropland that do not lead to satisfactory definitions for ecosystem types. If necessary and useful for more finely resolved analysis, structural and functional analysis could be done for each land ownership unit from other data.

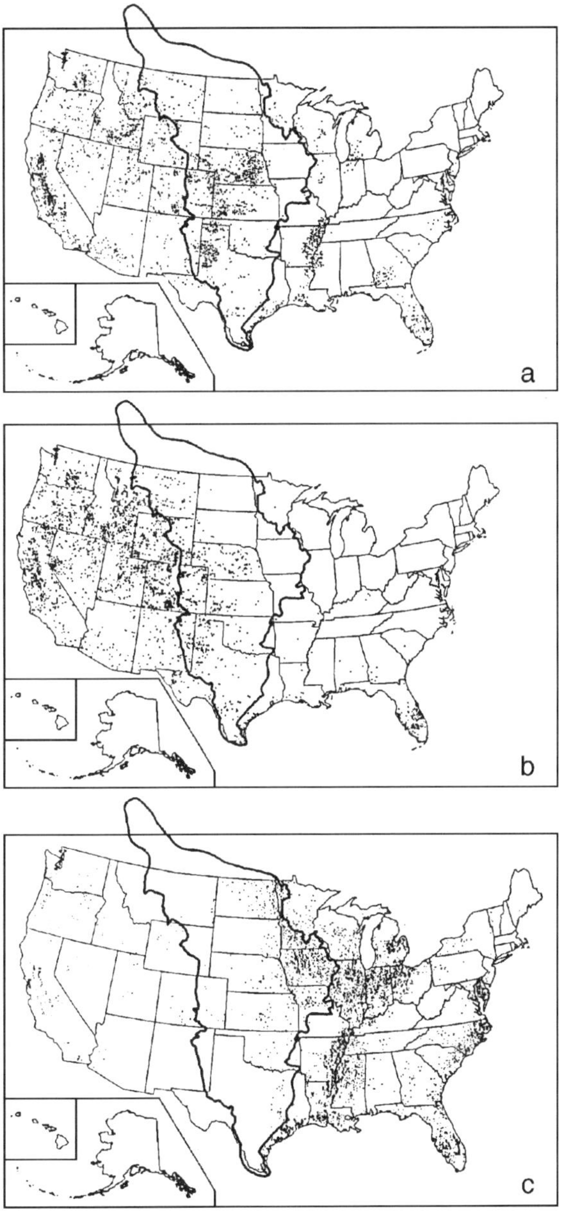

Source: U.S. Bureau of the Census 1990.

Figure 2.9a-c. Altered hydrologic regimes in the Great Plains

a. Irrigation land in farms (mostly groundwater) (each dot equals 10,000 acres)

b. Irrigation land in pasture and other land (mostly surface water) (each dot equals 1,000 acres)

c. Artificially drained agricultural land (each dot equals 20,000 acres)

The images of Great Plains ecosystems in Figure 2.10 are of such a scale that they do not include the very small remnants of original ecosystems embedded in the agricultural matrix making up most of the eastern half of the Great Plains. Those original ecosystem remnants are critical as habitat for species conservation. Their management, size, and potential augmentation with restoration projects are crucial for maintenance of the native biota of the region. To map these remnants of original ecosystems, however, would require more detailed work at subregional scales.

Riparian zones, wetlands, and aquatic ecosystems are mostly too small to be viewed on a regional scale like those in Figure 2.10. These wetland and aquatic systems include riverine floodplains, human-made reservoirs and catchbasin ponds, prairie potholes of the glaciated north, playa lakes of Texas, and ponds in the sand hills of Nebraska. The distribution of these, however, can be conveniently examined in the national wetlands map published by the U.S. Fish and Wildlife Service (1991). Clearly, different kinds of lowland systems are important in different parts of the Great Plains. This suggests that different kinds of ecosystem analysis must be directed to lowland systems in various areas. Regardless of the type, however, all these areas are extremely important to the collective Great Plains ecosystem and cannot be ignored, particularly from the viewpoints of wildlife, recreation, and hydrological resources (Van der Valk 1989; Great Plains Agricultural Council 1979; Richardson and Arndt 1989).

Fundamental Needs for Assessment and Planning

The many issues bearing on conservation of Great Plains ecosystems, in terms of both human-generated factors and natural ones, are addressed in this symposium's program. The issues are different for upland and lowland ecosystems. For example, some of the major issues for upland agro-ecosystems will revolve around sustainability of groundwater supplies, groundwater chemical quality, and soil qualities, while issues for upland natural ecosystems will revolve around destruction, fragmentation, and species loss. For lowland ecosystems, the issues will revolve around maintenance of flows, water quality, drainage plans and policies, and destruction and fragmentation of wildlife habitat.

The spatial relations of ecosystems with one another must be part of the assessment of ecosystem conservation. Lowland ecosystems are obviously and particularly related to surface and subsurface flows from upland ecosystems. Many other examples of neighborhood relationships deserve consideration. It will not be adequate to analyze ecosystems in isolation, since they are best analyzed as interconnecting networks with energy, matter, and biological information moving among them in critical ways.

To seriously approach the goals of this symposium, it appears that several organizational resources must be generated. One is a classification of regional ecosystems at appropriate scales. This must include agro-ecosystems as well as those composed of and supporting native biota—objects of concern from the viewpoint of the Endan-

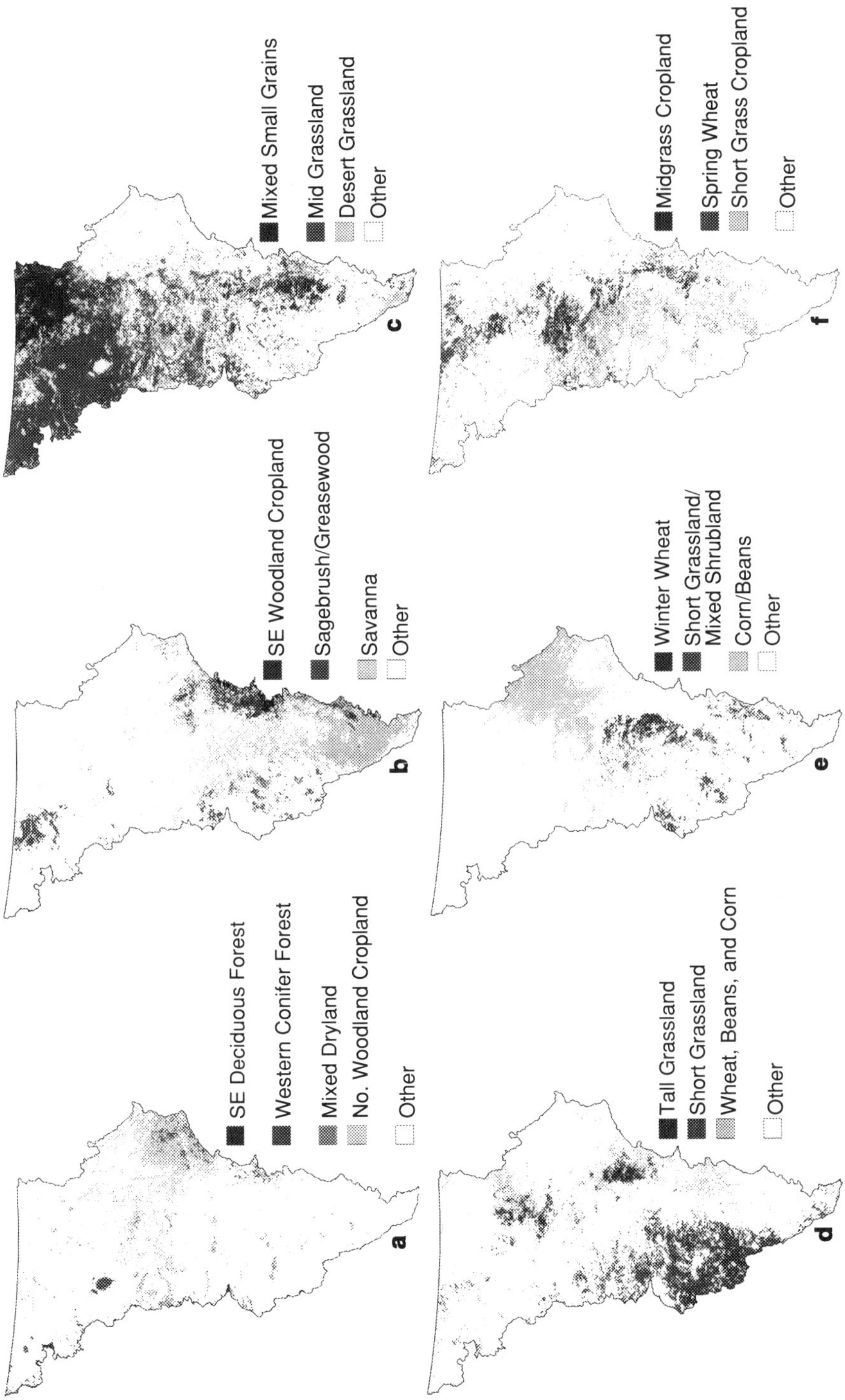

Source: Derived from data by Loveland et al. 1991.

Figure 2.10a-f. Land cover types of the Great Plains

Table 2.1. Types of land cover of the Great Plains

Land cover	Area (km^2)	Percent of Great Plains
Mid-grassland	507,236	23.21
Mixed small grains	282,352	12.92
Short grassland	204,861	9.37
Corn, beans	194,322	8.89
Savanna	193,142	8.84
Short grass cropland	136,239	6.23
Mid-grass cropland	129,219	5.91
No. woodland cropland	115,214	5.27
SE woodland cropland	71,105	3.25
Winter wheat	59,668	2.73
Sagebush-greasewood	52,752	2.41
Tall grassland	48,351	2.21
Wheat, beans, corn	34,823	1.59
Other	32,032	1.47
Mixed dryland	28,707	1.31
Spring wheat	22,710	1.04
Desert grassland	19,386	0.89
Short grassland-mixed shrub	18,966	0.87
SE deciduous forest	17,706	0.81
Western conifer forest	16,927	0.78
Total	2,185,718	100.00

Note: Types of land cover are listed in descending order of area (for the geographic area as defined by the symposium "Conservation of Great Plains ecosystems: Current science, future options"). These land covers are based on a simplification of the AVHRR-derived classified image for the 48 conterminous U.S. states by Loveland et al. (1991). See Loveland et al. (1991) for details on classes and their derivation.

gered Species Act. More refined classifications obviously will be needed for finer scale description and management.

A second necessary organizational resource is a network of regional spatial database centers to facilitate understanding and planning in a geographically explicit manner. Within the context of this geographic system, analyses of ecosystem distributions, their structural and functional properties, and spatial relationships within subregions could be better organized. Without this kind of geographic organization, it is difficult to see how the diversity and particularity of locations within this huge region can be addressed in a practical way, or how appropriate policies can be meaningfully developed.

Changes in atmospheric carbon dioxide levels effectively alter the climate for many kinds of plants by modifying their water use efficiency and carbon balance. Whether actual climate changes or not, these recorded changes in carbon dioxide concentrations may have already altered the potential distribution of plants and can be expected to alter such distributions even further in the future. In this sense, the future is already with us.

Acknowledgments

Preparation of this chapter was supported by the National Park Service's Global Climate Change Program through project number CEGR-R92-0195 under cooperative agreement number CA 1268-1-9009 with the University of Wyoming. I particularly wish to thank Ms. Marilee Doyle for assistance in finding materials and Mr. Gary Amende for development of the EROS imagery underlying the ecosystem classification in this chapter.

References

Akin, W. E. 1991. Global patterns climate, vegetation, and soils. Norman: University of Oklahoma Press.

Bolen, E. G., C. D. Simpson, and F. A. Stormer. 1979. Playa lakes: Threatened wetland on the southern Great Plains. In *Riparian and wetland habitats of the Great Plains: Proceedings, 31st annual meeting, Colorado State University, June 18-21.* Great Plains Agricultural Council Publication 91. Ft. Collins, Colorado: Rocky Mountain Forest and Range Experiment Station.

Botkin, D. B., and M. J. Sobel. 1975. Stability in time-varying ecosystems. *American Naturalist* 109: 625-646.

Cole, C. V., I. C. Burke, W. J. Parton, D. S. Schimel, D. S. Ojima, and J. W. B. Stewart. 1988. Analysis of historical change in soil fertility and organic matter levels of the North American Great Plains. In *Challenges in dryland agriculture: Proceedings of the International Conference on Dryland Farming, Amarillo, Texas, August 15-19.*

Cole, C. V., J. W. B. Stewart, D. S. Ojima, W. J. Parton, and D. S. Schimel. 1989. Modeling land use effects on soil organic matter dynamics in the North American Plains. In *Ecology of arable land,* ed. M. Clarholm and L. Bergstrom. Dordrecht, Netherlands: Kluwer Academic Publishers.

Espenshade, E. B., Jr. (ed.). 1986. *Goode's world atlas.* Chicago: Rand McNally & Company.

Frazier, I. 1989. *Great Plains.* New York: Farrar, Straus, and Giroux.

Great Plains Agricultural Council, Forestry Committee. 1979. In *Riparian and wetland habitats of the Great Plains: Proceedings, 31st annual meeting, Colorado State University, June 18-21.* Great Plains Agricultural Council Publication No. 91. Ft. Collins, Colorado: Rocky Mountain Forest and Range Experiment Station.

Harrington, J. A., and J. R. Harman. 1991. Climate and vegetation in central North America. Natural patterns and human alterations. *Great Plains Quarterly* 11:103-112.

Jenny, H. 1980. *The soil resource.* New York: Springer-Verlag.

Lewis, G. M. 1979. The cognition and communication of former ideas about the Great Plains. In *The Great Plains: Environment and culture,* ed. B. W. Blouet and F. C. Luebke. Lincoln: University of Nebraska Press.

Loveland, T. R., J. W. Merchant, D. O. Ohlen, and J. F. Brown. 1991. Development of a land-cover characteristics database for the conterminous United States. *Photogrammetric Engineering and Remote Sensing* 57:1453-1463.

Pielke, R. A., G. Dalu, T. J. Lee, H. Rodriquez, J. Eastman, and T. G. F. Kittel. 1993. The effect of mesoscale vegetation distribution on the hydrologic cycle and regional and global climate. In *Land-surface/atmosphere interactions on global and regional scales: Proceedings, AMS Conference on Hydroclimatology*. Boston: American Meteorological Society.

Richardson, J. L., and J. L. Arndt. 1989. What use prairie potholes? *Journal of Soil and Water Conservation* 44:196-198.

Riebsame, W. E. 1990. The United States Great Plains. In *The earth as transformed by human action: Global and regional changes in the biosphere over the past 300 years*, ed. B. L. Turner, W. C. Clark, R. W. Kates, J. F. Richards, J. T. Mathews, and W. B. Meyer. New York: Cambridge University Press.

———. 1991. Sustainability of the Great Plains in an uncertain climate. *Great Plains Research* 1:133-151.

———. 1993. Sustainable development questioned: The historical debate between adaptationism and catastrophism in Great Plains studies. Paper presented at the American Society for Environmental History Conference, Pittsburgh, Pennsylvania, March 5.

Rosenberg, N. J. 1987. Climate of the Great Plains region of the United States. *Great Plains Quarterly* 7:22-32.

Scott, J. M., F. Davis, B. Csuti, R. Noss, B. Butterfield, C. Groves, H. Anderson, S. Caicco, F. D'Erchia, T. C. Edwards, Jr., J. Ulliman, and R. G. Wright. 1993. Gap analysis: A geographic approach to protection of biological diversity. *Wildlife monographs* No. 123, supplement to *Journal of Wildlife Management* 57(1).

Stokes, W. L. 1966. *Essentials of earth history: An introduction to historical geology*. Englewood Cliffs, New Jersey: Prentice-Hall.

Tansley, A. G. 1935. The use and abuse of vegetational concepts and terms. *Ecology* 16:284-307.

Turner, M. G., and R. G. Gardner (ed.). 1991. *Quantitative methods in landscape ecology*. New York: Springer-Verlag.

U.S. Bureau of the Census. 1980. *1978 Census of Agriculture, Vol. 5, Special reports, Part 1. Graphic summary*. AC78-SR-1. Washington, D.C.: U.S. Government Printing Office.

———. 1990. *1987 Census of Agriculture, Vol. 2, Subject series. Part 1. Agricultural Atlas of the United States*. AC87-S-1. Washington, D.C.: U.S. Government Printing Office.

U.S. Department of Agriculture, Soil Conservation Service. 1967, reviewed 1985. Principal kinds of soils. In *National atlas of the United States of America*. Reston, Virginia: Department of the Interior, U.S. Geological Survey.

U.S. Fish and Wildlife Service. 1991. *Wetland resources of the United States*. Map sheet prepared under the direction of Thomas E. Dahl, National Wetlands Inventory, U.S. Fish and Wildlife Service, St. Petersburg, Florida.

U.S. Geological Survey. 1969. *Geology, national map atlas*. Sheet No. 74. Reston, Virginia: U.S. Department of Interior.

———. 1987. Soils, principal kinds of soils: Orders, Suborders, and great Groups. From the *National Atlas of the United States*, Map sheet 38077-BE-NA-07M-00. Reston, Virginia: U.S. Department of Interior.

Van der Valk, A. (ed.). 1989. *Northern prairie wetlands*. Ames: Iowa State University Press.

3. Valuing Ecosystems and Biodiversity

Jason F. Shogren
Iowa State University
Thomas D. Crocker
University of Wyoming

An Introduction to Nonmarket Valuation

Environmental resources in the Great Plains supply a flow of direct and indirect services to the citizenry. The services provided by these ecosystems and their corresponding levels of biological diversity are multifarious—ranging from basic life-support to the filtration of nonpoint source pollution from urban and rural runoff. While these resources provide a nearly limitless set of valuable attributes, many of their services remain unpriced by the market. The services are rarely bought and sold by the pound on the auction block, and therefore never enter into private markets and remain unpriced by the public sector. For example, the market price of cropland does not generally account for the nutrient filtration and wildlife habitat services provided by a Nebraska wetland. The market undervalues wetland services because the associated costs and benefits accrue to more than just the owner of the land. Water filtration benefits all those downstream; wildlife does not stay within the confines of one landowner's property. This inability to exclude others from enjoying benefits or suffering costs prevents the market price from sending the correct signal about the true value of the wetland (Crocker and Tschirhart 1992).

Recognition that ecosystem and biodiversity services are systematically mispriced by the market has forced policymakers to consider other means to assess the value of these resources. Nonmarket valuation, the economist's recommendation, uses the implicit and explicit trade-offs between conservation and development to assess the value of unpriced environmental resources. The trade-off between cropland and nutrient filtration reflects an economic value of a wetland, and the economist's job is to estimate its monetary value accurately. If an economist captures these trade-offs within a reasonable range of error, nonmarket valuation provides data to aid policymakers in their musings on how best to manage the natural resources of the Great Plains.

This paper examines some issues underlying the nonmarket valuation of ecosystem and biodiversity services provided by the Great Plains. Fueled by court cases over natural resource damage assessment of Superfund waste sites and the Exxon Valdez

Stanley R. Johnson and Aziz Bouzaher (ed.), *Conservation of Great Plains Ecosystems,* 33–46.

oil spill in Alaska, the past decade has been witness to an explosion in nonmarket valuation research. Probably more collective intellectual energy has been spent on valuing environmental resources than any other question in natural resources and environmental economics. The next section defines two measures of economic value—the willingness to pay for improved ecosystem and biodiversity services and the willingness to accept compensation for decreased services. Four techniques of nonmarket valuation are briefly discussed. A further section explores four debated issues in nonmarket valuation—biodiversity, total value and warm glows, unfamiliarity and learning, dynamic inconsistency and sustainable development, and endogenous risk. The final section offers concluding observations and sketches a four-step procedure for improved valuation of the ecosystem and biodiversity services provided by the Great Plains.

Economic Measures of Value and Nonmarket Valuation

Economists have a distinct and well-defined definition of value based on the ideals of rationality and consumer sovereignty—the rational consumer is purposive. With sovereignty, the consumer is best able to make the choices that affect his or her own welfare. Note that rationality in economics implies that the consumer's choices are consistent with purposes, however odd the choices may appear to the outsider. If consumers prefer improved wetland quality to a new John Deere tractor, rationality requires the choice of wetland quality. Based on rational choice, economists work within two general measures of value—the individual's willingness to pay (WTP) to secure an increase in ecosystem services and willingness to accept (WTA) compensation to allow a deterioration in these services. The value of an environmental service is the aggregation of these individual WTP or WTA value measures.

Although some economists have argued that the value measures should be nearly equal, a consistent pattern of empirical evidence reveals that WTA exceeds WTP, occasionally by a magnitude of 100 times (see for example Brookshire and Coursey 1987). The fear that this divergence is some form of cognitive mistake prompted Cummings et al. (1986) to recommend that only WTP measures be used in nonmarket valuation. But as Mäler (1974) argued and Hanemann (1991) demonstrated, WTA should exceed WTP if the environmental resource has limited substitution possibilities. For example, since the filtration services provided by a wetland have imperfect, if any, substitutes, an individual will require more compensation to give up access to this service than he or she is willing to pay to acquire access. Low substitutability drives the divergence in value measures, an argument supported by Shogren et al.'s (1994) laboratory valuation experiments on market and nonmarket goods. Therefore, the choice of value measure will depend on a clear definition of the technical and legal substitution possibilities for the resource or service in question.

Figure 3.1 illustrates the relationship between the WTP and WTA measures of value. The vertical axis represents the level of pollution, measured, for example, in an ambient concentration level. The horizontal axis reflects an individual's level of wealth. The set of curved lines that originate from the horizontal axis reflect the individual's preferences for income and for avoiding pollution—the set of utility functions. Each point on a utility function represents different combinations of pollution and wealth that leave the individual at the same level of satisfaction. The slope of a utility curve shows the willingness of the individual to trade between wealth and pollution while maintaining the same level of utility. The slope of the utility function is called the *marginal rate of substitution*—the willingness to trade pollution for wealth while holding utility constant. The flatter the slope the less effective is wealth as a substitute for pollution, and more wealth is necessary to compensate the individual for an increase in pollution. A steep slope implies the opposite—less wealth is needed to compensate the individual for an increase in pollution. Note that utility increases as one moves to the right, $U_{00} > U_0 > U_1 > U_2$ for a given level of pollution, P. If the individual has more wealth and less pollution the more satisfaction is received. Also note that although there are an infinite number of utility functions, only four are drawn on Figure 3.1.

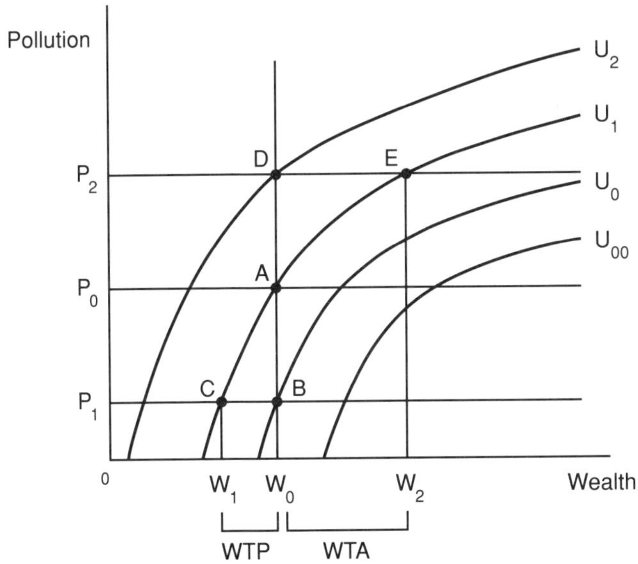

Figure 3.1. WTP-WTA measures of economic value

Consider point A: the initial condition of pollution level, P_0, wealth level, W_0, and utility level, U_1. If the level of pollution decreased to P_1 from P_0, then given the original wealth level, W_0, the individual's utility would increase to U_0, point B. Therefore, the maximum the individual would be willing to pay for this decrease in pollution is the amount of wealth needed to return to the original utility function, U_1, $WTP = W_0 - W_1$, point C. The individual would pay no more than this because to pay more would lead to a lower level of utility than possessed at the initial condition, U_1. The individual would surely pay less than this if possible, but this lower value is not the accurate economic value of decreased pollution. The accurate value is the maximum WTP.

If the level of pollution increased to P_2 from P_0, then given W_0, the individual's utility level decreases to U_2, point D. The minimum amount of wealth that the individual would be willing to accept in compensation for the increase in pollution equals $WTA = W_2 - W_0$, point E. This compensation restores the individual to his or her initial utility level given the increase in pollution. The individual would not accept less because this too would lead to a lower level of utility than possessed at the initial condition, U_1. Again the individual would gladly accept more than the minimum WTA, but this higher value does not reflect the accurate economic value to accept an increase in pollution. The accurate value is the minimum WTA. Finally, note that the slope of the function dictates the divergence between WTP and WTA. The more curved the slope, the more income is needed to compensate the individual for increased pollution, and consequently, the larger the WTP-WTA difference.

The goal of nonmarket valuation then is to estimate accurately the maximum WTP or the minimum WTA for a given change in ecosystem or biodiversity services. Four techniques are commonly used to value the consumption (as opposed to production) contributions of such services—the travel cost method, hedonic pricing, contingent valuation, and market experiments. Each is considered briefly (see Braden and Kolstad [1991] for detailed descriptions and evaluations of these and other techniques). First, the travel cost method is based on a model to predict the demand for a recreation area by visitors from surrounding areas. The use of a site is determined by the costs of access, including the direct travel costs, entry fees, and the opportunity cost of travel time and length of stay. The idea is that the higher the cost the less the use. These actual market costs are then used to estimate a demand function that can be employed to estimate indirectly an individual's WTP for improvements in the recreation site. The method is indirect in that the values of the ecosystem services are statistically teased out of the actual travel costs of the recreator. The economist's goal is to determine how a change in services affects the use of the area. While limited in scope, the travel cost method has been extensively applied over the years because it is relatively straightforward and it gives results.

The second, hedonic pricing, is another indirect approach that uses actual market prices to value an environmental asset. Hedonic pricing works as follows. Suppose one has a choice of two houses that are identical except for environmental quality. If environmental quality has economic value, the house with better quality should be more expensive than the other house. The difference in price can then be related to the economic value of the higher environmental quality. Given that the selling price of a house is determined by the market, the trick is to separate the fraction of the market price derived from environmental quality from that set of characteristics that define a house (e.g., square feet or number of bedrooms).

Third, the contingent valuation method (CVM) directly elicits value by constructing a hypothetical market for a nonmarket good through the use of a survey. The hypothetical market creates an opportunity for an individual to reveal his or her WTP or WTA for a change in the level of the good. The market is constructed so that features of actual markets and institutions are used to describe what the good is, how it will be changed, who will change it, how long the change will occur, and who will pay for the change. The major advantage of CVM is its flexibility to construct a market where no market currently exists. But flexibility is also the major weakness of CVM, as it allows ample opportunity for misperception. A researcher can specify a hypothetical good and elicit a value, but a respondent may perceive the good quite differently when providing the stated value.

Finally, experimental markets in controlled laboratory settings are a relatively new approach to value directly a nonmarket asset. Experimental markets are real markets selling real goods to real people, but within a highly stylized context. While using auctions whose efficiency properties are well understood, experimental markets can elicit values over several levels and types of market experience. Lab experiments can isolate and control how different auctions and market settings affect values in a setting of replication and repetition. The experimental approach can serve as both a complement and substitute for CVM surveys. Lab work complements CVM surveys by providing an opportunity for both extensive pretesting of the CVM design and evaluating survey respondents' values with repeated market experience; substitution occurs when the lab can auction a service that can actually be bought and sold but is currently not on the market.

Ecosystems, Biodiversity, and Economic Valuation

Nonmarket valuation has made significant advances over the past three decades, both in intensity and scope, as economists attempt to value an increasing number of environmental goods from around the world. Increasing sophistication, both in analytical structures and estimation procedures, has increased the optimism of economists about using nonmarket valuation as a viable tool to assist in decision making. But as nonmarket valuation advances so do new controversies and debates. Consider four topics—total value and warm glows, value formation and preference learning, dynamic inconsistency, and endogenous risk. These issues must be addressed if nonmarket valuation is to help guide the conservation strategy of the Great Plains.

Total Value and Warm Glows

The most difficult policy question confronting policymakers in the Great Plains is the management and valuation of regional ecosystem and biodiversity issues. This question involves open access and public goods that are not controlled or managed by any one county, state, or region. Therefore, given the dimensions of ecosystem and biodiversity services, the practice of nonmarket valuation faces a significant challenge in understanding how citizens perceive these services and how they value changes at the genetic, species, regional, and global scale. Although recent papers by Solow et al. (1993) and Weitzman (1993) are notable attempts by economists to provide a rigorous and workable definition of biodiversity that includes the costs of preservation, these papers are just as conspicuous for their lack of attention to a benefit dimension. A probable reason for this neglect is the difficulty in assigning economic value to goods that most people will never directly use. How can we attach an economic value to the mere existence of an environmental good that we may never even visit?

Following Krutilla (1967), economists have answered this question by proposing the concept of *total value*. Total value is the idea that consumers have both use and nonuse values for environmental resources. Use value is straightforward—the economic value of current use—but a nonuse value is more problematic and controversial. Option value is the economic value of potential future use of a resource, while existence value is the value of its mere existence, with no plans to ever use it. As academicians debated the theoretical justification, the U.S. District of Columbia Court of Appeals ruled in 1989 that nonuse value constitutes a valid representation of economic value. In *Ohio vs. U.S. Department of the Interior* (800f.2d 432) the court stated that "... option and existence values may represent 'passive use' but they nonetheless reflect utility derived by humans from a resource and thus *prima facie* ought to be included in a damage assessment." The view that nonuse values are bona fide is supported by a recently convened blue ribbon panel (including two Nobel Laureates) evaluating contingent valuation (Arrow et al. 1993).

But there is an opposing view—existence value is not really a measure of a value component of any particular environmental asset. Instead it is a surrogate measure of general preferences toward the environment—a "warm glow effect." Eliciting existence values with a contingent valuation survey provides the opportunity for respondents to state their general preference toward the environment rather than for the specific ecosystem or biodiversity service in question. This is probably the first, if not only, occasion they have been asked to reveal a public opinion on the environment. Therefore, the value revealed may reflect the warm glow of contributing to saving the general environment rather than the specific service in question. For example, Crocker and Shogren (1991a) find mixed evidence of surrogate bidding for atmospheric visibility in Oregon. They observed no significant difference in values for improved visibility in one specific mountain location as compared to the value for statewide improvements. In addition, Arrow et al. (1993) note that the bimodal distribution of value estimates in many CVM studies—zero or a positive value around \$30 to \$50—suggests that these values may serve a function similar to charitable contributions. Not only do the respondents want to support a worthy cause, but they also receive a warm glow from donating to the cause.

The recent debate between Kahneman and Knetsch (1992) and Smith (1992) further illustrates the debate. Kahneman and Knetsch observed that on average the WTP to clean up one lake in Ontario was not significantly greater than the WTP to clean up all the lakes in the province. They cite this as evidence that individuals are not responding to the good, but rather to the idea of contributing to environmental preservation in general, the warm glow. Smith questioned this view, arguing that incremental WTP should decline with the amount of the good already available, and as such the evidence is consistent with economic theory. But other reports such as Desvousges et al. (1992) support the warm glow argument, finding evidence that the average WTP to prevent 2,000 birds from dying in oil-filled ponds was not significantly different from the value to prevent 20,000 or 200,000 birds from dying. While accepting the argument that WTP for additional protection probably does decline, Arrow et al. (1993, p. 11) note that the drop to zero . . . "is hard to explain as the expression of a consistent, rational set of choices."

Separating total value from warm glows presents a challenge to the comprehensive monetary evaluation of the resources in the Great Plains. Total values are more accurately estimated for well-defined areas and well-specified resources. But a piecemeal resource-by-resource approach will overestimate economic value because it does not address substitution possibilities across the set of resources. For example, if ten resources across the Great Plains are valued, then the summed values of ten unique studies over each resource will exceed the value of one study over the ten resources. Hoehn and Loomis (1993) find that independent aggregation of the benefits of only two programs overstates their total benefits by 27 percent; the overstatement with three programs is 54 percent. But as we move toward one comprehensive valuation study to account for substitutions and complementarities, we increase the likelihood of warm glows as the resources become less tangible and more symbolic. The sheer size of the Great Plains and its numerous ecosystem and biodiversity services requires a valuation strategy that includes well-defined substitution possibilities and checks of internal consistency.

Unfamiliarity and Learning

Even beyond warm glows for Great Plains ecosystem services, policymakers must still appreciate that many individuals are simply unfamiliar with most of the services and functions that ecosystems and biodiversity provide. As an example, a survey of Scottish citizens revealed that over 70 percent of the respondents were completely unfamiliar with the meaning of biodiversity (Hanley and Spash 1993). Such levels of unfamiliarity are of concern if consumer sovereignty is to command respect in resource policy questions.

The question of unfamiliarity is central to understanding the values estimated with nonmarket valuation. Standard guidelines suggest that nonmarket valuation is more reliable if the respondent is familiar with the good. The person who is familiar with the good will be better able to value changes in its provision. For example, most U.S. respondents are familiar with the bald eagle and may be able to provide dollar values for increased levels of the species. But for many other environmental assets, such as wetland filtration, most respondents may be unfamiliar with the actual services provided. What does value mean when a person is unfamiliar with the asset being valued? There are two possible answers to this question—value formation and preference learning.

Hoehn and Randall (1987) define *value formation* as the process by which an individual assigns a dollar value to a good, given that he or she completely understands the relative ranking of the good relative to other goods. That is, a person knows that she or he prefers improved environmental quality to a new toaster but has not attached a monetary value to that preference. The formation of this monetary value is affected by the time and resource constraints inherent in any decision to allocate wealth. Time and resource constraints inhibit the individual's ability to comprehend the complex services provided by the good, thereby making the service appear unfamiliar. Hoehn and Randall examine how values are formed by comparing the values obtained under the ideal consumer problem to the values obtained given time and resource constraints. They argue that imperfect communication can cause an individual to undervalue the service relative to the same measure of value formed under ideal circumstances. This undervaluation problem can be alleviated if more time and decision resources are devoted to the value formation process.

Figure 3.2 illustrates the value formation issue. The revealed value is presented on the vertical axis, while time taken to learn is represented on the horizontal axis. Note that as time taken to learn about the good is constrained, the individual's valuation is low relative to "true" value. The individual did not have enough time to accurately translate preferences into a monetary value. But as the time to form a value is increased, the individual has more opportunity to translate preferences into a monetary manifestation, and therefore the revealed value will approach the true value from below.

However, the value formation argument presumes the individual understands her or his initial preferences for resources. In short, the argument assumes the individual's sense of well-being is static and invariant and that no doubt affects the preference for any outcome. This follows the standard view that preferences for goods are fixed and that changes in demand for an ecosystem service must occur because of changes in shadow prices, household technology, or resource constraints. Demand changes cannot arise because the individual does not understand how an unfamiliar service affects overall satisfaction or because of changes in preferences.

There are numerous goods and services involving environmental resources with which respondents are unfamiliar. Frequently the respondent has little day-to-day contact with the service, or if so, may view the efforts to alter these experiences as futile. The individual therefore has devoted little effort to understanding how these services affect his or her well-being. If this is the case, the individual may well need to form conjectures and accumulate experience with the resource in order to more accurately assess relative preference for the resource. Crocker and Shogren (1991b) demonstrate that if an individual does not know his or her preferences for a resource, the revealed value will be greater than when preferences are known. The individual is willing to pay extra to acquire information about the potential value the resource may provide in the future. In contrast to value formation, this result suggests that a person will overvalue an unfamiliar asset. In addition, overvaluation decreases as the individual becomes more familiar with the service. Figure 3.2 shows that as time increases, the revealed value approaches the true value from above.

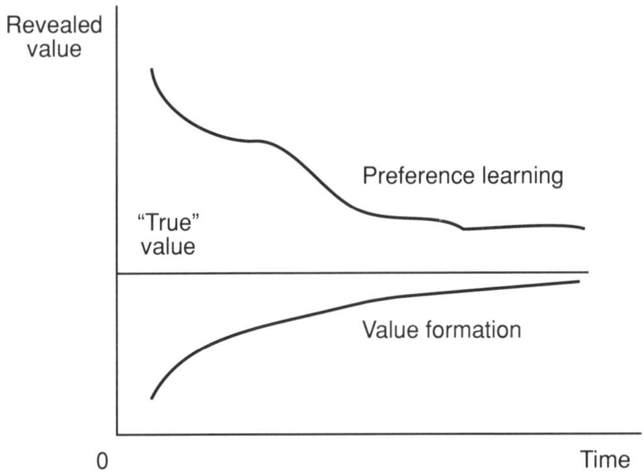

Figure 3.2. Preference learning and value formation

The two effects work in opposite directions—value formation implies undervaluation, while preference learning implies overvaluation. This suggests that nonmarket valuation efforts in the Great Plains should define explicit criteria that accurately specify the degree of value formation and preference learning. If such criteria can be defined, then an individual's values with situation, time, and resource decision constraints, and preference knowledge can be compared when the variables change simultaneously or sequentially. Otherwise, estimates of value for Great Plains resources will be inaccurate.

Dynamic Inconsistency and Sustainable Development

Sustainable development is the currently popular term symbolizing the belief that intergenerational equity must play an explicit and significant role in the management of environmental resources. Promoting sustainable development in the Great Plains requires changing how the region measures its economic progress including changes in the regional income accounts to include the stocks of ecosystem and biodiversity services and pricing the assets not currently sold in the marketplace. The concept will require regional managers to reflect on what exactly they want to define as "sustainable." Indicators of economic and environmental welfare will need to be established, as will the key issues likely to dominate the sustainable development debate—such as past trends, status quo, and likely development programs.

Demands for sustainable development of the region's resources should account for the fact that the resource allocation decisions each of us makes today generate costs and benefits that can accrue far off into the future. Unless society is shortsighted, current environmental policy must account for the temporal dimensions of these individual allocation decisions. Although scientists and policymakers have questioned the use of individuals' preferences to construct social discount rates, the policymakers nevertheless acknowledge the importance of preferences. By not understanding how individuals actually discount the future consequences of their choices, policymakers' understanding of behavior will be misspecified and their predictions about it will be inaccurate. Consequently, Great Plains policy that promotes sustainable development but ignores individuals' preferences toward the present guarantees unintended results.

Economists have stressed this point—unless we incorporate individuals' preferences into traditional benefit-cost analysis, policy predictions based on an independently derived social discount rate will be biased. One must account for the individuals' opportunities to reallocate consumption and investment decisions over time so as to smooth their satisfaction over time. Psychologists have pushed this a step further, arguing that not only might different individuals have different preferences, but that the same individual's preferences might differ depending on the situation. In contrast to the standard economic assumption of individually invariant time preferences, this dynamic inconsistency has been documented in several studies (see Lowenstein and Thaler 1989). This work suggests that an individual's preference for the present is situation-specific and influenced by opportunities to reallocate resources over time.

While employing the contingent valuation method, Crocker and Shogren (1993) explore how an individual's preference for the present is affected by the access conditions to an ecosystem. They ask whether an individual's time preference is inconsistent in the sense that she or he fails to discount the future in the same manner given alternative access conditions. The null hypothesis is that WTP to avoid delayed access to a desirable resource will exceed WTP to extend access, unless time and income are substitutes. Applying standard discounting formulae allowing calculation of an individual's preference toward the present, a dynamic inconsistency that is both individual and situation-specific was found.

These results suggest that although the change in value is positive for both the acquisition of extended access and the prevention of delayed access, the change in the incremental value for delayed access will exceed the change in the incremental value for extended access, provided that time and income are complements. These results point out an internal inconsistency in two common assumptions on preferences—an individual has an invariant rate of time preference and equally values a gain or loss from the status quo endowment of resources. Results show that if an individual has an invariant rate of time preference, his or her value of a gain versus a loss must be different. Alternatively, if the individual values a gain or a loss equally, his or her implied rate of time preference will differ. There is a dynamic inconsistency in time preference and valuation.

To illustrate, suppose an individual is willing to pay $40 to secure a status-quo level of access to an ecosystem service. She or he is then asked to change the bid given a change in the temporal conditions of access—either a ten-year extension or a ten-year delay. If the change in the WTP is zero for extended access and zero for delayed access, the implied rate of time preferences is infinite for the extension but zero for the delay. The change in value from the status quo is identical, but the implied discount rate is infinitely different. But if the change in value remains at zero for the extension but is $40 for the delay, then the rate in both cases is infinite. The rate is invariant, but the change in value is significantly different. While extremes are used to illustrate, the argument works equally well for intermediate examples. How an individual values a gain versus a loss can yield quite different implied rates of time preference and consequently influence the revealed value of conservation in the Great Plains.

Valuing Risk

Nonpoint source pollution is one of the biggest concerns in the Great Plains. Sources of nonpoint pollution include agricultural use of chemicals and sedimentation from urban runoff. Nonpoint pollution imposes a range of potential risks to individuals. Given the multiple pollutants arising from multiple pathways, a coherent Great Plains conservation plan requires accounting for the costs and benefits of reduced risks to human and ecosystem health.

Economists generally value the reduction of risk through some form of collective action or government intervention. But intuition and evidence both affirm that individuals take private action to influence the likelihood or the severity of an undesirable event. Resources are invested to change the probability that good things happen and that bad things do not. Individuals invest in water filters or bottled water to reduce the odds of becoming ill from tainted water supplies. These choices alter risk, and how one chooses to reduce risks depends on both one's preferences for risk and the technology to reduce risk. We define this ability to alter the likelihood and severity of a harmful event as *endogenous risk.*

Policies to reduce risk in the Great Plains must systematically incorporate the implications of endogenous risk into the decision-making framework. On the cost side, multiple and interdependent pollutants and pathways in designs for environmental policy can be incorporated. Advances in technology allow the use of computers as a creative research tool to integrate physical and economic processes into a unified system. An environmental economic modeling system that integrates physical and economic interactions can simulate the economic costs and benefits of alternative policy scenarios and patterns of activity. An example developed for the U.S. Environmental Protection Agency to evaluate the risk-benefit trade-offs of alternative pesticide policies is the integrated system CEEPES, which incorporates both the economic and environmental indicators of social costs and their interactions (see Bouzaher et al. 1993).

Estimating the benefit side is more difficult. A refocus of how researchers measure the value of reduced risk is needed to better account for endogenous risk in ecosystem management. Specifically, the assumption of exogenous risk in benefit-cost analysis can undervalue reduced risk and fail to identify those who value risk reductions most highly. There are several reasons for undervaluation, all involving the inability of an exogenous risk perspective to disentangle the relative values of private and collective contributions to risk reductions. When risk is considered exogenous to the individual, protection must be supplied collectively. But if self-protection is a viable substitute for collectively supplied protection, it expands an individual's opportunities to exploit personal gains from collective provision (see Shogren and Crocker 1991).

Resolving the undervaluation problem in the Great Plains must be explicitly addressed. Risk valuation should assess an individual's preference for alternative risk-reduction strategies. By allowing the individual to reveal whether he or she would prefer to reduce risk privately or collectively or both, or by reducing the probability or severity or both, the true value of risk reduction is better measured. Nonmarket valuation techniques such as the contingent valuation method or laboratory experiments can be used further to elicit preferences for alternative risk-reduction strategies. But without first understanding how people prefer to reduce risk, policy recommendations will be based on potentially precise but inaccurate and incomplete information. Again this produces a policy environment where a policymaker's prediction of the consequences of policies may well differ from actual consequences.

Conclusions

Consumer sovereignty is the foundation of the economic approach to value the ecosystem and biodiversity services provided by the Great Plains. If a person is best able to make his or her own choices regarding how to allocate resources, the WTP or WTA compensation reflects the value of a resource. But if the individual does not have the information revealing the actual services provided, the revealed values do not reflect the true value of the resource. This dilemma can be resolved by replacing consumer sovereignty with some form of institutional sovereignty that presumably allows better informed policymakers to make resource allocation decisions independent of individual valuations.

Second, consumer sovereignty may be preserved by acknowledging that nonmarket valuation has and will continue to advance if issues described in this chapter are addressed. Such issues must consider general preferences toward the environment, the impact of unfamiliarity and learning on values, preferences toward the present situation that differ for the same individual with differing terms of access, and risk management policies that undervalue risk reduction unless both private and collective means to adapt are included.

A four-step valuation process is set up to address these issues systematically for the Great Plains. Details will need to be defined for the relevant context. First, nonmarket valuation must begin by eliciting the knowledge of the services provided by the set of resources to be valued. Quantifying what a person knows about the benefits of an ecosystem and alternative definitions of biodiversity is essential. This will provide some measure of the precision and accuracy of any value estimates, especially to spot any warm glow values. Second, nonmarket valuation must provide opportunities for persons to reveal their preferences for the resource relative to other goods, both over spatial and temporal dimensions. Explicit criteria to identify preference learning, value formation, and temporal inconsistencies are needed. Third, hypothetical valuations over a range of levels and types of services can capture preferences given available substitution opportunities. The goal is to reflect the broad range of services provided in the Great Plains, and the substitution possibilities both within the Plains and between other regions, such as the Lake State region. Finally, a set of laboratory experimental auctions should be developed that allow a person to value a nonhypothetical set of services. Real money with real decisions over repeated market experience will improve the precision of nonmarket valuation of the Great Plains.

References

Arrow, K., R. Solow, P. Portney, E. Leamer, R. Radner, and H. Schuman. 1993. *Report of the NOAA Panel on Contingent Valuation.* Washington, D.C.: Resources for the Future.

Braden, J., and C. Kolstad (ed.). 1991. *Measuring the demand for environmental quality.* Amsterdam: North-Holland.

Brookshire, D., and D. Coursey. 1987. Measuring the value of a public good: An empirical comparison of elicitation procedures. *American Economic Review* 77:554-566.

Bouzaher, A., R. Cabe, A. Carriquiry, P. Gassman, P. Lakshminarayan, and J. F. Shogren. 1993. Metamodels and nonpoint pollution policy in agriculture. *Water Resources Research* 29:1579-1587.

Crocker, T., and J. Shogren. 1991a. Ex-ante valuation of atmospheric visibility. *Applied Economics* 23:143-151.

———. 1991b. Preference learning and contingent valuation methods. In *Environmental policy and the economy,* ed. F. Dietz, R. van de Ploeg, and J. van der Straaten. Amsterdam: North-Holland.

———. 1993. Dynamic inconsistency in valuing environmental goods. *Ecological Economics* 7:239-254.

Crocker, T., and J. Tschirhart. 1992. Ecosystems, externalities and economies. *Environmental and Resource Economics* 2:551-567.

Cummings, R., D. Brookshire, and W. Schulze. 1986. *Valuing environmental goods: An assessment of the contingent valuation method.* Totowa, New Jersey: Rowman and Allanheld.

Desvousges, W., F. R. Johnson, R. Dunford, K. Boyle, S. Hudson, and K. Wilson. 1992. *Measuring natural resource damages with contingent valuation: Tests of validity and reliability.* Research Triangle Institute, North Carolina.

Hanemann, M. 1991. Willingness to pay and willingness to accept: How much can they differ? *American Economic Review* 81:635-647.

Hanley, N., and C. Spash. 1993. *The value of biodiversity in British forests.* Report to the Scottish Forestry Commission, University of Sterling, Scotland.

Hoehn, J., and J. Loomis. 1993. Substitution effects in the valuation of multiple environmental programs. *Journal of Environmental Economics and Management* 25:56-75.

Hoehn, J., and A. Randall. 1987. A satisfactory benefit cost indicator from contingent valuation. *Journal of Environmental Economics and Management* 14:226-247.

Kahneman, D., and J. Knetsch. 1992. Valuing public goods: The purchase of moral satisfaction. *Journal of Environmental Economics and Management* 22:57-70.

Krutilla, J. 1967. Conservation reconsidered. *American Economic Review* 57:787-796.

Lowenstein, G., and R. Thaler. 1989. Intertemporal choices. *Journal of Economic Perspectives* 3:181-194.

Mäler, K-G. 1974. *Environmental economics: A theoretical inquiry.* Baltimore: Johns Hopkins University Press.

Shogren, J., and T. Crocker. 1991. Risk, self-protection, and ex-ante economic value. *Journal of Environmental Economics and Management* 20:1-15.

Shogren, J., S. Shin, D. Hayes, and J. Kliebenstein. 1994. Resolving differences in willingness to pay and willingness to accept. *American Economic Review* 84:255-270.

Smith, V. K. 1992. Arbitrary values, good causes, and premature verdicts. *Journal of Environmental Economics and Management* 22:71-89.

Solow, A., S. Polansky, and J. Broadus. 1993. On the measurement of biological diversity. *Journal of Environmental Economics and Management* 24:60-68.

Weitzman, M. 1993. What to preserve? An application of diversity theory to crane conservation. *Quarterly Journal of Economics* 108:157-184.

Community and Economic Resources

4. Population Dynamics and Their Implications for the Ecosystems of the Great Plains

Richard Rathge
North Dakota State University, Fargo
North Dakota Census Data Center

An important starting point in the examination of the Great Plains ecosystem is its population base. Residents play a dual role in the ecology of the region. First, they constitute a key resource base that contributes to the overall quality and conservation of the system. This resource is often referred to as *human capital.* Ironically, the second role they play within the ecosystem is that of consumers. People use and abuse the resources within the system. To a large degree, therefore, an understanding of the population size, distribution, and composition of residents in the region provides a useful context from which to assess and debate strategies aimed at appropriately managing the Great Plains ecosystem.

The purpose of this chapter is twofold. First, the changing nature of people in the region is addressed. One focus of the symposium was to identify at-risk components within the ecosystem that should receive priority and protection. Trends in residential movement in the Great Plains indicate that many communities in the region are, indeed, at risk. In addition, the future of many rural residents is dubious at best. The major thrust of this work, therefore, will be to provide a critical review of population settlement patterns within the region. This broad-level approach offers important insight into the human resources of the ecosystem. Although individuals as consumers play a significant role in utilizing resources, the focus of this chapter is solely on people as resources.

Second, this chapter outlines some management approaches aimed at revitalizing human capital. A theme emerging from the research regarding population movement in the Great Plains is the interplay between people and technology. Technology has been the major driving force responsible for the settlement trends in the region. The disparity between advancements in technology and those of human capital, however, has created tremendous challenges for policymakers and planners. Technological advances continue to change optimal patterns of settlement rapidly. However, the social institutions and cultural norms that bind residents together into a community react very differently to change. Continued clashes between the two have resulted in a dynamic population in the Great Plains.

Stanley R. Johnson and Aziz Bouzaher (ed.), *Conservation of Great Plains Ecosystems,* 49–62.

Population Shifts in the Great Plains

The movement of people in the Great Plains has been patterned. The general trend has been one dominated by urbanization. It is helpful, therefore, to review residential shifts within this context. This can be accomplished by subdividing the territory of the Great Plains into zones according to degree of metropolitan accessibility. Since the Bureau of the Census defines metropolitan from a county perspective, it is useful to subdivide the region into counties and categorize them by proximity to metropolitan centers. These become county types, or Beale codes.

Historical Overview

A review of the historical movement of people in the Great Plains is shown in Table 4.1. The period of peak population for each zone was determined using census data for 1990 and each previous decade. What is most striking about the data is that populations peaked before 1950 in nearly two-thirds of the counties in the region. Only 100 of the 870 counties (12 percent) continued to grow through 1990.

Closer inspection of the data in Table 4.1 confirms the linkage of population movement with metropolitan dominance. Nearly two-thirds of the metropolitan counties continued growing through 1990. In contrast, only 6.3 percent of the adjacent nonmetropolitan counties and 3.9 percent of the remaining nonmetropolitan counties had peak population in 1990. In fact, more than 70 percent of all nonmetropolitan counties in the Great Plains had fewer people in 1990 than they had before World War II.

The magnitude of population shifts in the region, which occurred during the past four decades, is shown in Table 4.2. Between 1950 and 1991, the total population in the Great Plains grew by slightly less than 10.2 million people, or 58 percent. However, almost all the growth that took place was concentrated in the metropolitan areas. For example, a review of Table 4.2 shows that metropolitan counties in the region expanded by 120 percent during this time period. In contrast, nonmetropolitan counties, in total, grew by less than 8 percent over the four decades. The relatively little difference in total growth rates that existed between adjacent and nonadjacent counties in nonmetropolitan areas during this period is due in part to a population revival that occurred during the 1970s. Demographers frequently referred to this period as the *population turnaround.* This was a unique period in the history of the United States when nonmetropolitan counties grew at a faster pace than metropolitan counties (see Sofranko and Williams 1980). However, this quirk in the historical pattern of settlement in the Great Plains was short-lived (Beale 1988).

Table 4.1. Peak population periods in counties of the Great Plains

Peak Population Periods	All Counties		Metro Counties		Nonmetro Counties					
					Total		Adjacent		Not Adjacent	
	Number	Percent	Number	Percent	Number	Percent	Number	Percent	Number	Percent
Before 1950	568	65.3	11	11.2	557	72.2	176	73.6	381	71.5
1950–60	152	17.5	10	10.2	142	18.4	32	13.4	110	20.6
1970–80	50	5.7	13	13.3	37	4.8	16	6.7	21	3.9
1990	100	11.5	64	65.3	36	4.7	15	6.3	21	3.9
Total	870	100.0	98	100.0	772	100.0	239	100.0	533	100.0

Table 4.2. Population change in the Great Plains by county types, 1950–91

County Type	Years 1950	Years 1991	Percent Change
Metro	7,837,775	17,254,141	120.1
Total	17,531,207	27,710,570	58.1
Nonmetro			
Adjacent	3,878,334	4,078,424	5.1
Not Adjacent	5,815,098	6,380,005	9.7
Total	9,693,432	10,456,429	7.9

Current and Future Trends

Estimates of the population since 1990 indicate that the historical patterns of residential movement described here are likely to continue. County population estimates for 1991 were combined with county-specific vital statistics regarding births and deaths to determine the rate of natural increase and net migration for each county.

Natural increase is a measure of population change that occurs in an area as a result of births and deaths. If an area were isolated from the rest of the world, changes in its population would be solely due to the difference in the number of people born in the area to those who die. Thus, natural increase offers insight into the natural population change that takes place in an area. Since few places are totally isolated from the rest of the world, a second component is needed to understand an area's total population: migration.

Net migration is a measure of the net difference in people leaving versus people coming into an area. It is calculated by first subtracting the population at a beginning time period from a later time period. The resulting number is the total population change over the two time periods. Next, natural increase is subtracted from total population change. The residual is net migration.

The components of population change for the Great Plains are reported in Table 4.3. As shown in Figure 4.1, nearly 60 percent of the counties in the region lost people through out-migration between 1990 and 1991. The greatest losses were among the nonmetropolitan counties not adjacent to metropolitan centers. Nearly two-thirds (66.4 percent) of these counties posted net losses (Table 4.3). A similarly high proportion of nonmetropolitan counties adjacent to metropolitan centers posted net migration losses (57.8 percent). In contrast, only 26.5 percent of the metropolitan counties had net out-migration between 1990 and 1991.

Table 4.3. Components of population change in the Great Plains by county type, 1990–91

Counties	Total	Natural Change				Migration			
		Increase		Decrease		Out		In	
		Number	Percent	Number	Percent	Number	Percent	Number	Percent
Total	870	635	73.0	235	27.0	515	59.2	355	40.8
Metro	98	97	99.0	1	1.0	26	26.5	72	73.5
Nonmetro	772	538	69.7	234	30.3	489	63.3	283	36.7
Adjacent	275	213	77.5	62	22.5	159	57.8	116	42.2
Not Adjacent	497	325	65.4	172	34.6	330	66.4	167	33.6

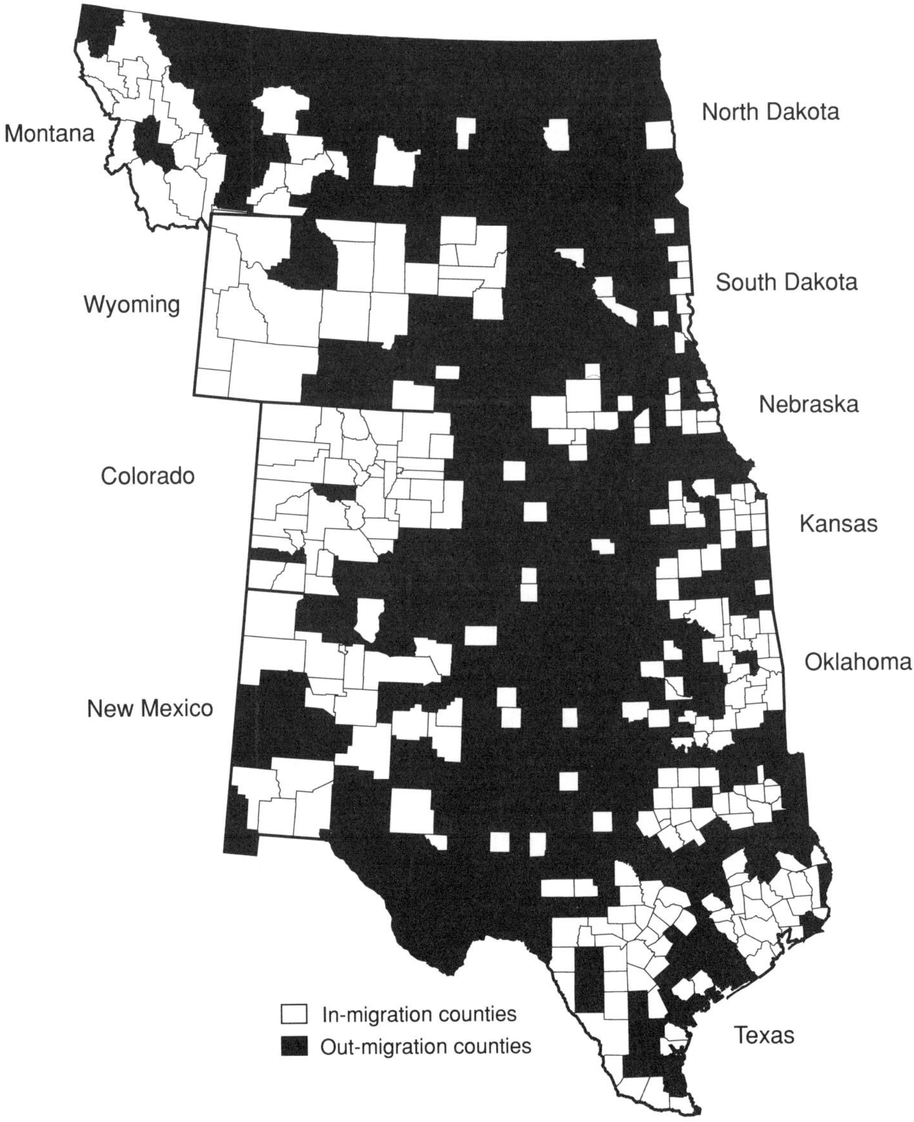

Figure 4.1. Net migration status of counties in the Great Plains, 1990–91

An important fact to consider when examining these data is that migration is typically selective. The majority of people who move from rural counties tend to be in their 20s and 30s. For example, in more than one-half of the nonmetropolitan counties in North Dakota between 1980 and 1990, the rate of out-migration of young adults between the ages of 20 and 29 exceeded 50 percent (Olson 1991). This selective migration distorts the age structure of a county by decreasing the number of young adults and enlarging the proportion of elderly. A deficit of young adults leads to the occurrence of natural decrease, when more people die in a county than are born. Unless the county can attract more people than it loses through migration, it is destined to decline.

Natural decrease is largely a nonmetropolitan phenomenon. Research by Johnson and Beale (1992) indicates that more than 95 percent of the counties experiencing natural decrease between 1950 and 1987 were nonmetropolitan. This is consistent with the data for the Great Plains. As shown in Figure 4.2, nearly one in three nonmetropolitan counties in the region revealed a natural decrease between 1990 and 1991. A review of Table 4.3 shows that among nonmetropolitan counties, those not adjacent to metropolitan centers were much more likely to have natural decrease. Nearly 35 percent of the nonadjacent counties had more deaths than births between 1990 and 1991 compared to 22.5 percent of the adjacent nonmetropolitan counties. In contrast, only one metropolitan county had fewer births than deaths during this period.

Prolonged out-migration, especially of the young, in many of the Great Plains counties is symptomatic of inadequate economic development programs. Unfortunately, as county populations fall below critical thresholds, they lose their ability to sustain needed local services such as schools, hospitals, and various service functions. In addition, business volume is reduced to a level at which retail and service competition can no longer be maintained. As a result, the local tax base is eroded, which places increased financial strain on residents. This, in turn, leads more people to leave and the downward spiral continues.

Economic Base

Although significant industrial and occupational restructuring of the rural economy has occurred throughout the United States, the Great Plains economy is still dominated by agriculture. The majority of nonmetropolitan counties in the region (Figure 4.3) are classified as farm-dependent, according to the latest classification scheme by the Economic Research Service at the U.S. Department of Agriculture (Hady and Ross 1990). Farm-dependent counties are those in which at least 20 percent of the total labor and proprietor income is derived from farming.

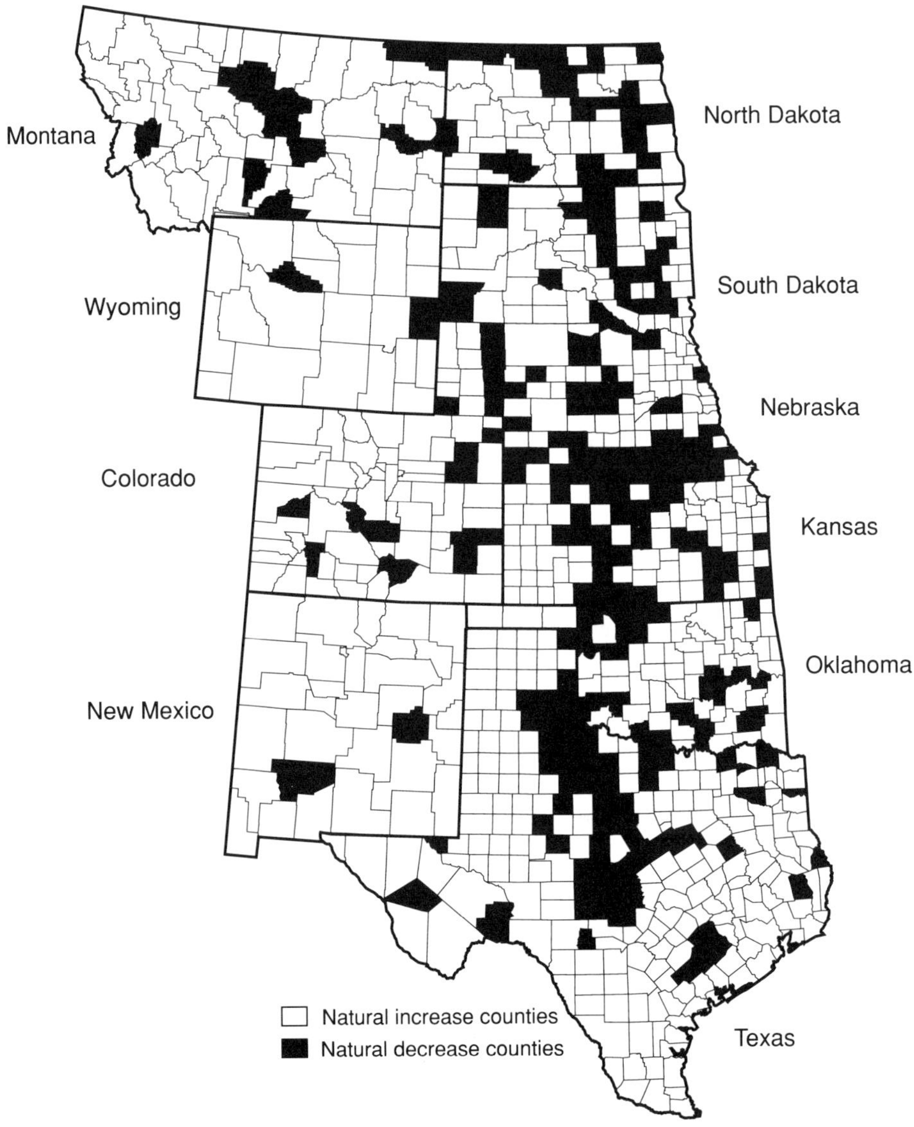

Figure 4.2. Natural decrease status of counties in the Great Plains, 1990–91

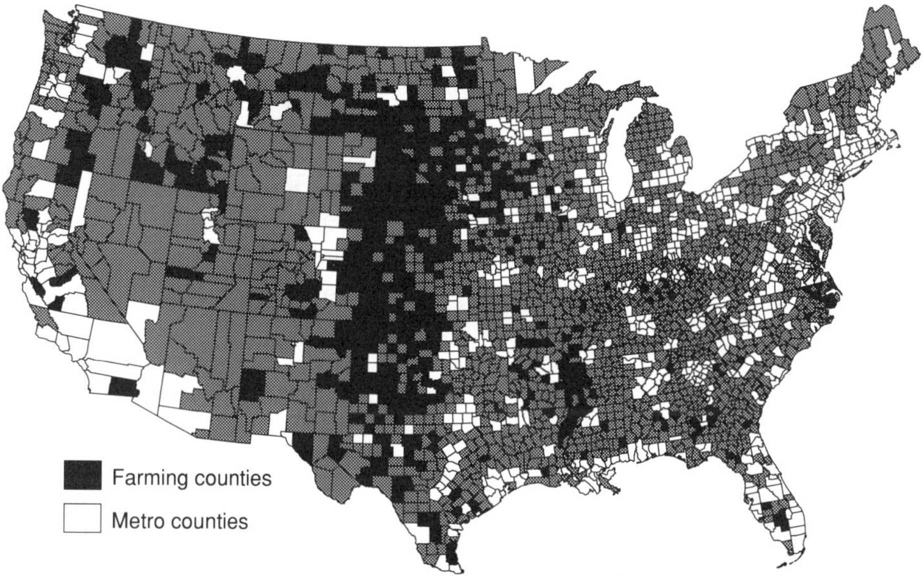

Figure 4.3. Nonmetropolitan farm-dependent counties, 1986

Factors Contributing to Population Loss

The loss of rural residents, especially in farm-dependent counties, is due to a complex set of circumstances. In one body of research, observers suggest that much of the problem can be attributed to shifts in the structure of agriculture (Leistritz and Ekstrom 1986). The basic argument from this theoretical approach is that technological advances in agriculture replace labor, thereby increasing farm size, which reduces farm numbers and farm population. This downsizing has a spillover effect in neighboring farm communities in terms of fewer demands for services. Less demand, in turn, reduces related employment opportunities in these communities and causes a spiraling population loss.

Evidence to support this argument is overwhelming. For example, agricultural advancements in equipment, chemicals, and seed varieties have greatly expanded the production capabilities of farmers. The index of agricultural output per hour of farm work rose about 1,300 percent between 1940 and 1989 (Beale 1993). Although harvested cropland has remained relatively stable over the past 45 years, productivity has more than doubled per acre (Albrecht and Murdock 1990). The use of irrigation, for example, increased yields two to six times over nonirrigated land (Casey et al. 1975). In short, economies of scale have encouraged farmers to continually expand their holdings and reduce labor.

In order to better appreciate the magnitude of farm population losses from agricultural restructuring, it is important to offer some comparative context. For example, in 1940 the farm population in the United States topped 30 million, or one-fourth of the nation's population. Currently, there are fewer than 6 million farm residents, representing only 2 percent of the nation's population (U.S. Bureau of the Census 1992). A similar free-fall decline occurred in neighboring rural towns. Population losses in rural communities between 1940 and 1970 exceeded 50 percent, caused by a lack of employment opportunities (Beale 1978, 1980; Larson 1981). Even greater losses occurred in the 1980s. Among the farm-dependent counties in the Great Plains, small towns declined in population by more than 80 percent between 1980 and 1990 (Beale 1993).

The dramatic changes in transportation and communications systems are a second factor contributing to rural population loss (Brown 1979). Brown contends that, prior to 1920, rural communities existed as relatively isolated and autonomous trade centers. Without effective transportation or communications systems, residents depended on local communities for goods, services, and their general livelihood. However, advances in transportation and communications shifted mobility, and residents no longer were restricted to local communities for either services or livelihood.

This perspective is supported by the settlement patterns of the Great Plains. The epoch of the railroad, for example, established many of the region's largest cities. These centers sprang up at strategic locations where important resources including water, coal, oil, and gas were located (Borchert 1981). Zones of influence around these cities grew in response to the expansion of minor rail corridors. As transportation systems were dominated by the automobile, residential movement shifted to reflect a new optimal pattern. Towns and service centers at strategic locations along major highways grew and prospered. With advancements in fuel economies, people increased traveling distances, thereby shifting the demand for services. Towns that once served as important service centers lost their viability. This is most apparent in the dramatic decline in retail and service establishments in rural areas. For example, there has been a net decline in retail and service establishments of more than 70 percent in rural places each decade since 1950 (Johansen 1993).

The interplay among people, environment, and technology with regard to population change is complex and not well understood. However, a common thread that ties many of the findings together is that residential change largely reflects the ability of people to adapt to their environment through technology. Human ecologists often refer to this as the *ecological carrying capacity* of an area (Duncan 1959).

Given this understanding of causes behind population shifts, one can begin to address adaptive strategies.

Recommended Management Approaches

The future of the Great Plains, like its past, rests in its ability to innovatively adapt to technology. A review of the current and historical settlement patterns of the region reveals the difficult challenges many rural counties face. In particular, farm-dependent counties in the Great Plains are likely to continue their historical decline as technological advances continue to reshape U.S. agriculture. Likewise, rural communities distant from metropolitan centers also will decline as their businesses lose market share and their employment opportunities shrink. In contrast, the larger urban service centers and metropolitan areas will continue to attract those leaving the farms and hinterland.

A major challenge that faces policymakers and planners is how to address the problem. One concern is how to prioritize programs or initiatives. Stabler (1992) contends that a common failure among some approaches is their assumption of community viability. He believes that an unfortunate reality that many policymakers wish to avoid is the fact that not all rural communities are capable of expansion or stabilization. As a result, programs do not employ selection processes that attempt to concentrate energies in areas most likely to succeed. This ultimately dilutes scarce resources and energies for revitalization.

A more pragmatic issue is one of community identity. As noted earlier, most communities in the Great Plains began as self-sufficient and autonomous entities. Limited communication and transportation systems made residents dependent upon communities for needed services. Concurrently, communities became focal points for social interaction (Hobbs 1992). This created a situation where residents became ingrained with a sense of belonging often referred to as *community identity.*

Community identity continues to influence the way residents think about their communities. Feelings of identity, however, can bar innovations. For example, one strategic action plan for rural revitalization is multicommunity collaboration. The underlying assumption to this approach is that independent, community-specific economic development efforts are increasingly inappropriate in an era of diminishing human and financial resources (Korsching 1992). Therefore, the solution to rural decline is cooperation through technology.

Some effective ways to accomplish this task have been debated at a recent conference on multicommunity collaboration (Korsching et al. 1992). Two of the more interesting areas of debate are highlighted here.

First, leaders and policymakers need to more aggressively explore policies that advance the development of information technologies for rural areas. Many of the obstacles that rural areas face can be overcome by technology. Powers (1992) offers several insightful solutions. For example, he suggests that information-based jobs may be one alternative for economic development in remote nonmetropolitan areas. Success in Nebraska, South Dakota, and North Dakota supports this alternative. Another suggestion is that interactive video or other forms of immediate response broadcast may serve rural students more effectively than long distance busing to consolidated schools. Similarly, data and resource sharing through computer technologies may allow small rural firms and businesses to be more competitive by keeping abreast of new advances and expanding to reach distant markets.

Second, leaders and policymakers need to more openly embrace the concept of interdependency and multicommunity collaboration. Cooperative ventures need to be explored along many dimensions and among various levels of government or organization. Powers (1992) also offers several suggestions along these lines. For example, he suggests reexamining the limits of traditional community boundaries that separate voluntary associations such as service clubs, churches, and fraternal organizations. Further, the advantages of community clusters to private business need to be examined in terms of the benefits as service centers. Similarly, the advantages of multicommunity participation in public service delivery or infrastructure need to be explored. Education, public safety, health care, emergency services, and basic infrastructure services may be more effectively delivered and financed by a cooperative of communities rather than by the traditional single community model.

The benefits of a collaborative approach to rural revitalization are many and varied. Four of the more salient ones are outlined by Schaffer (1992): (1) economic efficiencies arising from economies of size, (2) gaining access to more resources, (3) capturing the spillovers from collective actions, and (4) synergism. Cooperation among communities reinforces commonalities among localities and offers a new and unified approach to a shared problem.

Unfortunately, an equally impressive case can be made for the potential costs of a collaborative action agenda. Perhaps the greatest hurdle is cooperation itself. Issues of local pride and jealousy can easily undermine the best cooperative strategies, as illustrated by Salamon and Davis-Brown (1990) in their discussion of attempts by two communities in Illinois. Other costs identified by Schaffer (1992) include maintaining the collaborative efforts, the undermining of local organizations and voluntary efforts, and problems arising because of issues of political jurisdiction.

References

Albrecht, Don E., and Steve H. Murdock. 1990. *The sociology of U.S. agriculture.* Ames: Iowa State University Press.

Beale, Calvin L. 1993. Salient features of the demography of American agriculture. In *The demography of rural life,* ed. D. L. Brown, D. R. Field, and J. J. Zuiches. University Park, Pennsylvania: Northeast Regional Center for Rural Development.

———. 1988. Americans heading for the cities, once again. *Rural Development Perspectives* 4(3).

———. 1980. The changing nature of rural employment. In *New directions in urban-rural migration,* ed. D. L. Brown and J. W. Wardwell. New York: Academic Press.

———. 1978. People on the land. In *Rural U.S.A.: Persistence and change,* ed. Thomas R. Ford. Ames: Iowa State University Press.

Borchert, John. 1981. Geographical shifts in midwestern population in the twentieth century. In *Population redistribution in the Midwest,* ed. Curtis Roseman, Andrew Sofranko, and James Williams. Ames, Iowa: North Central Center for Rural Development.

Brown, David L. 1979. Farm structure and the rural community. In *Structure issues of American agriculture.* Economics, Statistics, and Cooperatives Service, Agricultural Economic Report 438. Washington, D.C.: U.S. Department of Agriculture.

Casey, James, Ronald D. Lacewell, and Lonnie L. Jones. 1975. *Impact of limited fuel supplies on agricultural output and net returns: Southern high plains of Texas.* MP-117. College Station: Texas Agricultural Experiment Station.

Duncan, Otis D. 1959. Human ecology and population studies. In *The study of human population,* ed. P. M. Hauser and O. D. Duncan. Chicago: University of Chicago Press.

Hady, Thomas F., and Peggy J. Ross. 1990. *An update: The diverse social and economic structure of nonmetropolitan America.* Staff Report AGES 9036. Agriculture and Rural Economy Division, Economic Research Service, U.S. Department of Agriculture.

Hobbs, Daryl. 1992. Costs and benefits of multicommunity collaboration. In *Multicommunity collaboration: An evolving rural revitalization strategy,* ed. P. F. Korsching, T. O. Borich, and J. Stewart. Ames, Iowa: North Central Regional Center for Rural Development.

Johansen, Harley E. 1993. The small town in urbanized society. In *The demography of rural life,* ed. D. L. Brown, D. R. Field, and J. J. Zuiches. University Park, Pennsylvania: Northeast Regional Center for Rural Development.

Johnson, Kenneth M., and Calvin L. Beale. 1992. Natural population decrease in the United States. *Rural Development Perspectives* 8(1):8-15.

Korsching, Peter F. 1992. Multicommunity collaboration: An evolving rural revitalization strategy. In *Multicommunity collaboration: An evolving rural revitalization strategy,* ed. P. F. Korsching, T. O. Borich, and J. Stewart. Ames, Iowa: North Central Regional Center for Rural Development.

Korsching, P. F., T. O. Borich, and J. Stewart (ed.). 1992. *Multicommunity collaboration: An evolving rural revitalization strategy.* Ames, Iowa: North Central Regional Center for Rural Development.

Larson, Olaf F. 1981. Agriculture and the community. In *Nonmetropolitan America in transition,* ed. A. H. Hawley and S. M. Mazie. Chapel Hill: University of North Carolina Press.

Leistritz, F. L., and Brenda L. Ekstrom. 1986. *Interdependencies of agriculture and rural communities: An annotated bibliography.* New York: Garland Press.

Olson, Karen. 1991. 1980-1990 migration in North Dakota. Population Bulletin 7(11). North Dakota Census Data Center, North Dakota State University, Fargo.

Powers, Ronald C. 1992. Why work together? In *Multicommunity collaboration: An evolving rural revitalization strategy,* ed. P. F. Korsching, T. O. Borich, and J. Stewart. Ames, Iowa: North Central Regional Center for Rural Development.

Salamon, S., and K. Davis-Brown. 1990. Rural communities in transition. In *National Rural Studies Committee: A proceedings,* ed. Emery Castle and Barbara Baldwin. Corvallis, Oregon: Western Rural Development Center.

Schaffer, Ron. 1992. Benefits and costs of multicommunity collaboration. In *Multicommunity collaboration: An evolving rural revitalization strategy,* ed. P. F. Korsching, T. O. Borich, and J. Stewart. Ames, Iowa: North Central Regional Center for Rural Development.

Sofranko, Andrew, and James Williams (ed.). 1980. *Rebirth of rural America: Rural migration in the Midwest.* Ames, Iowa: North Central Regional Center for Rural Development.

Stabler, Jack C. 1992. A perspective on multicommunity collaboration for economic development. In *Multicommunity collaboration: An evolving rural revitalization strategy,* ed. P. F. Korsching, T. O. Borich, and J. Stewart. Ames, Iowa: North Central Regional Center for Rural Development.

U.S. Bureau of the Census. 1992. Residents of farms and rural areas. *Current population reports,* Series P-20, No. 457 (March). Washington, D.C.: U.S. Government Printing Office.

5. Economic Prospects for the Great Plains

Mark Drabenstott
Federal Reserve Bank of Kansas City

The Great Plains economy slowed considerably during the 1980s. Two broad gauges of the region's economic performance—income growth and employment growth—both fell as the region adjusted to problems in its key economic sectors: agriculture, mining, and manufacturing. Although other regions also slowed during the 1980s, the slowdown in the Great Plains was more pronounced, and as a result, this region ranked among the slowest-growing regions in the nation during the 1980s.

The economic slowdown was especially pronounced for the rural Great Plains. Economic growth slowed in both metropolitan and nonmetropolitan parts of the region in the 1980s, but the metropolitan areas—and places adjacent to metropolitan areas—fared much better than rural areas. In fact, metropolitan counties in the region captured 97 percent of the overall growth that did occur in the Great Plains in the 1980s. Moreover, the poor performance in nonmetropolitan areas was not spread evenly across places with different economic bases. For example, natural resource-based rural counties fared much worse during the 1980s than those based on retirement.

In the decade ahead, the region will face four main economic challenges:

- A widening economic gap between urban and rural growth
- The lack of a university research powerhouse
- The need to help people and infrastructure as they adjust to new economic realities
- The need to diversify the region's economic base, a need felt most acutely in the rural Great Plains

Taken together, these challenges probably point to slower economic growth in the region than in the nation.

In this chapter the economic changes at work in the Great Plains are examined. First, a number of trends that swept the region in the 1980s are documented, with special emphasis on the relative performance of rural and urban areas. The second section investigates four factors likely to affect growth prospects in the Great Plains. Third, the region's economic challenges are summarized and the possible avenues for boosting growth in the rural Great Plains are highlighted. This chapter concludes that the region must face up to an enormous transition problem for some of its people

Stanley R. Johnson and Aziz Bouzaher (ed.), *Conservation of Great Plains Ecosystems,* 63–81.

and infrastructure and encourage development that adds value to the region's abundant natural resources.

The Great Plains Economy in the 1980s

The Great Plains economy encountered an economic slump in the 1980s, as key industries—agriculture, energy, and manufacturing—underwent dramatic downsizing and structural change. As a result, economic growth across the Great Plains was much slower than in the 1970s and, on average, trailed well behind growth in the nation.

Broad Indicators of Economic Performance

The slowing in the Great Plains region is reflected in broad indicators of the region's economic performance. Real income growth, one key indicator of overall pace of economic growth, slowed in the region from 4.2 percent in the 1970s to 2.0 percent in the 1980s. Employment growth, another indicator of the region's economic performance, slowed from 2.9 percent in the 1970s to 1.7 percent in the 1980s. Both are considerably larger decreases than the U.S. average. Of course, decade-long averages obscure the year-to-year variation in growth. Most notable is a modest rebound in income and employment growth at the end of the 1980s, resulting from a strong recovery in agriculture and a more stable energy sector. Broad economic indicators also mask variations in economic performance across individual states and specific localities within the region, but they provide a benchmark for comparing the average performance of the region with the average performance of other regions.

Economic growth in the Great Plains lagged considerably behind economic growth in most other regions of the United States during the 1980s. (The regions pictured in Figure 5.1 are as defined by the U.S. Department of Commerce.) Income growth in the Great Plains exceeded growth in only two of the Department of Commerce regions—the Plains and the Great Lakes. Job growth in the Great Plains lagged behind growth in most other regions (Figure 5.2). Only three regions—the Mideast, the Plains, and the Great Lakes—posted slower job growth during the decade.

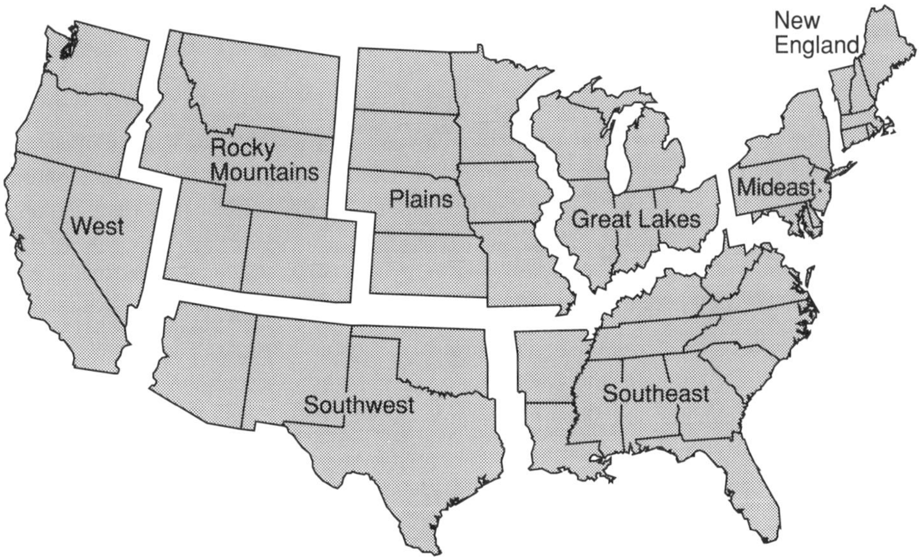

Figure 5.1. U.S. economic regions

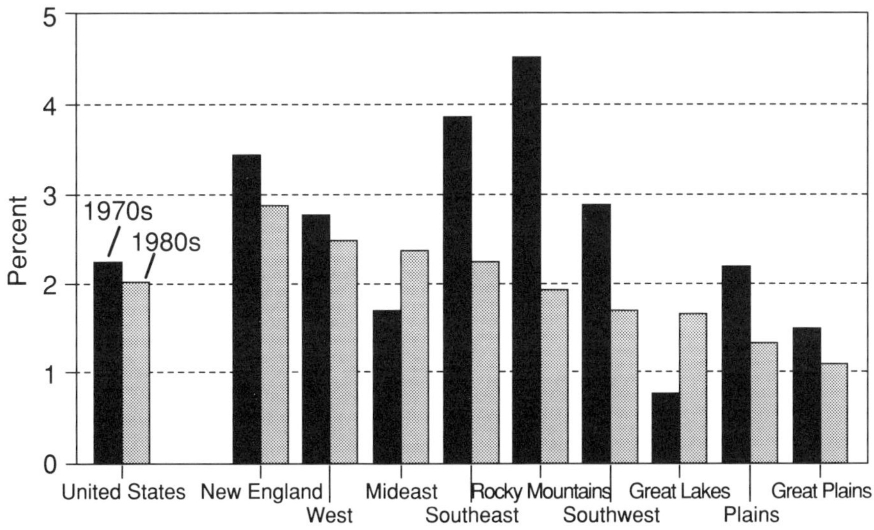

Figure 5.2. Comparison of employment growth for the 1970s and 1980s

Urban and Rural Growth

The lackluster performance of the Great Plains region during the 1980s is often linked to the slowing in its rural economy. Few would disagree that the region has a distinct rural character compared to most of the United States. Yet 70 percent of the region's population resides in metropolitan counties. Therefore, it is important to look at both urban and rural growth to understand the recent economic performance of the Great Plains.

Economic growth in the nonmetropolitan counties of the Great Plains was much slower than growth in metropolitan counties, a divergence in urban and rural performance that was not unique to the Great Plains. In virtually every region of the country, nonmetropolitan counties had slower economic growth than metropolitan areas in the 1980s (Figure 5.3); New England was the lone exception. In the remaining regions, as well as the Great Plains, rural growth in real income and employment underperformed metropolitan areas both within the region and in the nation. Rural counties in the central regions of the nation—including the Great Plains—performed worst. In contrast with rapid rural growth in New England, for example, rural counties in the Great Plains saw their incomes and job ranks grow by only 0.3 percent.

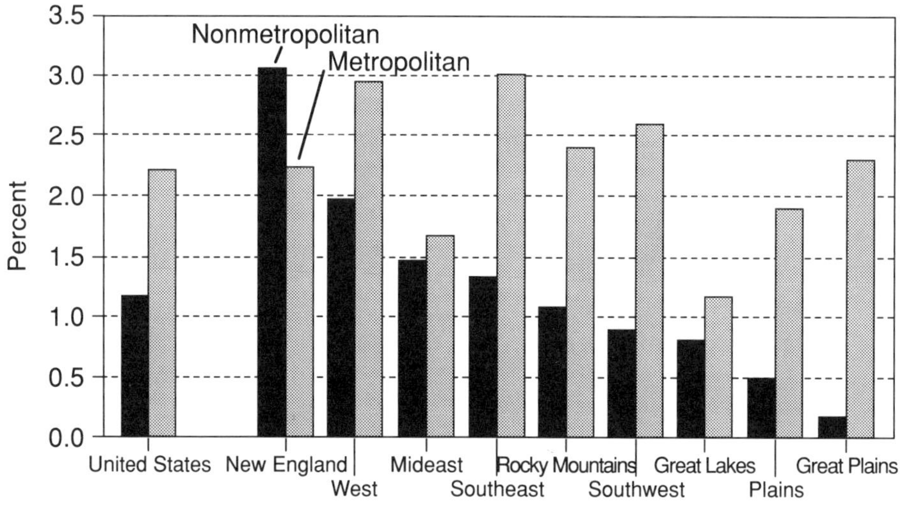

Figure 5.3. Average annual employment growth, 1980–89

Although the urban Great Plains outperformed the rural Great Plains in the 1980s, urban economic growth in the Great Plains was slow by comparison with most other regions of the country. The 2.6 percent growth in urban income in the Great Plains exceeded the growth rate of only three other regions—the Rocky Mountains, the Great Lakes, and the Plains (which includes the Great Plains). And the 2.3 percent growth in urban employment exceeded only three other regions—the Plains, Great Lakes, and Mideast. Thus, with the exception of the Mideast, the overlapping central regions were at the lower end of the regional distribution of urban growth in the 1980s.

The divergence between rural and metropolitan growth stands in sharper relief when shown as a ratio. The rural-urban income ratio, for example, represents rural income growth as a proportion of metropolitan income growth (Figure 5.4). When the ratio is 1, rural growth just matches urban growth. For the nation as a whole, the income ratio was 0.6 in the 1980s, suggesting rural places grew roughly half as fast as urban places. By contrast, rural places outpaced urban places in the 1970s, when the ratio was 1.4.

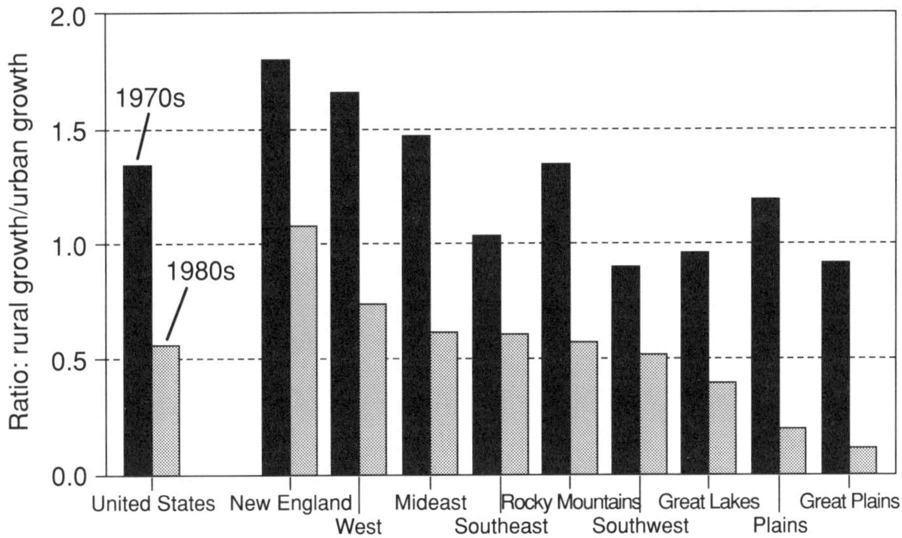

Figure 5.4. Rural-urban income ratios for the 1970s and 1980s

Rural Proximity to Metropolitan Areas

The rural Great Plains posted lackluster performance in the 1980s, but clearly not all rural places are alike. Some rural communities serve as bedroom communities to thriving metropolitan areas, while others are isolated from population centers.

Rural is a term that is often defined as not being urban. The difficulty comes in deciphering how far economic influence extends from a metropolitan area. To address this spatial uncertainty, the U.S. Department of Agriculture developed a series of ten categories, called Beale codes, to define the spectrum of counties from core metropolitan to absolutely rural (Table 5.1). The codes separate metropolitan counties into those in the center of large Metropolitan Statistical Areas (MSAs) and those on the fringe. The codes also identify which nonmetropolitan counties are adjacent to MSAs and which are not. The result is a useful taxonomy with central core MSA counties classified as 0 and completely rural counties classified as 9. In between lies a spectrum that gauges proximity to an MSA.

Table 5.1. Beale code definitions

Code	County Type
Metropolitan	
0	Central counties of metropolitan areas of 1 million population or more
1	Fringe counties of metropolitan areas of 1 million population or more
2	Counties in metropolitan areas of 250,000-1,000,000 population
3	Counties in metropolitan areas of less than 250,000 population
Nonmetropolitan	
4	Urban population of 20,000 or more, adjacent to a metropolitan area
5	Urban population of 20,000 or more, not adjacent to a metropolitan area
6	Urban population of 2,500-19,999, adjacent to a metropolitan area
7	Urban population of 2,500-19,999, not adjacent to a metropolitan area
8	Completely rural (no places with a population of 2,500 or more), adjacent to a metropolitan area
9	Completely rural (no places with a population of 2,500 or more), not adjacent to a metropolitan area

When rural counties in the Great Plains are grouped according to their respective Beale codes and their economic growth rates are compared, an unmistakable trend emerges for the 1980s. Growth declined—in nearly straight-line fashion—as the distance from a metropolitan area increased. With employment growth, for instance, the four metropolitan county codes (0 to 3) had the strongest growth and three of the four county codes in this group matched or exceeded the nation's growth in the 1980s (Figure 5.5). Meanwhile, counties adjacent to metropolitan areas (codes 4, 6, and 8) outperformed counties that were not adjacent (codes 5, 7, and 9). This last group ranked last in job growth. And within the nonadjacent group, counties with larger population centers outperformed completely rural counties. Exactly the same story holds for income growth.

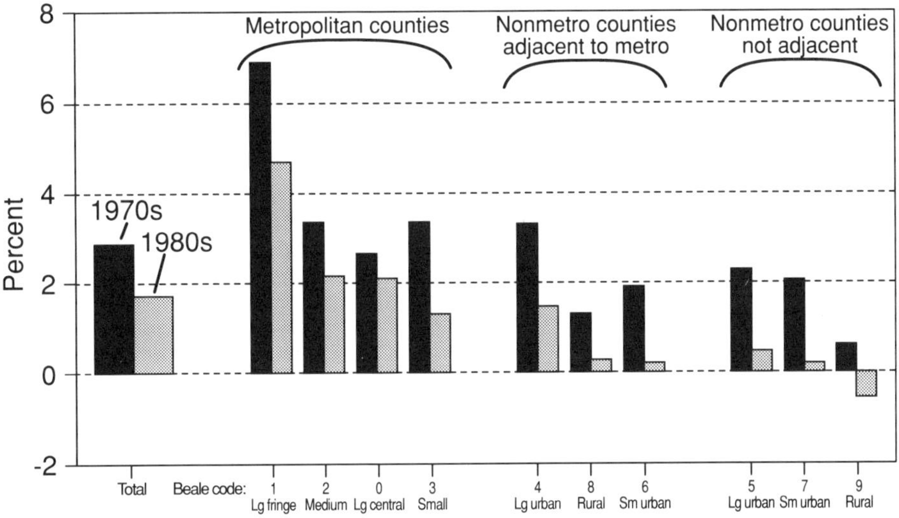

Figure 5.5. Employment growth by Beale code, Great Plains

A simple but stark conclusion arises from these data: Economic growth in the Great Plains (and elsewhere in the nation) migrated to the cities in the 1980s. Such was not the case in the 1970s, but that decade looks increasingly like an aberration. And while growth extends to suburbia and in some cases slightly beyond, economic growth in MSAs far exceeds growth in completely rural places. Indeed, metropolitan counties accounted for almost all of the region's new income and jobs in the 1980s. Moreover, if you were a resident of the rural Great Plains in the 1980s, the record shows that it was generally better to live in a county next to an MSA in the 1980s than in any other rural place in the region.

Economic Bases and Rural Growth

The rural Great Plains derives its economic growth from a variety of economic bases. Although the largest number of rural counties in the Great Plains are farming counties, most of the Great Plains rural population resides in trade counties. The share of rural population in farming counties fell from 32 percent to 28 percent between 1969 and 1989 while trade and other types of counties gained or were stable.

Surveying economic growth in the rural Great Plains across different types of economic bases offers useful insights into the rural slowdown of the 1980s. Figure 5.6 shows growth in employment for eight different rural economic bases. The performance data do *not* measure the performance of different industries themselves. Rather, they indicate the performance of rural places dependent on such industries.

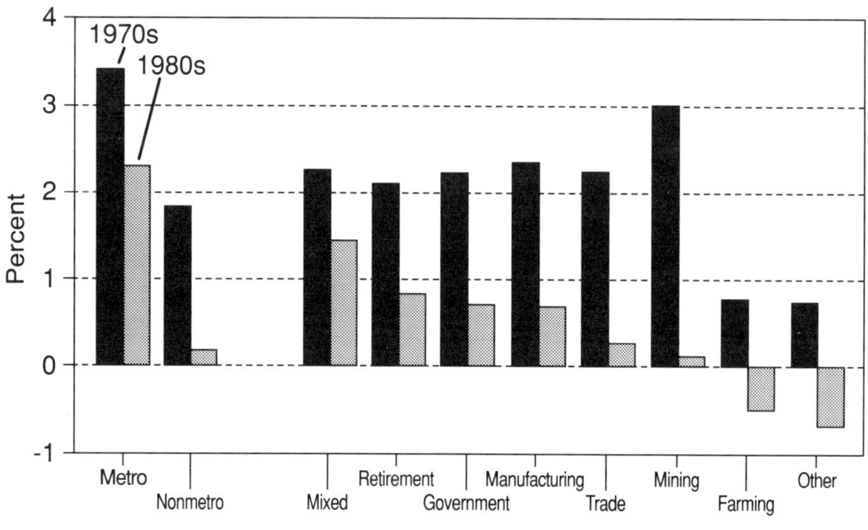

Figure 5.6. Employment growth for the Great Plains during the 1970s and 1980s

Rural counties dependent on retirement or mixed bases grew fastest in the Great Plains during the 1980s. These two county types were also the fastest-growing rural categories in the nation as a whole. Despite their strong performance among other types of rural counties, retirement and mixed counties (which have a diverse economic base) failed to keep pace with metropolitan counties. Growth in both jobs and incomes in these rural counties lagged well behind metropolitan counties in the Great Plains.

Both retirement and mixed counties can be viewed as emerging trade centers in the rural Great Plains, places that tapped into the growing U.S. service sector. Retirement counties thrived on transfer payments and a proliferation of services in the past decade. Retirement counties also benefited from growth in health care and financial services. Mixed counties, by becoming hubs for rural trade, experienced growth in a wide range of services.

Meanwhile, rural counties in the Great Plains dependent on industries that compete in global markets ranked last in economic performance. Income in farm-dependent and mining-dependent counties *declined* on average during the 1980s. (Employment fell in farming counties and edged up slightly in mining counties.) Agriculture and energy clearly underwent dramatic change in the 1980s because of the pressures of global competition. Manufacturing-dependent counties—where a much smaller share of the Great Plains rural population resides than in the nation (27 percent in the nation compared with only 5 percent in the Great Plains)—also did poorly, but their modest growth compared favorably with the income and job losses in the two county types dependent on natural resources.

Decade-long averages may overlook a possible rural recovery in recent years. Farm-dependent counties, for example, have been buoyed by a strong farm recovery. Yet annual data since 1987 suggest that even record farm incomes did not lead to a widespread rebound in economic activity in those counties. Anecdotal evidence from the seven-state Tenth Federal Reserve District (Colorado, Kansas, Missouri, Nebraska, New Mexico, Oklahoma, and Wyoming), for example, suggests that structural change in agriculture has led to fewer farms, fewer agribusinesses, and a weaker economic multiplier in farm communities. Economic activity appears to be migrating to farm trade centers that are prospering at the expense of surrounding communities.

Economic Winners and Losers in the Rural Great Plains

A useful way to highlight the fundamental transition facing rural places in the Great Plains is to examine closely those counties that did well throughout the past decade, both in the region and across the nation. These "winners" provide clear insights into what it may take for rural communities to prosper in the 1990s. To make the analysis manageable, this section draws on some research that looks at rural counties that did better than average in the nation (Smith 1992). The analysis divides rural counties into winners and losers. Winners are counties where employment and income growth exceeded average employment and income growth for all nonmetropolitan counties in the seven-state region. In other words, winners are counties that had average annual growth in employment of 1.1 percent and growth in income of 1.5 percent from 1979 through 1989.

A few qualifiers are in order. Past results are, of course, no guarantee of future performance. Agriculture, which is vital to the rural Great Plains, clearly went through some very wide swings in the 1980s. The farm recession was especially deep in the Great Plains, but agriculture also posted record or near-record incomes from 1986 through 1990. The 1990s may well be different. Nevertheless, some of the basic trends embedded in the 1980s will continue in the 1990s.

One-fourth of the nation's rural counties were winners in the 1980s. For the decade as a whole, 635 of the 2,357 nonmetropolitan counties had above-average growth in employment and income. In fact, the rural winners that met that test outperformed the nation's metropolitan areas, on average. Employment growth, for example, averaged 2.6 percent a year through the 1980s for rural winner counties, while real income growth managed 3.1 percent.

Who were the winners? Based on a breakout according to economic base, 25 percent were retirement counties, and another 21 percent were rural trade centers. The trade center counties often stole their growth from surrounding counties—the so-called Wal-Mart effect. About 20 percent were manufacturing-dependent counties, and only 2 percent were farm-dependent counties. Put another way, if you lived in a farm-dependent county, you had a one in 200 chance of living in a rural county that had above-average growth in the 1980s.

Geographically, the winners were clustered but appeared in many parts of the nation (Figure 5.7). Scenic amenities were important; there were clusters of winning counties in the Ozarks, the North Lake country, the Rockies, and the West Coast. The Eastern seaboard did well because it benefited from spillover from prosperous metropolitan areas. East Texas and the Sun Belt round out the list.

Equally revealing is an analysis of the one-half of all rural counties that made up the loser category. Of 1,261 counties with subpar employment and income growth, 28 percent were trade-dependent. These are the counties that did not get Wal-Marts. Fully 27 percent of the losers were farm-dependent counties, a rather sobering finding for this audience. Another 16 percent were manufacturing-dependent, suggesting that some rural manufacturers did quite well in the 1980s, while others did quite poorly.

Geographically, the losers were clearly clustered in the middle of the country. The Great Plains, the western Corn Belt, and the northern Rockies did poorly. So did farm-dependent parts of the Delta. In short, one can think of a "Heartland triangle" that takes in most of the losers. The Heartland is quite an amorphous term, but consider this definition: if a line is drawn from Chicago to Dallas, and from Dallas to Seattle, everything north of the two lines is, in fact, a sea of losing counties.

Based on Beale code classifications, one-half of the winning counties were directly adjacent to metropolitan areas. Still, one-sixth were completely rural with no urban center. While that suggests that completely rural counties can prosper, it appears to come mainly at the expense of other rural counties. Fully two-thirds of the rural losing counties, meanwhile, were not adjacent to metropolitan areas.

Within the Great Plains itself, nearly three-fourths of the region's 675 rural counties were losers—that is, growth in both jobs and income was below the national average

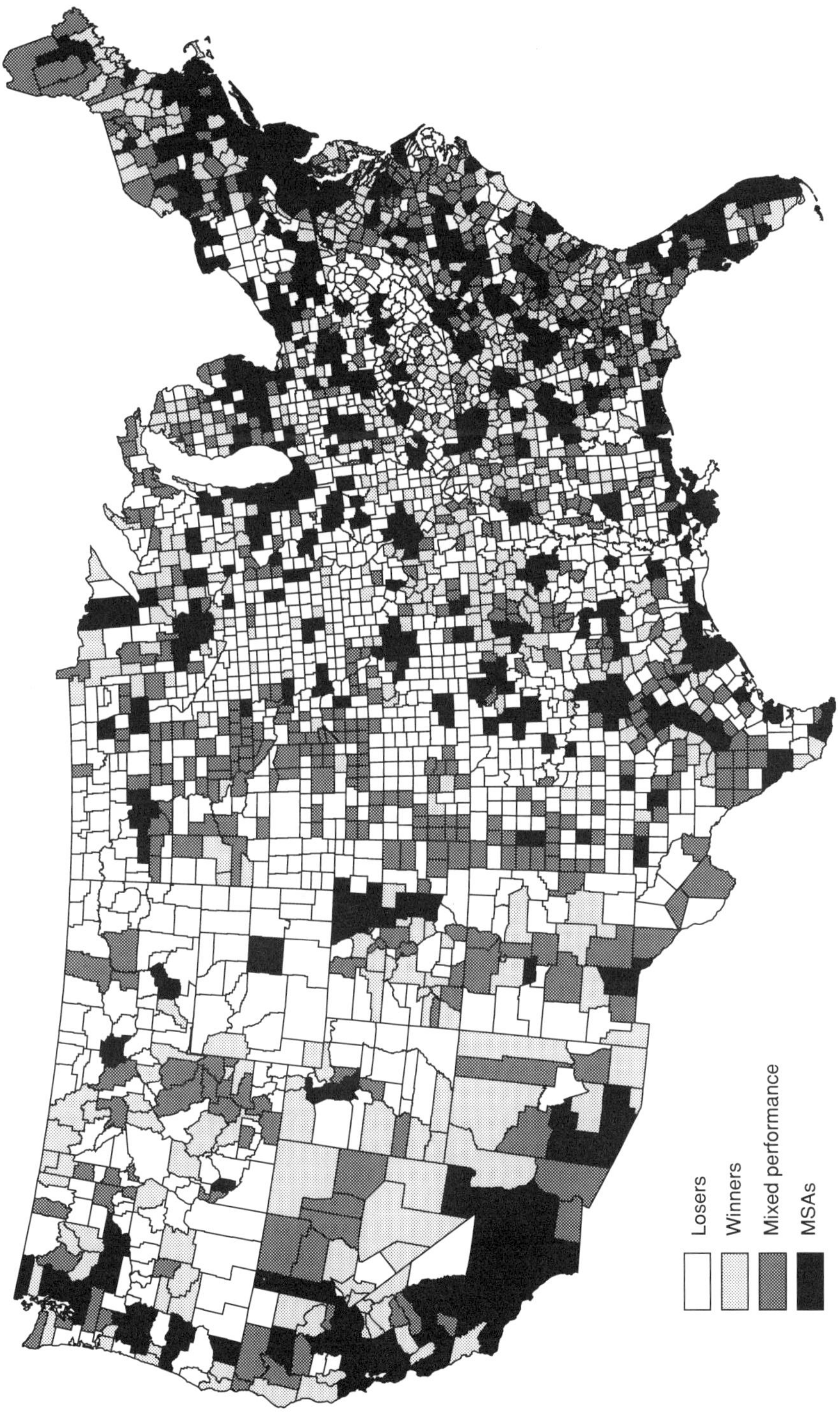

Figure 5.7. Rural winning and losing counties

for all rural counties. Only 27 counties, just 4 percent of the total number of rural counties, were winners. Put simply, the rural Great Plains is suffering from a massive consolidation in economic activity to trade centers. And while the region's metropolitan areas have been quite healthy economically, little of the growth has filtered down to the rural parts of the region.

Economic Prospects in the Great Plains

What are the region's long-term growth prospects? Economic models are weak instruments in forecasting long-term growth. A more useful approach may be to consider the region's principal growth assets and in the process uncover some economic challenges that must be addressed.

The region has five main assets: workforce, educational system, infrastructure, fiscal climate, and financial capital. In an analysis of whether these assets will be pluses or minuses for growth in the seven states of the Tenth Federal Reserve District, Drabenstott (1993) argues that the region's mix will lead to economic growth slower than the nation's.[1] The region is blessed with a good workforce, a plus for long-term growth. The region's schools and universities turn out quality graduates, but its research universities are mostly second-tier in size. In an economy where innovation is the fuel of progress, that tilts education to a slight minus for long-term growth. The region's infrastructure is extensive, but roadways need improving and the region lacks access to growth markets abroad, making infrastructure a slight minus for growth. The region generally has low taxes and low government spending, making fiscal climate neutral to long-term growth. Finally, the region has conservative banks, financial assets that are growing slower than the nation, and a fragmented venture capital market, making financial capital a minus for long-term growth.

That analysis suggests that the Great Plains will grow more slowly than the nation in the 1990s. Moreover, with the rural Great Plains lacking a new engine for growth, nonmetropolitan areas will likely grow even more slowly than urban centers.

This outlook for the Great Plains points to four challenges for the region in the decade ahead: addressing or accepting the widening gap between urban and rural economic growth, bolstering university research in the region, easing the economic transition for the region's population and infrastructure, and broadening the region's economic base, especially in rural areas where economic activity remains heavily tied to natural resources.

[1]While the Great Plains takes in additional states, a strong case can be made that prospects for the Tenth District are largely those for the Great Plains.

The Widening Gap in Rural and Urban Economic Performance

The Great Plains is commonly thought to be a predominantly rural region. While it has many wide open spaces, only 30 percent of the region's population now lives in nonmetropolitan counties. The urban economy of the Great Plains slowed in the 1980s, but incomes and jobs in the region's metropolitan counties still grew roughly 2.5 percent a year.

But the economy of the rural Great Plains was largely left behind in the 1980s. For every $1 of new income in Sioux Falls, rural incomes grew only 12 cents. Hurt by its overwhelming dependence on agriculture and energy, the rural Great Plains found no engine of growth in the 1980s. The Great Plains has nearly three-fourths of the nation's 428 farm-dependent rural counties, and nearly one-fourth of the nation's 106 mining-dependent counties.

Agriculture and energy may do reasonably well in the 1990s, but neither is likely to spur widespread economic growth in the region. Agriculture has undergone profound changes, characterized by new economic links among producers, processors, and consumers (Schertz and Daft 1994). The farm financial crisis of the mid-1980s brought an acceleration in farm and agribusiness consolidation. In fact, the industry played catch-up, offsetting the 1970s when consolidation slowed. The reality is that agricultural production now requires less rural infrastructure to support it, just as the 1960s required less than the 1930s. Put simply, there are fewer farms, fewer lenders, fewer suppliers, fewer buyers of farm products. The result is a consolidation of economic activity in farm-dependent rural places. Thus, only the strong communities survive.

As in agriculture, the energy industry seems likely to fuel economic activity in a much more focused way than in the past. In the 1970s, the energy industry was a strong source of rural growth, but in the 1980s disaster struck. The industry has downsized, forcing severe cost-cutting and a new search for productivity. The industry can now sink oil wells and mine coal with far fewer workers than it needed before. The Wyoming coal industry, for instance, has seen steady growth in coal output but has actually seen employment fall, which has had adverse effects on some rural Wyoming communities. In short, improvements in the energy industry will bring only muted gains for rural communities.

What, if anything, can be done about the widening gap in economic performance between rural and urban places in the region? Rural development will vex policymakers throughout the nation in the 1990s, but three things seem necessary to close the rural-urban gap.

First, acknowledge that there is no surefire path to rural prosperity. Rural counties with the best chance of prosperity probably have mountains or lakes or some other magnet amenity. But beyond drawing tourists and retirees, much of the rest of rural America is struggling with complex economic forces. Highways, fiber optic lines, and health care are all important but none is sufficient by itself to ensure rural prosperity.

Second, steps need to be taken to remedy rural remoteness. Throughout history, our nation has valued equal access for rural Americans. The postal system and rural electrification are but two examples. Today, rural citizens have mail service but they are removed from centers of innovation and culture. While that has always been true, the rural disadvantage today appears larger despite our mobile society and its world-class communications system. To prosper, rural businesses need an extension service aimed at a host of industries, not just agriculture, and they need a quality workforce that receives the social and cultural amenities they have come to expect. Is there a way to ensure access while allowing market forces to still operate? That is a thorny dilemma for public policy.

Third, rural places need options for achieving some agglomeration. Johnson County, Kansas—an affluent suburban county in the Kansas City metropolitan area—is successful because it has all the cultural amenities *and* the business networking that any service business could want. It will be impossible for rural counties in western Kansas to replicate that. But is it possible for rural communities to band together in innovative ways to produce some agglomeration, even if only on a small scale? Regional airports, shared public services, and regional community colleges are three possible examples.

Bolstering University Research

In an economy where information and services are critical and where productivity in base industries like manufacturing and agriculture is paramount, university research becomes an absolutely critical regional asset. Unfortunately, the Great Plains comes up short in this respect. In an analysis of research spending at universities in seven Great Plains states, Drabenstott (1993) found only three universities ranked among the nation's top 50 public research institutions, and only one among the top 20 (the University of Colorado at 19th).

What can the region do to spur more research and corresponding economic activity? There are three possible approaches. First, more states in the region could adopt the "Colorado model." Following the University of Colorado example, states might select research niches, invest some additional seed money, and then aggressively leverage the research effort with federal dollars. The region is not without its areas of research expertise. This strategy suggests that such expertise has not been fully exploited for the region's benefit. Some universities, such as Kansas State, receive more research dollars from the state than from federal sources. But federal research dollars are becoming more scarce, so the Colorado approach may offer only limited potential in boosting research.

Second, leaders could combine research programs from across the region into a de facto major-league research university. The creative energies of the region's universities have never been collectively harnessed, but under this approach they would be. Duplicate and competitive programs would be eliminated, and a new group of centers of research excellence could take their place. Strictly for purposes of illustration, the University of Nebraska might become the site for a Food Research Center, while de-emphasizing a few specific portions of its engineering research in favor of a new Engineering Research Center at Kansas State.

To be successful, this approach demands that rival states cooperate. The problem is that neither mechanisms nor incentives exist to channel the cooperation needed to create a regional research powerhouse. Can such a mechanism be created? It is worth noting that other regions are making gains in coordinating some development efforts through regional institutions such as the Great Lakes Commission and the Southern Growth Policies Board.

A third approach is consolidation. The region has more universities per capita than the rest of the nation. Some institutions might be eliminated and their teaching loads reassigned. The money left over could be spent to reinvigorate research programs. In a region like the Great Plains, which has a lot of space, new technologies might provide distance learning at lower cost. Notwithstanding the potential economic benefits, consolidation poses thorny political problems. In fact, most boards of regents in the region are debating whether to add, not subtract, institutions from their watch.

Easing Economic Transition in the Region

Regardless of how successful economic development efforts may be in the region, the legacy of the 1980s will hang on for many parts of the region. Put simply, there will be many communities in the region that will not be economically viable in market realities of the 1990s. People will migrate to other places in the region, or to places beyond the region. But infrastructure will be left behind: schools, roadways, and water systems, among other things. What, if anything, can public policy do to ease this painful transition?

The first task is to ease the human transition. Many of the people leaving the Great Plains are rural residents who are searching for a better livelihood. These economic pilgrims often lack the skills to be quickly assimilated into the economy elsewhere. In the past, farmers often moved to factories. But in the 1990s, those job openings no longer exist. Instead, the job market demands more technical and analytical skills.

The answer lies in better retraining programs. Many of the states in the Great Plains have extensive community college and university extension resources, which can be great assets in the retraining effort.

But public officials are asking whether the costs of retraining can be recovered if retrained workers leave the region. Their question has two answers. First, a case can be made that rural economic distress is a national problem that could justify federal involvement in rural retraining. Second, a case might be made that retraining is in the interest of Great Plains states even if workers move elsewhere. That is because the cost to the public of unemployment insurance and other social programs may more than offset the cost of retraining.

Easing the challenges left by public infrastructure will prove more difficult. Many states in the Great Plains will be left with critical decisions on infrastructure, such as highways, as well as how to deliver public services through a network of county governments that may be facing fiscal deficits. Choosing which public infrastructure to maintain will be very difficult in the Great Plains, where distance between cities is great. Nevertheless, careful comparisons of costs and benefits will be necessary in making investments in facilities like roadways, bridges, and water systems. New analytical techniques that use digital geographical information systems hold the promise of better informed decisions (Henry et al. 1990).

Making public service delivery more efficient and less costly in rural areas will require innovations in public policy. County governments are slow to cooperate with neighboring counties because there are few incentives to do so. State governments may want to consider making grants to counties to encourage cooperation in delivering local services. Such incentives would encourage efficiency and phasing out of duplicate facilities.

In general, there will be value in Great Plains governments cooperating more fully in the future to provide public services more efficiently. That includes shared goals in universities and physical infrastructure. With limited resources, the region will have a better chance to achieve excellence if it cooperates more than it has in the past (Fosler 1991).

Broadening the Rural Economic Base

The cities of the Great Plains are already well along in broadening their economic bases. The time is long past when cities such as Omaha and Denver depended heavily on agriculture and energy. But the rural areas of the Great Plains remain tied to natural resource industries. Of the region's rural population, 28 percent live in counties that depend on agriculture while 5 percent live in mining-dependent counties (Table 5.2). For the U.S. rural population, the corresponding figures are 6 percent and 4 percent. Thus, the rural Great Plains is more than three times more dependent on agriculture and mining than the rest of rural America. Meanwhile, only 5 percent of the region's rural population lives in counties dependent on manufacturing compared with 27 percent for rural America.

Table 5.2. Population statistics, Great Plains

		Share of Population			Share of Rural Population		
	Counties	1969	1979	1989	1969	1979	1989
	Number			Percent			
Metropolitan	80	64	66	70			
Nonmetropolitan	595	36	34	30	100	100	100
Farming	306	12	10	8	32	30	28
Government	37	4	4	3	10	11	11
Manufacturing	16	2	2	2	5	5	5
Mining	24	2	2	1	4	5	5
Mixed	25	1	1	1	3	3	3
Other	2	0	0	0	0	0	0
Retirement	19	2	2	1	4	5	5
Trade	166	15	14	13	41	42	43
Total	675						

Source: Bureau of the Census 1994.

To improve economic prospects in the rural Great Plains, it will be necessary to broaden the economic base. In the 1990s, both agriculture and energy will remain commodity industries driven by international markets, requiring Great Plains producers to operate on thin margins in both cases.

The key will be to add more value to the resources before products leave the region. For agriculture, adding value will require more food processing or new industrial uses for agricultural commodities. The food processing strategy is being promoted in many states: Nebraska, Kansas, and Iowa, in particular. But the prospects for the region appear mixed. Those three states seem to be in a good position, while Northern Plains states face a tougher challenge because they are so far from major consumer markets (Barkema and Drabenstott 1990).

In energy, the region seems to have few options in adding value. The region has large reserves of natural gas, and production could rise if environmental concerns raise the demand for natural gas, a fuel that burns cleanly. The region also has large deposits of low-sulfur coal; production is increasing steadily and could increase even more rapidly if the nation tries to limit the burning of high-sulfur coal. Finally, the region might find some of its agricultural products in greater demand if the nation chooses to encourage fuels from biomass, such as ethanol.

From the point of view of the rural Great Plains, research is needed on new technologies that can keep rural industries, like agriculture and manufacturing, world class while at the same time stimulating the broader rural economy. For more than 125 years, the public has made big investments in research at land grant universities, an investment that has paid a steady stream of dividends in increased productivity. There may be value in asking whether future research should focus strictly on productivity, or whether the impact on rural economic activity is also important. With citizens growing more concerned about the social problems that now plague some cities, and with growing concern for responsible stewardship of the rural environment, there may be a place for research that makes rural industries more competitive while also providing greater economic impact for rural communities.

Conclusions

The Great Plains economy slowed sharply in the 1980s, and compared with other regions of the country, the Great Plains was one of the slowest growing portions of the country. Within the region, the decade was marked by a widening gap between rural and urban areas. With diverse economic bases, cities in the region maintained solid, steady growth in the 1980s. However, the rural Great Plains remained heavily dependent on agriculture and energy and fared poorly in the 1980s, with little if any gains in income or employment.

These economic trends pose stiff challenges for the region in the decade ahead. To close the gap between rural and urban economic performance, steps must be taken to remedy rural remoteness, while allowing rural communities to achieve some economies of scale. To bolster university research, the region must consider ways to pool its resources to build a world-class research program. To help rural people and rural infrastructure adjust to the economic trends underway, public policy will need to provide better retraining programs, as well as innovative programs to encourage more efficient delivery of public services. To broaden the rural economic base of the region, new ways will need to be found to add value to the region's abundant base of natural resources. In particular, value could be added to Great Plains agricultural products either through food processing or new industrial uses.

References

Barkema, Alan, and Mark Drabenstott. 1990. Processing food in farm states: An economic development strategy for the 1990s. *Economic Review* (Federal Reserve Bank of Kansas City) 75(July/August):5-23.

Butler, Margaret A. 1990. *Rural-urban continuum codes for metro and nonmetro counties.* Staff Report AGES 9028. U.S. Department of Agriculture, Agriculture and Rural Economics Division.

Drabenstott, Mark. 1993. The district's long-term growth prospects. *Economic Review* (Federal Reserve Bank of Kansas City) 78(1):55-65.

Drabenstott, Mark, and Tim Smith. 1992. The changing Great Plains economy: New directions for economic policy. Policy paper (March), Center for the New West.

Fosler, R. Scott. 1991. Economic development: A regional challenge for the heartland. In *Regional economic development and public policy.* Kansas City: Federal Reserve Bank of Kansas City.

Henry, Mark, K. Brooks, M. Roche, and L. Singletary. 1990. Modeling impacts of new highways on regional economies with the aid of geographic information system. In *Application of geographic information systems, simulation models, and knowledge based systems for land use management.* Proceedings of a conference at Virginia Polytechnic Institute and State University (Nov.).

Schertz, Lyle, and Lynn Daft (ed.). 1994. *Food and agricultural markets: The quiet revolution.* National Planning Association Report no. 270. Washington, D.C.: U.S. Department of Agriculture, Economic Research Service; Food and Agriculture Committee, National Planning Association.

Smith, Tim R. 1992. Long-term growth in rural places: Winners and losers in the 1980s. Paper presented to the North American Meetings of the Regional Science Association International, Chicago (Nov. 14).

U.S. Bureau of the Census. 1994. Regional economic information system. CD-ROM database. Washington, D.C.: U.S. Bureau of the Census.

6. Sustaining the Great Plains Ecosystem: Integrating People, Economics, and the Landscape

Robert D. Sopuck
Sustainable Development Coordination Unit,
Province of Manitoba, Canada

The Great Plains is one of the most precious resources of the North American continent. Not only does the area produce most of our food, but the Great Plains is also a unique and productive ecosystem and is home to some of the most industrious people and communities in North America.

It is a tribute to those people that agriculture and the Great Plains developed the way they did. The short growing seasons and drought make Great Plains agriculture a risky venture, and it was for good reason that the explorer John Palliser dismissed much of the prairies as being unfit for farming. Unfortunately, while the ability of the Great Plains to produce food remains strong, the economy, the landscapes, and the people are all at risk. Only by dealing with all three components at the same time will society be effective in the goal to which it aspires, and that goal should be the conservation of rural life, with rural life being defined in the broadest possible terms. Sustainable community development is the intersection of the people, economy, and environment. That is where society wants to be. Most of the information presented in this chapter is from Canada, but similarities between rural life on the Great Plains in both countries far outweigh any differences.

The Great Plains Environment—Soil, Water, and Landscapes

Soil

Prairie soils have been affected by agriculture but are still capable of producing agricultural commodities in the foreseeable future. Formerly, soil degradation was seen as a threat to the long-term survival of agriculture and, ultimately, the food supply. This apocalyptic vision is no longer accepted. To date, modern agriculture has adapted well to any loss of soil quality that has occurred, primarily through the addition of chemical and biological inputs that have in effect replaced such losses. Crosson (1991) referred to research that showed that if U.S. erosion rates remained at the 1992 level for 100

Stanley R. Johnson and Aziz Bouzaher (ed.), *Conservation of Great Plains Ecosystems*, 83–95.

years, corn, soybean, and wheat yields would be 4.6 percent, 3.5 percent, and 1.6 percent less, respectively, than they would be in the absence of erosion. Therefore, Crosson concludes that there is every reason to assume that technological advances will continue to buffer agriculture against any yield losses due to soil degradation. This chapter agrees with that assessment, maintaining that the effects of agriculture "off the farm" are the most important.

Water

On the prairies, surface water receives pesticides and nutrients carried by soil particles from eroded land. There are cases where prairie lakes experience algal blooms due to excess phosphorus that results in fish kills. As well, intensive livestock operations can affect local water quality because of runoff from feedlots. To a greater or lesser extent, water quality has been negatively affected in most agricultural areas, especially in areas with intensive cultivation of land, wide-row monoculture, and intensive livestock production.

Landscapes and Biodiversity

Intensive agriculture, if it is to survive, requires the removal of most natural habitats such as wetlands, forests, and grasslands. Often the removal of these "impediments" has been an explicit goal of public policy. The monoculture and lack of diversity on the Great Plains are evidence that these policies have worked well.

The loss of wetlands has been felt in every region of Canada and the United States as more than one-half of the potholes on the prairies have been lost. The greatest single threat to wetlands historically has been drainage for agricultural purposes, accounting for 85 percent of known conversions. Prairie wetlands provide critical habitat for more than 60 percent of the waterfowl of North America; as prairie wetland areas have declined, so have waterfowl populations.

More than 80 percent of the prairie landscape has been transformed by agriculture. In Canada, all of the tallgrass prairie is gone and 90 percent of the fescue grassland has been plowed. Approximately 24 percent of the mixed prairie and 25 percent of the aspen parkland remain in the native state. Consequently, a disproportionate number of Canada's threatened and endangered wildlife species inhabit the prairie ecozone.

The Great Plains Economy

Overview

The rural economy differs from the urban economy in a number of important ways. Apedaile (1992) describes rural economies as:

> Rural economic systems are above all, human systems. They are characterized by activities which transform natural resources into intermediate products for further processing by other systems. Forestry, agriculture, mining, and fishing are typical rural activities. These are carried out by human interventions in the ecosphere, principally the biosphere. Humans are an integral part of the biosphere.

In areas with less than 5,000 population, 41 percent of the residents are employed in the goods producing sector and 59 percent in the service sector. Services are the dominant source of employment in cities, accounting for between 71 and 76 percent of the urban labor force (Bollman and Biggs 1992). The heavy rural reliance on the production of goods puts rural economies at risk in a number of important ways, not the least of which is that the use of primary products is increasingly marginal to the economies of developed industrialized economies (Freshwater et al. 1992).

Canada was largely a rural country in the 1880s, but rural people became a minority in 1931. The urban component of Canada's population has risen significantly since the early 1950s. Farm residents represent a small and declining proportion of the Canadian population and in 1991 were 3.2 percent of the total population (Statistics Canada 1992). Within rural Canada, farmers became a minority in 1956. The trend for North America's agricultural landscapes has been to have fewer and fewer people living on them.

Agriculture

Despite the declining rural population, agriculture remains a dominant activity on the Great Plains. Because of Canada's small population relative to its food production base, well over 60 percent of its grain production is exported, with grains and oilseeds being Canada's largest agricultural exports. The prairie region is very dependent on international agriculture and is thus vulnerable to changes in world commodity prices.

The trend in Canada has been toward fewer and larger farms. The highest number of farms was estimated in 1941, but beginning in the mid-1940s the number declined significantly. The number continues to decline—the 1991 census of agriculture reported a 4.5 percent decrease in the number of farms from 293,089 in 1986 to 280,043 in 1991 (Statistics Canada 1992). However, the farmers continue to provide cheap and abundant food and raw materials for the rest of society.

Due to increases in farm productivity, world food price trends have shown a consistent annual decline of about 0.5 percent for much of this century. The grains and oilseeds sector in Canada and the United States is most affected by world food prices, and it has shown a decline in net income and corresponding increases in public subsidies in the last decade.

High levels of production are encouraged by direct commodity payments and subsidies as well as technological improvements. Unless drastic and unforeseen weather or political events intervene to inhibit worldwide agricultural production, abundant and cheap supplies of agricultural commodities should be available in the foreseeable future. It may be time to accept that prices at the farmgate will never increase.

Other Resources

In many areas of the Great Plains, mining and energy (especially oil and gas) have been important components of the family farm income. However, the price outlook for all primary resource commodities that are traded internationally is negative. This downward trend may occasionally change for some commodities, but the overall decline is inevitable, as commodity producers around the world continue to improve efficiency. For many rural and farm residents, seasonal and temporary employment in other resource sectors is an important source of income that buffers the farm against low market prices. However, it appears that this safety valve is increasingly at risk.

Services

The service industries constitute the largest employment sector in rural and small town areas and account for the largest growth in employment. The future of the service sector in rural areas is by no means secure, and the widely acknowledged growth of the service sector has been largely an urban-based phenomenon. Furthermore, as rural population diminishes, there is less need for service providers.

Manufacturing

Manufacturing provides jobs for farmers and other rural dwellers and contributes to the growth and stability of the larger rural economy. In rural and small town Canada, manufacturing accounts for about 15 percent of employment. Stabler and Molder (1992) noted that there are no compelling reasons of a technical nature that constrain rural-based manufacturing industries. The quality of the rural workforce will be a factor in retaining and attracting manufacturing.

Interrelationships within the Rural Economy

Agriculture may be the dominant land use in the agricultural regions of Canada, but it is no longer the dominating feature of the rural economy. The level of nonagricultural activities in rural areas exceeds agriculture in all regions. Thus, the fate of agriculture is not necessarily the fate of rural areas.

In the United States, a major event in the 1980s was the decoupling of agricultural conditions from rural conditions (Freshwater and Deaver 1992). This has major implications for public policy because agricultural policy is no longer necessarily synonymous with rural policy.

This phenomenon is not restricted to North America. In Europe, agriculture is no longer the driving force behind rural development, and in most villages farmers have become a minority. However, agriculture and forestry still use more than 80 percent of the land area of the European Community, a distinction that is becoming, like in North America, increasingly significant.

The farm family's linkage to the nonfarm sector has increased in variety and importance. Farm families are obtaining a growing share of their income from off-farm sources, and more than one-half of farm family income in Canada comes from off-farm work. These multiple activities are becoming the norm in rural areas. Maintaining off-farm job opportunities for farm people is essential.

The People of the Great Plains

An understanding of rural and farm people is vital if policies are to be designed that will address the issue of conserving rural life. To what extent do rural and farm people differ from urban people?

The rural world view has a unique character and history and is still important to rural residents. The rural milieu is composed of cultural practices rooted in the agrarian tradition and a moral and ethical consensus about acceptable behavior (Apedaile 1992). McKie (1992) notes that rural residents continue to value the style of life and the social institutions that are characteristic of rural areas. Urban residents as well idealize the rural lifestyle as an attractive alternative to urban living.

The agricultural way of life produces people who, by and large, have a work ethic. One reason that farmers are such reliable workers is because, paradoxically, they get their satisfaction from their farms rather than from off-farm jobs. They are less interested in office politics and more interested in keeping the off-farm jobs that allow them to hold onto their farms.

The attachment that farmers and rural people have to private property is often underestimated. The private, independent farmer is the repository of rural values articulated in the Jeffersonian ideal of yeoman farmers. Private farms are the essential component of the agricultural way of life, and an understanding of private land ownership is a prerequisite for implementing conservation programs on private land.

Farm and rural people tend to have a utilitarian, stewardship view of the environment. To many rural people, wilderness preservation without use, or at least management, is a waste. Conversely, urban people tend to prefer preservation as a means to protect the environment. Because the vast majority of urban people are employed in the service sector, it is easy for them to demand a stop to agricultural activities in faraway places that affect faraway people.

Many environmental conflicts are really urban versus rural conflicts. Rural people are known to have difficulty appreciating the fact that their private landscapes support significant public resources such as water and wildlife, and that the larger society, to a point, has legitimate interest in how those resources are managed, especially given the level of farm subsidies. Urban people, on the other hand, must realize that resources such as wildlife on private land is a public resource produced at private expense. All of these differences often make rural versus urban conflicts over resource use and environmental management a dialogue of the deaf.

These lifestyle differences have led to extreme differences in world views and on public policy, and these differences become more profound as the population urbanizes. Gomez-Pompa and Kaus (1992), in a perceptive and thoughtful paper, state:

> The perspectives of the rural populations are missing in our concept of conservation. Many environmental education programs are strongly biased toward elitist urban perceptions of the environment and issues of the urban world. This approach is incomplete and insufficient to deal with the complex context of conservation efforts at home and abroad. It neglects the perceptions and experience of the rural population, the people most closely linked to the land, who have a firsthand understanding of their surrounding natural environment as a teacher and a provider. It neglects those who are most directly affected by the current political decisions made in urban settings regarding natural resource use. It neglects those who feed us.

Sustainable Community Development and Agricultural Policy

Overview

Agriculture stands out as the most distorted market in the world economy. The total cost of distortion to world consumers was about $260 billion in 1990 and without reform it will climb to $300 billion by the year 2000. Although agriculture is a relatively small part of the overall world economy, it is significant that the entire global trading system is currently being held hostage pending the resolution of agricultural trade policy issues. Subsidies have kept agriculture alive and Lerohl (1990), for example, states that "crude measures suggest that all or most of the producer surplus enjoyed by the agricultural sector in the Great Plains in several of the past years has come from public programs." It is a truism in ecology and perhaps life itself that "you can never do just one thing." Commodity price supports and export subsidies may have goals such as the preservation of agriculture and the improvement of rural life in general, but they have profound effects on biodiversity, economic diversification, and the social structures of rural communities.

Effect of Farm Subsidies on Biodiversity

Most of the commodity support and farm marketing policies in western Canada promote the cultivation of land or crops produced under cultivation as opposed to perennial crops. Cultivation of land requires the replacement of native or perennial vegetation with monoculture, resulting in essentially zero biodiversity. Moreover, the disturbance of the soil by cultivation, especially on light and fragile soils, causes soil erosion, degradation, and off-farm runoff. Only 13.2 percent of improved prairie farmland has a low soil erosion risk; the other 87 percent has medium to high erosion risk (Crerar 1991).

Direct support for the grain industry not only affects land use but discourages mixed farming and livestock operations. Livestock, especially beef cattle but other ruminants as well, contribute to farm diversification, the diversification of the landscape, and a decrease in soil erosion because of their reliance on perennial forage.

Effects of Agricultural Subsidies on the Rural Economy

In addition to affecting land use practices, current farm support programs distort the rural economy and can actually prevent the very economic developments that rural areas need. By encouraging the export of the unprocessed grain, value-added processing of grain and livestock production are discouraged. What link is there between agricultural diversification, or lack thereof, and community economic stability? To answer this question, a study in Saskatchewan compared the farm structure and viability of 598 communities in the southern, transitional, and northern zones within the agricultural region (Stabler and Olfert 1993). The authors note that, "A distinct positive association is observed between agricultural diversification, smaller farms, higher population density, and community viability."

Effect of Agricultural Subsidies on the Rural Community

The rural community can be thought of as a horizontal entity caught in a vertical world. The goal of the community is to grow, prosper, and provide a good life for its members. However, the consequence of agricultural policies that inhibit diversification is the diminishment of communities. Obviously, the decline of farming communities cannot totally be the result of agricultural subsidies since any rural community that sells resources on the global market is at risk. Nonetheless, agricultural policy must help and not hinder the adaptations that are needed.

Regarding community viability and vitality in Saskatchewan, Stabler and Olfert (1993) state: "Consolidation and lack of diversification (of farms) even within family ownership is seen to be sufficient to produce community decline." They go on to say, "It seems reasonable to infer, however, that a loss of social vitality accompanied the loss of economic viability since many social structures depend upon a number of people in the same way that commercial activities depend upon population size to satisfy market-threshold requirements."

Integrating the People, Economy, and Landscape

Overview

Around the world the distorting effect of outdated agriculture policies is affecting the sustainable development of rural communities. Von Meyer (1990) provides a European perspective:

> Due to an agricultural policy strongly biased toward the regulation of product markets and prices, European agriculture has run into serious crisis, which is at the same time economic, social, and environmental. The crisis not only affects agriculture but has also become a threat to the future development of many rural areas. Positive perspectives for the future development of agriculture and the rural areas in Europe can only be found if these economic, social, and environmental problems are solved simultaneously.

It is tempting to portray agriculture as the villain in the game of landscape and biodiversity conservation, but working with agriculture and the rural community is vital if such efforts are to succeed. Real progress is made when landscape and biodiversity programs contribute to the economic development of rural areas.

In the past, biodiversity conservation efforts tended to view farm people and agriculture as necessary evils to be dealt with in an adversarial fashion. After all, if they are responsible for the decline of biodiversity, how can we work with them to conserve the landscape? The "we-they" approach only alienates farm people and ensures that any successes will be small, grudging, and of little significance in the larger scheme of things. A much better approach is to ensure that biodiversity conservation enhances agriculture and the communities that depend on agriculture. In other words, conservation programs must be designed within and not outside the agricultural way of life. Farm people are not antagonistic to landscape and biodiversity conservation; they just do not feel that the burden of landscape and biodiversity conservation should be borne by them alone. In this day and age, farmers absolutely cannot bear additional costs, but skillfully designed programs that contribute to a farmer's bottom line and the regional economy as a whole will be widely supported by farm, rural, and urban people alike.

Private and Public Goods

The agricultural landscape (i.e., private land) produces public goods, private goods, and goods with both public and private qualities. Public goods from the countryside include an attractive landscape, biodiversity and wildlife, and water management, while private goods are crops and livestock sold from the farm. An owner of farm property has a mixture of complete, partial, and nonexistent property rights. The specific mix of property rights varies across North America, but the principle remains the same. Landowner have every incentive (some provided by government) to convert the public resources for which they receive no benefit with private resources that pay their own way. Agricultural policy has attempted to maximize the return from private goods (such as wheat or corn), but this has been accomplished at the expense of the public goods. Public goods belong to everyone or no one and cannot be bought and sold; therefore, they are often replaced on private land in favor of private goods.

The reason that farmers choose private goods over biodiversity and other public goods is that public goods, historically, have not paid their way. Furthermore, with a few rare exceptions such as hunting rights, it has been difficult or impossible to establish markets for public goods such as aquifer recharge, biodiversity, water management, duck production, or water quality. Somehow, there must be transactions in these public environmental resources if they are to be valued by farmers. The obvious solution is to use public policy to create incentives encouraging the provision by farmers of ecological services to a market that demands them.

Ecological Services: Newest Product from Farms

It is likely that under a new General Agreement on Tariffs and Trade (GATT) for agriculture, commodity and export subsidies will be reduced. Such reformed GATT, however, will allow programs that support rural infrastructure, family farm income, and the environment. Governments should investigate redirecting that portion of farm production subsidies, which will be removed at any rate under a reformed GATT, to create incentives for farmers and rural landowners to deliver ecological services from private land as the newest farm product. These services would include biodiversity conservation, wetlands for flood control and water management, trees, endangered species habitat, buffer zones along waterways, and wildlife habitat. What would be some of the benefits of such an approach?

Benefits for the Environment

Obviously, the spending of these farm subsidy dollars on environmental improvement will conserve and enhance biodiversity, improve water quality, and improve rural landscape amenities. This is already happening as part of the North America Waterfowl Management Plan (NAWMP) in the prairies of Canada. Under the Prairie Habitat Joint Venture of the NAWMP, 398,743 acres of waterfowl habitat in the agricultural region of western Canada have been secured as wildlife habitat. An additional 228,870 acres of habitat have been restored and 703,932 acres are being managed for wildlife. Agricultural considerations are also factored in, and much of this land produces farm commodities that are compatible with wildlife. It must also be emphasized that although the NAWMP is primarily a waterfowl program, it is estimated that at least 168 prairie wildlife species are expected to benefit from habitat projects of the NAWMP.

Benefits for the Economy

The provision of ecological services as a rural "industry" may be a stable source of rural income that farm communities need. Given that many farm subsidies will be lost under a new GATT, using what would be lost to pay farmers for ecological services would provide support for agriculture in a very cost-effective and environmentally sound manner. Actually, all governments are trying to make current spending more efficient, and addressing multiple objectives with the same spending will appeal to them.

The economic benefits from landscape changes are difficult to quantify. An example is the U.S. Department of Agriculture's Conservation Reserve Program (CRP). By 1988, CRP had 25.5 million acres of erodible land sown down to permanent cover (Crerar 1991) with 11 million acres on the Great Plains in Montana, Minnesota, North Dakota, South Dakota, and Nebraska (Luttschwager and Higgins 1991). An estimate the economic benefits of the CRP found an overall natural resource benefit of $10 billion per year for the entire country (Ribaudo et al. 1989). For the Great Plains the annual benefits were $216 million in improved soil productivity, $306 million in water quality benefits, and $148 million in air quality benefits.

It is assumed that a landscape that has been diversified with ecological services would provide broader economic opportunities such as tourism, guided hunting, agroforestry, trapping, and extensive livestock production. The land would generate a base income from ecological services leaving the landowner free to pursue other income opportunities on that land provided that the contracted services are maintained. Given the many activities of most farm families, an additional source of income from ecological services would be welcome.

Social and Political Considerations

The incentive approach, as opposed to the regulatory approach, is within the agricultural way of life. Once farmers get accustomed to the strange idea of ecological services, landscapes and biodiversity will become crops like any other. There is evidence that on-farm conservation programs, like the NAWMP, are acceptable to the farm and rural public. The Manitoba Habitat Heritage Corporation, the Manitoba delivery body for NAWMP, determined attitudes to the NAWMP in the Manitoba target zone—an area of primarily agricultural land with a strong agrarian tradition. When farm and nonfarm rural residents were surveyed, 89 percent of respondents registered "strong support" or "somewhat support" of the NAWMP. Keeping in mind that this was a farm and rural audience, compare this with the problems caused by attempting to regulate private land use to preserve wetlands, for example. Actually, the ecological services approach is best suited to wetlands because of society's general agreement on value of wetlands, and because wetlands on private agriculture land are usually nothing but a cost to the individual farmer.

No matter how small the basket of agricultural subsidies becomes due to general budget reductions, a portion should be reserved for ecological services. Indeed, this should be done even if the GATT is not reformed. A proactive approach by farmers, farm groups, and other supporters of rural life will be a show of goodwill to the nonfarm, nonrural public that is increasing in number, influence, and political power. The farm and rural community and, indeed, the entire agricultural establishment on the Great Plains has a window of opportunity in which to take the initiative to design a program of ecological services that will appeal to a broad range of individuals. This window exists for several reasons:

- Farmers are still well thought of by the rest of society.
- Concern for the environment is still high.
- If a GATT agreement is imminent, many millions of dollars of farm support will be available for other purposes.
- The farm and rural community is urgently looking for other sources of income.

Stated more bluntly, unless the farm and rural communities take this opportunity to design a system they can live with, one may be designed for them. And it is apparent from the experience of the NAWMP that systems compatible with the agricultural way of life are possible. For governments, the ecological services approach should be attractive for a number of reasons:

- It provides support for the farm and rural community in a manner regardless of what is in the final GATT agreement.

- It "greens" agriculture.
- It solves difficult land use conflicts.
- It appeals to the urban majority.
- It is cost-effective.

Implementation

Implementing a system of programs to improve the rural environment, economy, and community that is based, in part, on the delivery of ecological services must fit within the agricultural way of life to be successful. Specifically, the system must be locally driven because rural residents know the characteristics and, of great importance, the history of the landscape. Also, the system must fit within the local agricultural economy and be driven largely by incentives as opposed to regulation. Finally, rural development agencies should develop the capabilities to assist rural communities in taking economic advantage of the landscape changes that will occur.

Conclusions

The Great Plains is a complex system where social, landscape, and economic factors are constantly interacting. In order to achieve truly sustainable community development on the Great Plains, public policies must achieve more than one objective. If sustainable community development, or the "conservation of rural life," is the ultimate goal, this can only be reached through policies that understand the complexity and simultaneously advance the social, landscape, and economic agendas.

This chapter supports the redirection of a portion of current agricultural production and export subsidies to assist farmers with providing ecological services to the rest of society. This new approach will have multiple benefits.

References

Apedaile, L. P. 1992. The economics of rural-urban integration: A synthesis of policy issues. In *Rural and small town Canada,* ed. Ray D. Bollman. C92-094720-4. Statistics Canada.

Bollman, R. D., and R. Biggs. 1992. Rural and small town Canada: An overview. In *Rural and small town Canada,* ed. Ray D. Bollman. C92-094720-4. Statistics Canada.

Crosson, P. 1991. Sustainable agriculture in North America: Issues and challenges. *Canadian Journal of Agricultural Economics* 39:553-565.

Crerar, A. 1991. Prairie agriculture: The real options. *Policy Options*, Sept.

Freshwater, David, and K. Deaver. 1992. Falling further behind: Current conditions in rural America. In *Rural and small town Canada,* ed. Ray D. Bollman. Statistics Canada.

Freshwater, David, P. Ehrensaft, and L. P. Apedaile. 1992. The implications of global restructuring of primary industries for rural and single industry regions. Paper presented at the National Seminar on the Development of Rural and Single Industry Regions, Annual Policy Seminar, Coaticook, Quebec (April 1991).

Gomez-Pompa, A., and A. Kaus. 1992. Taming the wilderness myth. *BioScience* 42(4):271-279.

Lerohl, M. L. 1990. The western Canadian farm policy dilemma. *Canadian Journal of Agricultural Economics* 38:557.

Luttschwager, K. A., and K. F. Higgins. 1991. Some sociological and ecological effects of the Conservation Reserve Program in the north Great Plains. Paper prepared by South Dakota Cooperative Fish and Wildlife Unit, South Dakota University.

McKie, C. 1992. Does rural matter? In *Rural and small town Canada,* ed. Ray D. Bollman. C92-094720-4. Statistics Canada.

Ribaudo, M. C., S. Piper, G. D. Schaible, L. Langer, and D. Colacicco. 1989. CRD—What economic benefits? *Journal of Soil and Water Conservation* 44(5):421-424.

Stabler, C., and J. Molder. 1992. *Rural manufacturing industry: Products, markets, and location requirements.* Department of Agricultural Economics, University of Saskatchewan, Saskatoon.

Stabler, C., and M. R. Olfert. 1993. *Farm structure and community viability in the northern Great Plains.* University of Saskatchewan, Saskatoon.

Statistics Canada. 1992. Trends and highlights of Canadian agriculture and its people. In *The 1991 census of agriculture.* Statistics Canada, Ottawa, Ontario.

von Meyer, H. 1990. From agriculture to rural policy in the economy. Paper presented at a seminar, Europe 1993: Implications for rural areas, Aberdeenshire, Scotland (Mar. 19-23).

7. Sustainable Rural Economic Development in the Great Plains

Margaret G. Thomas
Midwest Research Institute

The goals of rural economic development and natural resource conservation are mutually supportive when both are pursued in a context of sustainability. Being able to visualize a sustainable society is the first step toward building one, but old paradigms limit the ability to create a new vision of the future and move toward it. Transportation systems, water resource and flood control projects, mining activities, utilities, and settlement patterns have all played a role in changing the landscape and degrading ecosystems of the Great Plains. Probably, however, the most critical force has been that of conventional agriculture. Indeed, the same agricultural forces that have diminished ecosystems have indirectly diminished many Great Plains communities.

New ways of viewing agriculture and rural development to make each more sustainable will benefit farmers, ecosystems, and communities. The natural resource community needs to become actively involved in helping farmers diversify to new types of sustainable growing operations that would provide more wildlife habitat and additional opportunities for entrepreneurship on farms. Many ideas for rural enterprises that are based on stewardship rather than consumptive use of resources lend themselves to a regional strategy that could both improve the quality of life in an area and promote unique approaches for tourism. Governments can help guide and smooth the process of change toward a future that is both environmentally and economically sustainable.

Environmental Sustainability

The goals of rural economic development and natural resource conservation are mutually supportive when both are pursued in a context of sustainability. The goal of environmental sustainability is necessary not only because it ultimately protects us all from threats to life, health, and quality of life but because it creates the setting in which long-term, rewarding employment is possible. Ultimately, a livable environment makes employment possible. In fact, even in the short term, emphasizing sustainability principles makes good business sense because it can create competitive advantages and new opportunities for local areas.

Stanley R. Johnson and Aziz Bouzaher (ed.), *Conservation of Great Plains Ecosystems*, 97–106.

Likewise, the awareness that we must become stewards, not consumers, of our resources will only come from areas where residents have reached some level of economic security, where communities are stable, leadership is enlightened, and where there is sufficient economic opportunity to permit a diversity of thinking and talent among residents. It is hard to imagine how ecological sustainability could be achieved in the absence of economic sustainability and well-being. The protection of our ecosystems requires long-range perspectives on life, perspectives that are not possible for those who are living on the edge in our society.

A holistic approach to improving the future of our rural communities and the future of our natural resource base is needed. This is true in every rural region of the country. *The key is to educate both the economic development community and the environmental community to work together to develop strategies that are mutually supportive.* These strategies for a sustainable future will vary somewhat according to the natural resource and economic issues of an area, but there are several general observations that can be made that would assist any region's communities in developing such strategies.

Visualizing Sustainability

Being able to visualize a sustainable society is the first step toward building one. A *sustainable society* has been defined as one that satisfies its needs without diminishing the prospects of future generations. Pursuing sustainability requires taking into account the full range of costs, including opportunity costs, that are at stake in our present activities. It means making fundamental changes in the way products are produced, used, and disposed of, and developing more benign production technologies. It means reducing the activities that are directly or indirectly degrading the environment. And it means developing plans to guide future decisions—plans that take into account the values of local residents, what the community needs to produce to sustain those values, and the steps needed to sustain the natural resources and landscape values that will support what they wish to produce.

Within a community, the vision of sustainability includes:

- Maximizing energy conservation and energy use efficiency
- Placing greater reliance on renewable energy and less on fossil fuels
- Maximizing reuse and recycling while minimizing waste
- Emphasizing environmentally benign production technologies and minimizing the use of hazardous materials
- Expanding human resource capabilities through lifelong education and training
- Using small-scale, appropriate technology where possible and producing community goods and services locally

It should be noted that adopting these strategies will not mean massive layoffs in any existing industries. In fact, avoiding energy use via improved efficiency is cheaper and promises more job creation than supplying energy from either conventional or renewable energy sources. For example, jobs will be created in home insulation, carpentry, and sheet metal work as homeowners seek to acquire more weather-tight homes.

As another example, recycling already is an important source of jobs. Compared with incineration and landfilling, recycling creates more jobs and is still the cheaper alternative because of its lower capital requirements. Renewables create more jobs than conventional energy industries because their capital requirements, with the exception of photovoltaic cells, are much more modest and their labor needs are greater.

These sustainability strategies will actually create jobs because the best aspects of the human propensity to buy, sell, and produce can be an engine for change. The demand for more efficient autos, stoves, refrigerators, air conditioners, heaters, dryers, and lights and for more cogeneration plants, including those that use biomass to generate power, will all stimulate entrepreneurial activity. Similarly, several activities will not only create jobs but also new technologies with export potential: (1) comprehensive, systematic recycling of metal, glass, paper, and other materials; (2) eliminating waste flows by restructuring industrial processes; (3) recycling nutrients in sewage to aquaculture; and (4) composting food and yard wastes. Likewise, designing and producing materials that are durable and repairable will offer entirely new production processes and products.

And what of the outlying rural areas beyond the community—what is the vision of sustainability in the undeveloped lands where ecosystem conservation is a real concern? It would include:

- Embracing sustainable agriculture and forestry principles
- Emphasizing the identification and preservation of biodiversity, functioning ecosystems, wildlife habitat and wetlands, recreational opportunities, prime farmland, and scenic landscapes as unique resources with many short-term and long-term values
- Extracting resources only to the degree that they can be renewed
- Emphasizing compact development so that homes, jobs, and shopping are close together and encouraging rail lines, bike paths, electronic home offices, and solar power to reduce dependence on the automobile

Outmoded Paradigms

Old paradigms are limiting the ability to create a new vision of the future and move toward it. One such outmoded paradigm is the concept of growth. Many in society are still reluctant to come to grips with a truth—limits on some forms of growth, such as the growth that consumes physical resources, are inevitable. In the Great Plains, economic growth that is achieved through spending the primary assets of the region—landscapes, soils, wildlife and fisheries, wetlands, clean air and free-flowing rivers, groundwater—is not growth at all but a shameful depletion of the region's (and the nation's) capital assets. And in communities, the economic growth that is achieved through technologies that actually reduce the number of jobs, or public subsidies to new industries that reduce funds for schools or infrastructure, or shopping malls that drain the Main Street merchants, is not truly growth but a reallocation of resources from "old" residents to "new" residents.

In short, possibly the greatest hurdle before Great Plains residents will be confronting the reality that increased growth is no longer a sufficient goal for society. The goal needs to be an improvement in human progress, an articulation of deeper values, and a greater commitment to improving the long-term quality of life in an area.

The environmental community has a great stake in promoting the conservation of ecosystems within the context of sustainable development plans. The remainder of this chapter looks specifically at the Great Plains, the major threats to sustaining ecosystems and rural communities, and how sustainable development strategies can be used to conserve both.

Problems of Conventional Agriculture

Degraded Ecosystems

Our transportation systems, water resource and flood control projects, mining activities, utilities, and settlement patterns have all played a role in changing the landscape and degrading ecosystems on the Great Plains, but probably the most critical force has been that of conventional agriculture. The separation of grain and livestock production and the monoculture cropping systems that developed following World War II led most American farmers to rely heavily on chemical fertilizers, pesticides, and energy, as well as increased use of credit. The intensive chemotherapy that has been built into the business of farming—the use of pesticides, herbicides, and artificial fertilizers—combined with extensive tillage and excessive water use has adversely affected natural resources across much of the Great Plains. Wetland, terrestrial and aquatic habitats, and soil, water, and groundwater resources have all declined in quantity and quality, and much of this decline has been caused by national agricultural practices.

Indeed, it is hard to think of a single ecosystem of the Great Plains that has not been adversely affected by agricultural practices. The majority of wetland and terrestrial habitats have probably vanished under vast fields of row crops. The current feed grain and feedlot system for producing beef that is characteristic of much of the Great Plains agriculture is perhaps the most resource-intensive and polluting portion of U.S.

agriculture. The majority of the corn raised in this country is used to feed cattle and the erosion created in this process is immense. The crops take large quantities of water, much of which is mined from aquifers at a rate greater than recharge. Heavy applications of fertilizers, herbicides, and pesticides are necessary, leading to widespread pollution problems and higher production costs. The feedlots themselves contaminate groundwater and surface water.

In addition, the single cash crop system has also robbed farmers of their acclaimed independent lifestyles. Conventional farming has trapped farmers—once the very symbol of self-reliance in this country—into a spiral of borrowing to meet rising production costs; planting increasing acreage in hopes of reaping a large enough crop to make enough money to cover costs, interest, and livelihood; depending on world markets to support the existing farm structure; seeing prices fall because of fluctuations caused by the overproduction of all other single crop farmers who are also overplanting to increase their revenues; netting a return lower than needed to pay costs; and being forced to borrow again. And, ironically, the outcome of all this grain-fed beef for human consumption has also contributed substantially to the high fat content of the American diet, which is a major health concern.

Diminished Communities

The same agricultural forces that have degraded ecosystems have indirectly diminished many Great Plains communities. Many, perhaps most, of the once agriculture-based communities in the Great Plains have been "simplified" and homogenized because of the decrease in the outlying farming population, the increase in farm size, and the growing dominance of large, high-technology agribusinesses. As the farming operations in the areas surrounding agriculture-based communities have become more efficient, labor has become less important and jobs have declined. Thus, as the size of farms has increased, rural towns have been further diminished, providing fewer opportunities for employment, entertainment, and social commitment. Young people have been drawn to urban areas and rural communities have been weakened.

As the technology of agriculture has become more complex and expensive, fewer businesses have been able to participate in developing and producing desired products. Many small companies that were once suppliers have disappeared, merged with large companies, or gone out of business. Thousands of small agricultural communities may become ghost towns in this decade. In the United States, it is estimated that one Main Street business closes and three to five workers in the community lose their jobs for every five farmers who go out of business. In 1982 there were 2.2 million farms in the United States; by the year 2000 there will probably be only 1.2 million.

The increase in farm size has also worked to isolate the farm owner from his or her land—the fundamental source of wealth for that farmer—and from the surrounding human community.

Benefits of New Approaches

New ways of viewing agriculture and rural development to make each more sustainable will benefit farmers, ecosystems, and communities. The truth of the matter is that the bulk of farming operations that produce one or two cash crops or types of livestock will have a tough time surviving in the future. They will face a future of commodity gluts, volatile prices, rising production costs, and other financial uncertainties. The cost of farming has increased at a rate greater than the income received. The disparity would be even greater without the direct and indirect subsidies provided by government. Yet while costs have gone up, the benefits of higher retail prices have not gone to farmers. Under the present system, only 3 to 27 percent of the retail dollar goes to those who grow the crop. The rest goes to those who process, wholesale, and retail the farm product.

Thus, in the future many farmers will have to make fundamental changes in the way they farm. They will need to produce a greater variety of crops and services, supply more of their own raw materials, institute natural pest control and soil conservation techniques, and become entrepreneurs. The Great Plains farmers are already exploring ways to stay in business while reducing their dependence on off-farm inputs (and thereby lowering operating costs). They are looking at local or direct marketing, value-added processing, and niche marketing. In the future, they will consider production—such as ethanol from biomass, solar, or wind power technologies—on their farms. Just as organically grown products are finding their niche, "sustainable farm" regions and products will be identified by logos. More communities will work with farmers to identify "win-win" projects such as community-supported agriculture programs, whereby a farmer sells shares in his or her crop to local households who in turn get a say in the food crops that are grown.

Indirect impacts of these changes are certain to benefit terrestrial and aquatic ecosystems, particularly as chemical inputs and the frequency of tilling are reduced. Therefore, any activities by the rural development and the environmental communities to aid the Great Plains farmers practice more sustainable agriculture are going to benefit the ecosystems of the Great Plains.

It is worth noting that trends in the rural economic development field are consistent with the trends toward sustainable agriculture. Current strategies for rural economic development have moved beyond the industrial recruitment focus of the past and can be characterized by:

- A focus on retaining existing businesses, encouraging the formation of new small businesses, stimulating entrepreneurial and new venture creation, developing microenterprises, and creating job training and education programs for the current labor force
- Increased emphasis on adding value to existing products that are locally produced, finding new niche markets that can utilize existing resources, and tapping import substitution opportunities

In other words, the economic development community has moved toward a new examination and appreciation of the resources, be they natural, human, business, or capital, that are unique to each area or community. Moreover, the major thrust of current economic development thinking is in looking afresh at the uniqueness of rural areas and their competitive advantages.

All of this is encouraging news for those concerned with environmental issues. Even more important, new opportunities now make wildlife habitat and wetland conservation and management and the protection and enhancement of biodiversity part of the "new" goods and services that farmers might market.

Diversification through Natural Resources

The natural resource community needs to become actively involved in helping landowners diversify to new types of sustainable growing operations that would provide more wildlife habitat and to other new opportunities for entrepreneurship on farms. Indeed, the special knowledge of natural resource specialists may be the key to success for some new ventures. Examples include:

- Extracting essential oils from native trees and plants for unique aromatics
- Reintroducing native fruit trees that can be harvested for their fruits or for flavorful cooking wood
- Growing and harvesting native trees that provide decorative wood (e.g., diamond willow, juniper, walnut, chestnut)
- Reintroducing edible greens, roots, tubers, and forest botanicals that have value as medicinals or pharmaceuticals
- Growing plants for greenery, landscaping, or use in the floral industry
- Growing crops for honey production
- Developing recreation and wildlife enterprises

All of these enterprises have the potential to benefit both landowners and rural residents, in general, while changing agricultural activities in ways that would benefit wildlife.

The last category, recreation and wildlife recreation enterprises, includes both consumptive and nonconsumptive wildlife-based enterprises—such as commercial hunting preserves, fishing—as well as appreciative use activities such as wildlife observation, bird-watching, photography, the formal or informal field education of schoolchildren or adults, and even scientific research. Many other recreational enterprises also would be more successful in the marketplace when combined with access to a high-quality wildlife habitat, such as bed and breakfasts, conference facilities, farm and ranch vacation homes, guide services, and organizational camps.

Benefits of Stewardship

Many ideas for rural enterprises that are based on stewardship rather than consumptive use of resources lend themselves to a regional strategy that could both improve the quality of life in an area and promote unique approaches for tourism. As Great Plains residents work to create a new vision of sustainable communities and sustainable farms, they should also be looking for ways to pull together the community and the region's landowners and resource managers. Many rural enterprises would also offer at least seasonal employment for area youth. Regional strategies would tie together an area's communities and natural resources so that all the region's resources might be appreciated and marketed.

A three-county area of north central West Virginia has taken a regional approach: Back Roads Adventures (BRAD), an innovative, new travel business developed as a way to match visitors and tourists with the unique scenic, historic, cultural, and natural resources of a rural area. The service was designed primarily for individuals, families, small groups, and visitors to travel through the area to prearranged locations for personal encounters related to the area's natural resources, traditional culture, craftsmanship, and rural life. Some 150 persons were identified who (1) possessed skills that reflected unique aspects of rural life in the area or had special knowledge and experience with interesting features of the natural environment, (2) were entertaining communicators and interpreters of their skills and expertise, and (3) welcomed the opportunity to interact with individuals or small groups of visitors for short periods of prearranged time in their shops, homes, or outdoors. Examples of their activities are shown in Table 1.

The initiators of BRAD are confident that they have hit upon a very important strategy that reaches several objectives:

- To provide activities that allow interested residents and visitors to learn about a rural area and the values embodied therein
- To provide employment and sources of additional income to those who make a significant contribution to protecting and preserving the rich and diverse cultural heritage of a region
- To increase financial support for important natural, cultural, and historic features that may be threatened by development
- To open new areas of tourist activity and promote a positive image of life in rural areas

Table 1. Promoting a rural region's unique natural, cultural, and historic resources

Exploring the Outdoors with a Personal Guide
Edible wild plants and roadside herbs
Hunting with a camera
Hiking a gentle mountain trail
Mountain wetlands adventure
Collecting wild mushrooms
Stargazing
Mountain fly-fishing
Seining for minnows

Unique Features of Rural Mountain Life
Folk medicine and herbal remedies
Making home brew
Sheep, wool, and the spinning wheel
Growing shiitake mushrooms
Mining for coal

Outstanding Local Craftspeople
Basket weaving—white oak, wild vines
Hand weaving—traditional and contemporary
Quilt-making and patchwork
Muzzle loader rifle construction
Original design pottery
Botanical designs in clay
Deep bas-relief in wood
Hand-braiding of rugs
Silversmithing

Mountain Music
Mandolin, guitar, and fiddle
The hammer dulcimer
Saturday night country music show
Bluegrass in the mountains

Outdoor/Nature Study
Winter botany and nature study
Identifying wild birds of spring
Seining for minnows
Butterflies and insects
Exploring the unusual courtship of woodcock and snipe
Tree-tapping and making maple syrup
Growing culinary herbs
Hunting wildlife with a camera
Growing shiitake mushrooms
Identifying fall insects by sound

Back Roads Adventures shows how sustainable use of natural resources can be effectively woven into a broader rural economic development strategy that can provide extra money to many different residents in a rural region. Those involved with BRAD see many opportunities for fee-generating activities related to nature study and outdoor recreation. Successful recreation and wildlife recreation enterprises are, ultimately, businesses based on both human and natural resources. Any enterprise that can tap grass roots resources has important implications for natural resource, cultural, and historic preservation as well as for economic development.

Eventually, farmers will be weaned from their dependence on federal crop subsidies. Perhaps those subsidies could reimburse landowners for stewardship that maximizes the ecological services that wetlands and wildlife habitat provide—such as endangered species habitat, observation of habitat areas, green belts around urban areas, flood water retention, research and education, groundwater recharge, wind erosion control, soil formation, energy production, air quality enhancement, scenic riverways, water supply, waste removal and assimilation, chemical recycling, wave erosion control (large waterways), and maintenance of global chemical balance. The environmental community, the agricultural community, and the economic development community should all work together to develop innovative ways to capture these values in the marketplace among the products of the "new farm."

Role of Government

Governments can help guide and smooth the process of change toward a future that is both environmentally and economically sustainable. Community leaders will need technical assistance to guide residents and local governments in weaving principles of sustainability into economic development plans. Governments can help by providing materials that show clearly how closely tied our natural resource management policies are with the long-term stability and economic health of our rural communities. These materials need to answer several questions:

- What is happening to our communities, our natural resource base, and our ecosystems in the Great Plains?
- What changes are necessary to move to a concept of sustainable rural development?
- What will be the entrepreneurial opportunities and the job prospects with sustainable community development, agriculture, and resource management?
- How can we identify and strive for "win-win" strategies that improve both our land resources and the future of our rural communities?

Overall, the protection of principal ecosystems and natural resources in our Great Plains requires a multifaceted approach. Traditional tools such as education, regulation, and financial incentives are needed. But pursuing this protection through community-based sustainable rural economic development strategies is a key approach that should be used as well.

8. Public Infrastructure: Highways

Bruce Cannon
Federal Highway Administration

The Intermodal Surface Transportation Efficiency Act (ISTEA), a landmark act of 1991, will have a major impact on the economic and environmental health of the Great Plains states, as will other ground transportation issues of the future.

The federal/state cooperative relationship relative to highways was primarily defined by the Federal-aid Road Act of 1916 and continues today. The states' role is to select, plan, design, and construct highway improvements and to maintain and operate the highway system. The federal responsibility is to promulgate standards, review and approve state proposals and project actions, ensure compliance with federal laws, provide technical assistance, distribute federal funds, and reimburse the states for previously approved eligible expenditures.

Overview of ISTEA

The ISTEA of 1991 expanded the funding and, of more importance, substantially modified the nature of the federal-aid highway programs. This legislation covers not only highways, but also highway safety and transit programs. To grasp the impact of this massive and complex legislation on the operation of the federal-aid highway programs, it is helpful to look at ISTEA's prevailing themes:

- **Intermodalism.** The new program directs that one transportation mode must interface with other modes. This means that all different modes of transportation must work together to form a seamless transportation network that will allow people and products to move from one mode to another smoothly, with minimal congestion or interruption.
- **Flexibility.** The state and local communities are the best architects of programs to meet their transportation needs, so they are given choices about the use of federal funds.
- **Efficiency.** The effective and efficient use of limited resources is reflected in the use of performance-oriented life-cycle cost principles and in six required management systems. It is also promoted in the assistance given to states to develop uniform commercial motor vehicle registration and fuel tax reporting agreements and in preserving land for future transportation facilities.

Stanley R. Johnson and Aziz Bouzaher (ed.), *Conservation of Great Plains Ecosystems*, 107–118.

- **Mobility.** This nation's demand and need for improved access and for safe, comfortable, convenient, and economical movement of goods and services is best met by a National Highway System (NHS), designed to identify the most important roads to receive special emphasis.
- **Innovation.** Epitomizing this best is the substantial funding to develop the Intelligent Vehicle Highway Systems.
- **Environment.** In emphasizing improvement and enhancement of the environment, ISTEA considers land use, environmental and social effects, and coordination with the Federal Water Pollution Control and Clean Air Acts. New programs—Congestion Mitigation and Air Quality Improvement, Scenic Byways, and Congestion Pricing Pilot programs—are established. New uses of funds—transportation enhancements, wetland banking, and mitigation of adverse impacts to wildlife, habitat, and ecosystems—are permitted.
- **Highway Safety.** New initiatives include an emphasis on work zone safety and a new program encouraging the use of seat belts and motorcycle helmets.
- **Investment.** A number of new activities are designed to enhance investment and encourage innovative financing approaches, such as cost-sharing partnerships among the private sector, state transportation agencies, and state and federal revenue agencies.
- **New Partners.** The traditional highway agencies and interests are mandated to reach beyond their boundaries and create new or enhanced partnerships (e.g., a state working in a pro-active partnership with air and water quality agencies).
- **Planning.** Solid, comprehensive planning is strengthened and expanded through revisions to the existing metropolitan transportation planning process and establishment of statewide transportation planning.

In addition to restructuring the surface transportation programs, the ISTEA authorizes funds for fiscal years 1992 to 1997. The total authorized for highway, highway safety, and transit programs is over $155 billion, which is 50 percent more on an annual basis than the amount authorized by the previous multiyear legislation. Clearly, Congress recognizes serious needs in transportation infrastructure. The highway funds are entirely supported by the Highway Trust Fund from fuel taxes and heavy-truck user fees.

The federal funds are delivered through categorical programs on a cost-reimbursable basis, not as one lump sum. The Great Plains receives nearly 20 percent as its share of the U.S. total.

ISTEA and the Great Plains States

As the ISTEA critical issues relating to the Great Plains states are considered, we must understand the area's statistical setting. Although the Great Plains encompasses 37 percent of the nation's land area, only 17.6 percent of America's people live there.

The total gross state product (GSP) of the Great Plains states in 1989 was $856.7 billion—almost 17 percent of the U.S. gross domestic product. These states contribute more than 45 percent of total U.S. production in farming. Within the region, however, mining and farming together contribute less than 8 percent of the region's total GSP, and the three primary industries in the region (manufacturing, services, and finance, insurance, and real estate) each contribute between 16 and 17 percent of the region's total GSP.

The Great Plains's 1,408,244 miles of public roads represent 36 percent of the total road mileage in the United States. By far, most of these roads (82 percent) are under the jurisdiction of local governments. The higher order roads (15 percent of the total), those that carry more traffic and are more important to interregional and interstate traffic, are primarily under the jurisdiction of the states. The small remainder of the roads are owned by the federal government and are on federal land.

In the Great Plains states, more miles were traveled per unit of population than in other states (19.1 percent vehicle miles of travel versus 17.6 percent of U.S. population), which is typical of rural areas.

Total spending for highways in the Great Plains states was $15.9 billion in 1991 for capital improvements and administrative and maintenance expenses of highways. As in the other parts of the nation, the Great Plains receives about 41 percent of its funds for capital highway improvements from federal government programs, primarily through the federal aid highway programs.

Because of ISTEA, and other recent landmark legislation such as the Clean Air Act Amendments of 1990, transportation service now goes well beyond the transport function. It now includes many other societal concerns such as wetlands, clean air, congestion damage to urban space, and much more.

ISTEA-produced changes will clearly have a major impact on transportation in the Great Plains region; its mandates and choices are creating opportunities for new approaches to transportation.

Several ISTEA activities will play a major role in the economic health and well-being of the Great Plains states: establishing the National Highway System (NHS), developing planning and management systems, coordinating the federal-aid highway program and the environment, and identifying major existing and emerging trade corridors.

National Highway System

Right now the Federal Highway Administration (FHWA), states, and local officials are participating in defining the NHS. As envisioned by ISTEA, the NHS will be the major focus for the federal-aid highway programs into the twenty-first century. In the words of the law, the NHS will "provide an interconnected system of principal arterial routes which will serve major population centers, international border crossings, ports, airports, public transportation facilities, and other intermodal transportation facilities and other major travel destinations; meet national defense requirements; and serve interstate and interregional travel" (Public Law 102-240 1991).

Clearly, the system will define the most important roads in the nation. The first step, as outlined by the ISTEA, is to functionally reclassify all roads in the states and propose which roads will be included in the NHS. Getting to this point will involve trade-offs between urban and rural areas.

The proposed NHS must include certain routes—Interstate, Strategic Highway Network (STRAHNET), major STRAHNET connectors, and high priority corridors—and mileages are required as part of the NHS in each of the Great Plains states.

The required mileage is the base from which the states started when developing an illustrative 150,000-mile NHS that was submitted to Congress last year. States will have considerable flexibility to propose routes other than the required mileage, or those included on the illustrative NHS, as long as the routes are functionally classified as principal arterials or provide major intermodal connections to the NHS, and the total mileage is within the mileage targets based on the 150,000-mile illustrative NHS. Variation from the target mileage is allowed if there is consistency with the objectives of the NHS and specific criteria established by the ISTEA. (For further information see FHWA 1992.)

It is important to note that the number of miles that a state has on the final NHS does not affect the amount of funding that the state will receive from the NHS program established by the ISTEA. However, this major initiative is crucial to the future surface transportation programs of the Great Plains.

Planning and Management Systems

The underpinning for transportation in the future will be comprehensive multimodal planning, not only at the metropolitan level but also on a statewide basis. The ISTEA enhances the planning process by bolstering it with new requirements for management systems. Effective planning will be essential for the health and well-being of each of the Great Plains states.

First, metropolitan planning is expanded and strengthened. A long-range transportation plan will include not only the current urban area but the area that is expected to be urbanized in 20 years. In areas that cannot meet Clean Air Act standards for ozone and carbon monoxide, the plan must encompass the area within the boundaries of the nonattainment area. In that way, the plan will truly consider the future. Also, the transportation improvement program that flows from the plan must be realistic rather than a wish list of projects. The transportation program must prioritize a list of projects and provide a financial plan to demonstrate that resources are available to carry out the long-range plan.

Second, although many states have been carrying out statewide planning, intermodal statewide planning is now a congressional requirement. As in metropolitan planning, the state is required to be involved in long-range planning and to produce a comprehensive state transportation improvement program containing a list of prioritized projects that are financially realistic.

The two planning processes do not exist in isolation but must be coordinated. This will obviously require trade-offs between the rural areas and the urban areas of the states. The statewide plan must also be coordinated with the appropriate Indian tribal governments and the Secretary of the Interior. In addition, the statewide transportation improvement program must fit with the Clean Air Act State Implementation Plan. Working with partners across agency boundaries will be absolutely essential.

The plans must also be multimodal. This means more cooperation and coordination among the metropolitan planning organizations, the state, transit operators, airport and waterport operators, and railroads. This linkage to commerce and trade necessitates coordination with economic development, resource management, and various industries such as tourism.

Each state must develop, establish, and implement six management systems:

- Highway pavement of federal-aid highways
- Bridges on and off federal-aid highways
- Highway safety
- Traffic congestion
- Public transportation facilities and equipment
- Intermodal transportation facilities and systems

In addition, each state must have a traffic monitoring system for highway and public transportation systems. This legislation clearly requires that states look at transportation outcomes rather than just the amount of money spent or miles put into service.

Federal-aid Highway Programs and the Environment

The environment receives top priority under ISTEA. For more than 100 areas that fail to meet air quality standards, the Clean Air Act requires them to give first priority in funding to projects that will reduce mobile source emissions.

Overall, the planning processes will ensure consideration of environmental issues, since the emphasis is on the integration of transportation planning objectives with sustainable development decisions, particularly the environmental consequences of those decisions. But more than requiring the states and locals to seriously consider the effect of their decisions on the environment, the ISTEA broadens the choices and increases the funding available to carry out these decisions.

A hallmark of the ISTEA is the wide range of options that state and local decision makers are given in the use of highway funds. The underlying philosophy is that the best transportation solutions do not necessarily emanate from Washington, D.C., or even from the states' transportation agencies. This flexibility seeks to put decision-making responsibility and money in the hands of those who are really on the "firing-line": the states, metropolitan areas, and local communities. Again, making these decisions will require cooperation with many different partners who have an interest in transportation outcomes. Some of these provisions give transportation decision makers choices for using highway funds to improve the environment.

- Transportation control measures identified in the Clean Air Act may be federally funded under ISTEA, in order to reduce single occupant vehicles and air quality emissions.
- Wetland banking and other wetland conservation planning measures are eligible for federal assistance, and funding for wetland banking may occur in advance of project construction. The FHWA has begun working on an agreement with The Nature Conservancy and Ducks Unlimited for technical assistance on wetland banking.
- Mitigation of adverse effects to wildlife, habitat, and ecosystems caused by projects under the Surface Transportation Program (STP) are eligible for funding.
- Federal-aid highway funds may be used for acquiring and removing nonconforming outdoor advertising signs.
- Federal funds can now be used to acquire land to preserve corridors for long-range (up to 20 years) transportation projects. Federal highway funds may be used in the cost of rights-of-way acquired in advance of federal approval.
- Federal-aid highway funds may be used to reduce congestion. This includes operational improvements such as traffic surveillance and control equipment, computerized signal systems, motorist information systems, integrated traffic control systems, incident management programs, and transportation demand management facilities, strategies, and programs. The ISTEA specified certain funds eligible (subject to a two-year limit) for start-up costs.
- A state may choose to use highway funds for transit and capital rail projects and vice versa. During the first year of ISTEA, this meant more than $300 million of federal-aid highway funds.

The ISTEA also established several programs whose only purpose is to ameliorate the environmental effects of highways or to make a positive contribution to the area surrounding the highway. The program funds may be used only for environment-related programs such as:

- Congestion Mitigation and Air Quality Improvement, which provides $6 billion over six years to reduce transportation-related air quality problems.
- Scenic Byways, through which states receive $50 million of funding for planning, design, and development of scenic byways. An additional $30 million in interim grants is provided to states with scenic highway programs to carry out eligible projects that include construction of pedestrian and bicycle paths along highways, rest areas, turnouts, highway shoulder improvements, passing lanes, overlooks, and interpretive facilities. Grants are available only when the scenic, historic, recreational, cultural, natural, and archaeological integrity of the highway and adjacent area are protected.
- Recreational Trails, for the purpose of maintaining and improving recreational trails. Projects must implement trail plans included or referenced in a statewide comprehensive outdoor recreation plan required by the Land and Water Conservation Fund Act. Permissible uses of the funds are administrative costs, environmental and safety education programs, development of urban trail linkages, maintenance of existing trails, restoration of areas damaged by trail use, trail facilities development, provision of access for people with disabilities, acquisition of easements, and fee-simple title for property and construction of new trails.
- Transportation Enhancement Program, which is funded through a 10 percent earmarking of the STP. Unlike the programs listed above, the 10 percent is a minimum; spending above the 10 percent level is permissible. The amount earmarked for transportation enhancements from this program will be around $2 billion over 6 years (over $100 million for the Great Plains states in 1993 alone). The transportation enhancement may include: facilities for pedestrians and bicycles; acquisition of scenic easements and scenic or historic sites; scenic or historic highway programs; landscaping and other scenic beautification; historic preservation; rehabilitation and operation of historic transportation buildings, structures, or facilities (including historic railroad facilities and canals); preservation of abandoned railway corridors (including the conversion to pedestrian or bicycle trails); control and removal of outdoor advertising; archaeological planning and research; and mitigation of water pollution caused by highway runoff. The Congress intended transportation enhancements as a means of stimulating additional efforts beyond what is customarily provided as environmental mitigation. States may not use transportation enhancement funds to finance normal environmental mitigation work.

In summary, the highway and the surrounding environment are a predominant concern of ISTEA. The legislation emphasized this by expanding the list of things for which federal-aid highway funds can be used and by dedicating some programs solely to addressing environmental matters. Again, in making decisions on the use of federal-aid highway funds, state and local officials will have to make trade-offs between many competing interests. Win-win scenarios are certainly possible for environmental, economic development, and transportation programs.

Trade Corridors and Border Crossings

An issue that could have profound impact on the Great Plains states involves the possibility of expanded trade with Canada and Mexico. The links between the United States and our neighboring countries were addressed by Congress in the ISTEA. The North American Free Trade Act (NAFTA) reinforces the emphasis on international trade, as does the Clinton Administration's support for economic expansion with Mexico and Canada.

The designation of the NHS highlights these concerns. Major international trade routes must be considered in the development of NHS alternatives and route selections. The ISTEA also requires the Secretary of Transportation to determine the need for a discretionary border crossing program for the U.S.-Mexico border and to conduct an assessment of existing and emerging international trade corridors.

International trade corridors exist among the United States and several other locations: Canada, Mexico, South America, western Pacific Ocean, and eastern Atlantic nations. Studies will identify these corridors and the associated major modal (highways, railways, waterways, airways) and intermodal facilities (ports, airports, and other interfacing facilities). These collectively form the nation's trade corridors that contain key transportation facilities serving freight and tourist trade. Preliminary analysis indicates that about 25,000 miles nationwide of the National Highway System's mileage will serve the trade corridor network.

Within existing trade corridors, the overall goal is to maximize the efficiency and effectiveness of the transportation system. While serving economic, environmental, social, and community interests, the highway programs on these existing trade corridors should immediately address:

- Lack of, or weaknesses in, intermodal connections
- Transportation performance weaknesses caused by recurring or nonrecurring delays such as maintenance and reconstruction; accidents, weather, and the like; and capacity deficiencies
- Highway pavement and bridge condition weaknesses associated with poor or structurally weak pavements or bridges for national and international loads

In addition, a long-range program should be developed that is based on forecasts and focuses on deficiencies in intermodal connections and transportation capacity, as well as pavement and bridge conditions.

Emerging trade corridors have similar overall goals and immediate action programs with one added complexity. More multistate cooperation will be necessary where a freeway does not presently exist. This is essential to ensure coordinated design standards, including access management and design lives of pavements and bridges. Coordinated capital improvement programs are also desirable.

Future Issues in Surface Transportation

Even though ISTEA has set the course for surface transportation, at least through 1997, it is not a rigid path with predictable actions and results. The essence of the legislation, after all, was that transportation should be responsive to many customers.

The question then arises, "What additional factors will influence transportation decision making in the future?" The issues identified here tend to overlap, bearing out the truth of ISTEA that transportation is not simple and the trade-offs are many.

Needs and Federal Funding

The perennial problem with the transportation infrastructure is that the needs are great and the funding is insufficient. The 1993 version of the *Status of the Nation's Highways, Bridges, and Transit Systems: Conditions and Performance* (USDOT 1993) report found that funding necessary to maintain U.S. highways in acceptable condition and to meet capacity demands far exceeds the resources that are expected to be dedicated to them from all levels of government. To illustrate the enormity of the problem, just maintaining the 1991 condition and performance of the nation's arterials and collectors required an annual investment of about $47 billion—about $20 billion above the actual 1991 investment. This is equivalent to a 20-cent-per-gallon increase in gasoline taxes.

At the same time, needs for dollars are increasing, the pressure to control the nation's deficit is growing, and the federal-aid highway program is affected by deficit-cutting measures. For example, in fiscal year 1993, the states have been limited by the U.S. congressional appropriations process to using only 80.5 percent of the federal-aid highway funds indicated by ISTEA. Until the deficit situation improves, or the view of the highway program in the overall context of government spending and taxes changes, constrained federal-aid highway spending will probably continue. However, the seriousness of the needs and funding dichotomy provides a strong basis for funding above the ISTEA level.

Efficiency

The public's demand for "smarter" spending by the government will drive investment decisions in the future. Especially in the face of insufficient federal dollars, more efficiency is encouraged.

Pavement quality is, and will continue to be, a major issue. Pavements are not lasting as long as is desirable. The design life is too short in urban areas, and the result is that road-user costs skyrocket due to motorist delays caused by having to reconstruct or repair pavements frequently.

Traffic congestion—one of the top issues that concern Americans—is expected to continue its high level of importance. This requires maximizing the use of our transportation facilities through traffic control and transportation demand strategies. Beyond these relatively simple ways to reduce congestion, a further look is required at root causes of congestion, such as inefficient land use patterns like urban sprawl. These are only indirectly related to transportation but must be managed in the long run in order to ultimately control the demand for more transportation infrastructure.

Parochialism

With tight spending a reality, attention naturally focuses on the best way to channel federal funds to benefit overall transportation. The trend toward parochial distribution of funds, as opposed to broad-based programs such as the NHS and the STP, is becoming stronger in Congress.

Parochialism can take several forms in the future. More demonstration projects are likely to show up in any legislation (e.g., regular appropriations, economic stimulus) that opens up the doors for new highway spending. If parochialism raises its head during the designation of the NHS, a cohesive system of the nation's most important roads could turn into an unsystematic assignment of routes based on political considerations. Earmarking certain administrative, research, or discretionary funds could also subvert the competitive process. Demonstration projects, designation of routes, and earmarking of funds are in conflict with the philosophy of "smart spending."

Highway Safety

The incidence of highway traffic accidents is a serious problem that continues to plague the nation. In 1991, approximately 41,500 people were killed in traffic-related accidents; deaths, injuries, and property damage from motor vehicle accidents cost the nation approximately $150 billion annually. A 10 percent savings would be equivalent to the total federal-aid highway funding in 1991.

The ISTEA provides for safety funding and initiatives like seat belts, motorcycle helmets, and substance abuse programs, but the public is likely to demand more improvement in this tragic safety situation. More interagency and public partnering are necessary as is greater use of advanced technology.

Environment

On the legislative horizon, the anticipated reauthorization of the Clean Water Act is expected to bring about a shift in the regulations affecting water quality. Instead of looking at actions on a project-by-project basis, the view of program and project managers will probably be broadened to watershed management. Already, the FHWA has been working with other agencies to decide on the best approach to watershed management.

Finally, in the overall move toward more thorough consideration of environmental effects, secondary and cumulative effects of highway projects will be evaluated. The FHWA has begun to more effectively incorporate such effects into highway development. Unlike evaluating direct effects, which are more measurable and easier to verify, the consideration of these secondary and cumulative effects will be more difficult. However, the FHWA is committed to minimizing future social, economic, and environmental effects and will be developing methods to consider all impacts—direct, secondary, and cumulative.

The National Highway System

The NHS will serve as the major focus of federal highway investment for the future. Although this system essentially exists, it needs major improvement, because it is expected to carry the bulk of interstate and interregional travel and commerce and will be the backbone of the U.S. transportation system as we enter the twenty-first century. In order to successfully compete in the global marketplace, the United States must begin now to develop that system to move goods and people efficiently. Linked with the NHS process is the identification of major trade corridors, trade facilities, intermodal facilities, and associated border crossings.

Conclusions

The ISTEA, which includes the financing, federal-aid highway funding program restructuring, and its associated themes, provides new direction or emphasis to surface transportation.

Of particular relevance to the Great Plains states are the National Highway System, planning and management systems, the federal-aid programs concerning the environment, and trade corridors. Future initiatives that merit watching by the Great Plains states are the needs versus funding dichotomy; parochialism (demonstration projects and earmarking); efficiency related to pavement and bridge quality and traffic congestion; highway safety; new environmental activities; and congressional action on the National Highway System.

The program changes in ISTEA go beyond providing funding to mitigate the environmental effects of transportation improvements. This legislation recognizes that in many situations it is simply not possible to build enough new roads or to add enough new capacity to existing roads to meet the forecast of unconstrained demand for transportation infrastructure. While recognizing that highways will continue to play a critical role in providing national mobility, ISTEA requires that options be considered to reduce trip demand and provides incentives to consider alternative transportation modes for meeting mobility needs. This shift in focus from meeting unconstrained highway demand can produce a positive situation for both highways and the environment that goes beyond simply minimizing the impact of highways.

It is imperative that solutions be found for the various problems, although some answers may often seem at odds with each other. In satisfying multiple goals, no solutions will be perfect for all program areas; however, in total everyone benefits. Win-win scenarios that advance economic development, as well as environmental, social, community, and transportation initiatives are certainly possible. That is what the ISTEA legislation is all about.

References

Federal Highway Administration (FHWA). 1992. *Federal-aid highway program obligations—Fiscal year 1993*. Notice N4520.122. Washington, D.C.: U.S. Government Printing Office.

Public Law 102-240. 1991. Intermodal Surface Transportation Efficiency Act of 1991. Washington, D.C.: U.S. Government Printing Office.

U.S. Department of Transportation (USDOT). 1993. *The status of the nation's highways, bridges, and transit: Conditions and performance* (Report to the U.S. Congress). Committee Print 103-2. Washington, D.C.: U.S. Government Printing Office.

Climate and Biological Resources

9. Conserving the Biotic Integrity of the Great Plains

Fritz L. Knopf
National Biological Service
Fred B. Samson
U.S. Forest Service

In the United States, a legal mandate for the conservation of biological diversity can be found in approximately 30 federal laws (Rein 1991). The mandate has now evolved into promoting a global strategy for conservation of diversity (World Resources Institute, The World Conservation Union, and United Nations Environment Programme 1992). The loss of biological diversity is recognized as a major national and global concern with potentially profound ecological and economic consequences (Council on Environmental Quality 1993). As a contribution to the multidisciplinary Great Plains Initiative, this chapter provides perspectives on the biological diversity of the Great Plains by defining (1) the concept, (2) landscape-level threats to its conservation, and (3) the conceptual approach to maintaining the biological integrity of the region in sustainable, yet economically developed, ecosystems.

Native grasslands cover almost seven billion hectares, more than one-half of the earth's terrestrial surface (Imboden 1988). Grasslands historically covered 17 percent of the American continent from south central Saskatchewan into northern Mexico and were surrounded by grassland and shrub associations on the west and south or fire-maintained woodland savannas on the north and east. The grasslands, originally named the Great American Desert, were passed over by early travelers who sought arable lands in Oregon, gold in California, and religious freedom in Utah. Ironically, however, the grasslands represent much of what is unique about the biology of North America—a native fauna with many forms that lack close relatives elsewhere in the world. Most notable of the unique species is the pronghorn, the symbol of the American Society of Mammalogists. In addition, the Great Plains area is instrumental in the conservation of biological diversity of the continent as stopover habitats for species migrating between more northern and southern regions.

This chapter illustrates how the biotic integrity of the Great Plains can be protected by focusing management on less than 10 percent of the fauna. Diversity conservation demands (1) protecting the biotic integrity (endemic species) of a biogeographic region, (2) understanding the ecological processes that support endemic species, and (3) developing dynamic management programs to ensure that biotic integrity is sus-

Stanley R. Johnson and Aziz Bouzaher (ed.), *Conservation of Great Plains Ecosystems*, 121–133.

tainable through time. A cost-effective and objective strategy for conserving biological diversity demands conservation of biogeographic regions irrespective of political boundaries. To this end, the Great Plains Initiative is the prototype for proactive conservation of the biological diversity of North America.

From Biological Diversity to Biotic Integrity

Historically, the landscape and biological diversity of the Great Plains can be envisioned from descriptions of the megafauna of the early 1800s. The region included shortgrass and tallgrass prairies intergrading just east of an irregular line from El Reno (Oklahoma), through Fort Hays (Kansas), North Platte (Nebraska), and northwestward into the west central Dakotas. The landscape of the eastern section included heavily wooded stream bottoms in uplands of fire-maintained grasses a meter or more tall. Wapiti and white-tailed deer were abundant. The grasses, however, do not cure well (i.e., they lose their nutritive value when dried), and these ungulates survived the season of vegetative dormancy by grazing or browsing in the wooded stream bottoms. The western landscape usually included treeless stream bottoms and uplands dominated by blue grama and buffalo grass, two warm-season grasses that flourish under intensive grazing pressure by reproducing both sexually and by tillering. Unlike the more eastern species, shortgrass species retain their digestible protein when dormant. This character supported the evolution of a major herbivore assemblage dominated by bison, pronghorn, and two species of prairie dog and their associated large carnivores, the gray wolf and grizzly bear.

Contemporary definitions of biological diversity are about as diverse as the biological resource itself. As definitions become broader and less specific, conservation becomes increasingly more difficult (Knopf 1992a). In simple form, biological diversity is the variety of life and accompanying ecological processes (Keystone 1991). What is meant by biological diversity, however, is not necessarily the number of species but rather the quality of those species relative to natural ecological processes of a region. An arboretum or zoo has more species per unit area than a native landscape but contributes little to natural biological diversity. Such quality-over-quantity decisions about biological diversity are developed by viewing diversity across biological, spatial, and temporal perspectives (Table 9.1). It is the complexity of these multiple scales that has fostered much operational disillusion in attempts to define and conserve biological diversity or to develop arguments to counter advocates who argue that "every species counts everywhere."

Thus, describing biological diversity of the Great Plains is not an exercise in listing all the living organisms that occur in the region. Because only 1.4 million of the estimated 10 million species on earth have been described to date (Swaminathan 1992), a description of all the species of the Great Plains will probably never be completed

Table 9.1. The scales of biological diversity

Biological	Spatial	Temporal
Genetic	Locale	Instantaneous
Population	Between locales	Daily/seasonal
Community	Regional	Generation
Ecosystem	Continental	Evolutionary

anyway. Using birds as an archetypal taxon, 330 of the 435 species that breed in the United States occur on the Great Plains (Johnsgard 1979); a simple list of just the birds would exceed the limitations for this chapter. However, birds are an excellent group for focusing a conversation about biological diversity because they are the best known taxon, have the ability to disperse widely (illustrating how regional losses in biological diversity originate with localized decisions), and are highly visible (ensuring excellent public support for conservation actions that simultaneously favor other biological groups).

Although 330 species of birds breed on the grasslands, conserving the avifaunal diversity of the Great Plains does not necessarily require all 330 species. The Great Plains's avifauna is relatively depauperate, since only 5 percent of all North American bird species are believed to have evolved in the region (Udvardy 1958; Mengel 1970). Mengel listed 12 species of birds endemic to the grasslands; 20 others evolved there but range more widely (Table 9.2). Protecting the avian diversity of the Great Plains demands that only these 32 species (and especially the 12 endemics unique to the region) receive priority. These species plus a list of endemic counterparts for plants, invertebrates, fish, amphibians, reptiles, and mammals would identify the precise focus for conservation of biological diversity of the Great Plains. The remaining species (298 species of birds, for example) will either receive special consideration in contiguous biogeographic provinces or will not need special consideration because they are ecological generalists that occur in a variety of settings across the continent.

Two further points need to be made about the significance of focusing conservation of biological diversity on endemic species of the Great Plains. First, it is precisely these organisms (because of their limited distributions and narrower ecological requirements) that either are, or ultimately will be, declining to threatened and endangered population levels regionally and nationally. Second, despite the relatively simplistic endemic avifauna on the Great Plains, this group of species is currently showing steeper, more consistent, and more geographically widespread declines than any other group of North American species. Of the 32 grassland species, 10 declined at a statistically significant rate of 1 to 3 percent per year from 1966 to 1991 (Knopf 1993) and conservation of these 10 species is the greatest concern for the preservation of biological diversity. Unlike birds that migrate into southern Mexico or Central America to winter in subtropical and tropical areas, grassland birds also winter mostly within the Great Plains. These endemic species are the biotic integrity of the region, the unique offerings by the Great Plains to the biological diversity of North America.

Table 9.2. The endemic avifauna of the Great Plains

Nonpasserines	Passerines
Primary Species (endemic)	
1. *Buteo regalis* (ferruginous hawk)	1. *Anthus spragueii* (Sprague's pipit)
2. *Charadrius montanus* (mountain plover)	2. *Calamospiza melanocorys* (lark bunting)
3. *Numenius americanus* (long-billed curlew)	3. *Ammodramus bairdii* (Baird's sparrow)
4. *Limosa fedoa* (marbled godwit)	4. *Aimophila cassinii* (Cassin's sparrow)
5. *Phalaropus tricolor* (Wilson's phalarope)	5. *Calcarius mccownii* (McCown's longspur)
6. *Larus pipixcan* (Franklin's gull)	6. *C. ornatus* (chestnut-collared longspur)
Secondary Species (more widespread)	
1. *Ictinia mississippiensis* (Mississippi kite)	1. *Eremophila alpestris* (horned lark)
2. *Buteo swainsoni* (Swainson's hawk)	2. *Sturnella magna* (eastern meadowlark)
3. *Circus cyaneus* (northern harrier)	3. *S. neglecta* (western meadowlark)
4. *Falco mexicanus* (prairie falcon)	4. *Spiza americana* (dickcissel)
5. *Tympanuchus cupido* (greater prairie chicken)	5. *Passerculus sandwichensis* (Savannah sparrow)
6. *T. pallidicinctus* (lesser prairie chicken)	6. *Ammodramus savannarum* (grasshopper sparrow)
7. *T. phasianellus* (sharp-tailed grouse)	7. *A. henslowii* (Henslow's sparrow)
8. *Bartramia longicauda* (upland sandpiper)	8. *Pooecetes gramineus* (vesper sparrow)
9. *Athene cunicularia* (burrowing owl)	9. *Chondestes grammacus* (lark sparrow)
10. *Asio flemmeus* (short-eared owl)	10. *S. pallida* (clay-colored sparrow)

Note: Modified from Mengel (1970). Mengel's list included five additional widespread species: *Centrocercus urophasianus* (sage grouse), *Oreoscoptes montanus* (sage thrasher), *Pipilo chlorura* (green-tailed towhee), *Amphispiza belli* (sage sparrow), and *Spizella breweri* (Brewer's sparrow). These species are really more closely tied to the Great Basin shrub steppe than to the Great Plains and have been deleted from this discussion.

Threats to the Biotic Integrity of the Great Plains

The principal ecological processes that shaped endemic plants and animals on the Great Plains over evolutionary times have been drought (Wiens 1974), grazing (Hobbs and Huenneke 1992), and wildfires (Zimmerman 1992). These ecological forces favored grassland homogeneity locally and heterogeneity regionally (Collins 1992; Glenn et al. 1992).

Securing the future of endemic grassland species on the Great Plains requires conservation that protects their native habitats and ecological niches. Wetland birds such as the marbled godwit and Wilson's phalarope obviously occur locally at predictable moist-soil sites. Birds of taller grasses, such as the greater prairie chicken and dickcissel, nest in habitats of residual vegetation standing from a preceding growing season and are dependent on the rejuvenation of stands by periodic fires (e.g., Kirsch 1974). Many endemic birds of shortgrass and mixed-grass landscapes such as the Baird's sparrow and the McCown's and chestnut-collared longspurs coevolved with grazing ungulates, whereas others such as the ferruginous hawk, prairie falcon, and burrowing owl are strongly associated with prairie dog towns.

The landscape of the Great Plains has undergone significant alteration since historical times. Developments that have had primarily local effects on the native species are ignored here. Activities with more universal effects on native biological diversity have included (1) fragmentation of grasslands, (2) draining of wetlands, (3) invasions or introductions of alien and exotic species, and (4) water development activities.

Fragmentation

The consequences of landscape fragmentation to local and regional faunas are well documented (Harris 1974; Martin 1992; and others) and have profound implications for conserving the biotic integrity of grasslands (Samson 1980; Johnson and Temple 1986; Gaines et al. 1992). All living organisms have some minimum area requirements for survival. Limited space for organisms in a region as expansive as the native grasslands of North America is initially difficult to comprehend. Probably the best illustration, however, is from human history—the 0.65-square-kilometer parcel size of the 1862 Homestead Act. Simply stated, when the parcel size becomes too small to provide necessary resources, the individual either dies or moves elsewhere. Moreover, the minimum area requirements of species vary; larger species require more room than smaller species. A viable population of the ferruginous hawk requires a much larger area than does the Great Plains skink. One approach is to focus conservation on species such as the hawk that have large minimum-area requirements because the space requirements of species with smaller area needs are met simultaneously. In this context, the hawk is an umbrella species (Wilcox 1984).

Fragmentation or insularization of the grasslands has been a consequence of cultivation for cereal grain production, eradication of prairie dogs, and augmentation of woody vegetation (Knopf 1993). The eastern Plains have been virtually obliterated for grain (primarily corn) production. Only 10.4 square kilometers of the original 103,600 square kilometers of native prairie survive in the state of Illinois (Mlot 1990). Alternatively, the western grasslands are less disturbed, with more than 70 percent still being managed as rangeland (Knopf 1993); 17 areas are protected as national grasslands (USDA 1990)

A second agricultural practice fragmenting the native grasslands has been the intensive eradication of prairie dogs. Prairie dog populations have been reduced 98 percent in the last 100 years (Marsh 1984). Prairie dogs were poisoned to reduce competition with cattle for grass forage, although the need for poisoning is supported neither by scientific data (O'Meilia et al. 1982) nor by inferences that ungulates and prairie dogs are actually symbiotic foragers (Krueger 1986). Prairie dogs support biological diversity because they are an ecological "keystone" species (Gilbert 1980; Terborgh 1986) that, by their presence, favor the continued existence of other endemic species such as the ferruginous hawk, mountain plover, and black-footed ferret. The management of prairie dogs is an example of a major biological diversity issue that can be easily resolved at reduced costs by updating government subsidy strategies (Miller et al. 1990).

Even where native grasslands remain somewhat intact, fragmentation has occurred in the form of woody vegetation encroachment on the northern and southern grasslands due to control of natural fires (Bird 1961; Pulich 1976). Almost 3 percent of the Great Plains is now forested by shelter belts planted to reduce wind erosion (Baer 1989). Equally significant has been the ecological development of streamside forests of alien and exotic tree species favored by water management practices.

Loss of Wetlands

The drainage of wetlands has profoundly altered the phytogeography of grasslands locally (Dahl 1990). Losses in Illinois and Iowa equal 86.3 percent of an estimated 49,421-square-kilometer historical wetland area (Knopf 1993). Comparable values in Colorado, Wyoming, and Montana show only a 40.0 percent loss of 20,838 square kilometers, and losses in the midplains states are an intermediate 56.6 percent of 70,478 square kilometers. These numbers support the intuitive relation between the loss of wetlands and intensity of cultivation across the Great Plains. Because a variety of wildlife depends upon wetlands and because of the potential for physical, chemical, and biological alteration of remnant sites, the conservation of wetlands has been a special concern in North America for decades (The Conservation Foundation 1988).

Besides the contribution of its own endemic species, the Great Plains contributes to overall continental biological diversity by providing habitats for migratory wetland species (such as the whooping crane) that cross the province semiannually. Many shorebirds require a regional complex of wetlands when migrating from the Gulf of Mexico coast to northern Alaska breeding grounds (Skagen and Knopf 1993). The wetlands provide food resources needed to complete migration since flight ranges of birds are determined by fat reserves. The highly dynamic nature of historic wetlands on the Great Plains (Fredrickson and Reid 1990) undoubtedly had a major influence on the evolution of shorebird migration, and coordination of wetlands conservation across the Great Plains is a regionally unique challenge for the Great Plains Initiative.

Exotics and Aliens

Agriculturally cultivated species such as corn and domestic livestock are usually aliens (i.e., native to the continent but not to the Great Plains) or exotics (species introduced from foreign continents). The continued presence of these species on the Plains is a requisite for economic vitality and viability of the region. Yet a second group of aliens and exotics are also present. These include plants such as purple loosestrife and Russian olive that enhance habitats for some wildlife species but are capable of widely displacing or significantly altering native plant communities and creating new, monotypic landscapes dominated by ecological generalist species and devoid of endemic species (Olson and Knopf 1986; Thompson et al. 1987). The issue of displacing native faunas with exotics and aliens is most severe among fishes, with entire ichthyofaunae being destroyed nationwide (Courtenay and Moyle 1992, Wilcove et al. 1992). Even among wildlife, species that are popular with sportsmen, such as the mallard and ring-necked pheasant (an alien and an exotic), can displace closely related endemic species regionally (Cade 1983; Westemeier 1988).

Water Management

Besides the significance of their wetlands to wildlife, the Great Plains are characterized by a major network of flowing water. This water historically has been impounded for agricultural and industrial purposes. Conservation of biological integrity in running water represents the maximum conflict between economic development and environmental conservation. Besides faunal mixing by introducing exotics and aliens, losses of species in lotic water can be attributed to five additional factors: habitat loss and degradation, overexploitation, secondary extinctions, chemical and organic pollution, and climate change (Allan and Flecker 1993).

Changing the timing and magnitude of flows in major river systems has also influenced the distribution of many terrestrial species. On the shortgrass prairie, woody vegetation in the form of streamside forests has colonized at many locales. Such forests have fragmented the native sea of grasses and facilitated movements of alien and exotic tree and wildlife species across the Great Plains (Knopf and Scott 1990; Johnson 1994). This fragmentation of the grasslands has had more severe consequences for the native flora and fauna of the western Great Plains than has the plowing of sod for grain cultivation.

Biotic Integrity Versus Biotic Enhancement

The synthetic landscapes along rivers and elsewhere (suburban settings, reservoirs, etc.) have resulted in major enhancements of biotic assemblages at many locales throughout the Great Plains. Easily 90 percent of the breeding bird species in northeastern Colorado in 1980 did not breed there at the turn of the century (Knopf 1986). Such enhancements have numerous benefits, not only for promoting awareness of natural resources but also for augmenting local economies. A staggering benefit of an estimated $15 million to the local economy of Grand Island, Nebraska, accrued from 80,000 visitors to the large staging area for sandhill cranes in 1991 alone (Lingle 1992). Ecotourism can offset costs of conservation and can be exploited for the enhancement of the quality of life as it has for years in towns like Jackson, Wyoming. Such benefits may be expected from the reintroduction of bison onto a native Oklahoma prairie in October 1993. Traditional fish and wildlife management promotes species of recreational interest, also providing major economic incentives while simultaneously promoting appreciation of natural resources. All of these enhancements from landscape evolution or management are compatible with the conservation of biotic integrity unless aliens and exotics usurp endemic species.

Conservation into the Next Century (Back to the Future)

Just as habitats are fragmented, so too is the structure of political bodies to protect the continent's biological diversity. Over 120 offices in seven state and federal natural resource agencies make daily land-use decisions in the relatively small Platte River headwaters (Knopf and Scott 1990). Simply stated, political and biological provinces are drastically askew and the splintering of biological provinces across political entities is a major hurdle to conservation of the biotic integrity of North America (Knopf 1992b).

Many activities and initiatives are already under way to focus management more specifically on biological diversity. Such programs require inventories of flora and fauna such as those being conducted by the Fish and Wildlife Service's Gap Analysis (Scott et al. 1987) and the Environmental Protection Agency's EMAP (USEPA 1990; National Research Council 1992). These efforts are coordinated with The Nature Conservancy's Natural Heritage Program in many states; all are working to identify and document species distribution patterns across the Great Plains. The next step will be to overlay these databases with biotic integrity filters to extract species-specific information for focusing conservation. Such filters will drastically enhance cost effectiveness by narrowing the focus for protecting the biotic integrity (endemic species) of the Great Plains.

Conservation of the biotic integrity of the Great Plains also demands refocusing agency actions. Within a given agency, the most glaring need is to consolidate the national grasslands under one autonomous administrative unit based in a central location such as Chadron, Nebraska. Ultimately, however, protecting biotic integrity demands a larger forum for working across political entities. Evolving natural resource programs with interagency coordination include:

- North American Waterfowl Management Plan (USFWS 1986) with its joint ventures defined along ecological units
- The recently established "Partners in Flight" program (Information and Education Working Group 1992) developed to pool resources and information relative to species of birds that breed in North America but winter in Latin countries to the south
- The "Bring back the natives" strategy of the Bureau of Land Management, Forest Service, and Fish and Wildlife Foundation (Williams and Williams 1992)

This final strategy is designed specifically to protect and restore endemic fish species (and, thus, stream health) in major lotic systems that cross agency boundaries.

Finally, and somewhat as a summary, the recent collaborative effort of Samson and Knopf (1993) is adapted as a philosophical guide to conserving biological diversity on the Great Plains. This work identifies fundamental, central approaches to guide daily decisions to protect native biological diversity. The four directions are:

Understand alpha and beta diversity. Alpha (α) diversity is the variety of life that occurs locally. Beta (β) diversity is the change in diversity that occurs across a landscape gradient, or between disjunct landscape parcels. Alpha diversity increases with the number of species in a given area, such increase being strongly correlated with landscape fragmentation and the subsequent invasions of edge or weedy species at a locale. Beta diversity is high when two sites share few species and the contribution of ecological generalists is reduced. An α-diversity activity is akin to placing a bird feeder in your backyard and has dominated species-diversity thinking in traditional wildlife enhancement activities. A β-diversity emphasis is the fundamental approach for identifying what is unique bioregionally, the quantitative template for developing sound conservation decisions.

Emphasize biotic integrity. Aliens and exotics pose a significant threat to endemic species. The concept of diversity has value only when applied to the native species of a particular landscape, and diversity is conserved only when the uniqueness of the native biota of each biogeographic province is secured. The number of species in an area is irrelevant to biological diversity just as the quality of care and display in a zoological park has nothing to do with the number of confined species. Focusing conservation on endemic species is to focus conservation on those elements most susceptible to landscape fragmentation.

Restore ecological processes. The loss of a species, community, or ecosystem can be traced to the loss of an ecological process that favored it. Historical perspectives are critical to understanding how endemic species evolved, especially in an environment like the Great Plains where climate is periodically harsh and always unpredictable at local and regional levels. Dynamic ecosystems require dynamic management. Restoring natural ecological processes champions conservation of regional diversity.

Promote ecological sustainability. The Homestead Act encouraged rapid settlement of the Great Plains, but its early versions ignored the issue of the settlers' ability to survive periodic, harsh conditions. Similarly, protecting the biotic integrity of the Great Plains dictates management that sustains viable populations of endemic species through ecological crunches—those natural, unpredictable, periodic moments when critical resources needed for survival become limited (Wiens 1977).

Taken together, these approaches form a paradigm for conserving the biological diversity of the Great Plains while simultaneously maintaining the long-term ecologic and economic vitality of the region (Samson and Knopf 1994).

References

Allan, J. D., and A. S. Flecker. 1993. Biodiversity conservation in running waters. *BioScience* 43:32-43.

Baer, N. W. 1989. Shelterbelts and windbreaks in the Great Plains. *Journal of Forestry* 87:32-36.

Bird, R. D. 1961. *Ecology of the aspen parkland of western Canada in relation to land use*. Branch Publication no. 1066. Canadian Department of Agricultural Research, Ottawa, Ontario.

Cade, T. J. 1983. Hybridization and gene exchange among birds in relation to conservation. In *Genetics and conservation,* ed. C. M. Schonewald Cox, S. M. Chambers, B. MacBryde, and W. L. Thomas. Menlo Park, California: Benjamin/Cummings Publishing Company, Inc.

Collins, S. L. 1992. Fire frequency and community heterogeneity in tallgrass prairie vegetation. *Ecology* 73:2001-2006.

Council on Environmental Quality. 1993. *Incorporating biodiversity considerations into environmental impact analysis under the National Environmental Policy Act.* Washington, D.C.: Council on Environmental Quality.

Courtenay, W. R., Jr., and P. B. Moyle. 1992. Crimes against biodiversity: The lasting legacy of fish introductions. *Transactions of the North American Wildlife and Natural Resources Conference* 57: 365-372.

Dahl, T. E. 1990. *Wetlands losses in the United States, 1780s to 1980s.* Washington, D.C.: U.S. Department of the Interior, Fish and Wildlife Service.

Fredrickson, L. H., and F. A. Reid. 1990. Impacts of hydrologic alteration on management of freshwater wetlands. In *Management of dynamic ecosystems,* ed. J. M. Sweeney. West Lafayette, Indiana: North Central Section of The Wildlife Society.

Gaines, M. S., J. Foster, J. E. Diffendorfer, W. E. Sera, R. D. Holt, and G. R. Robinson. 1992. Population processes and biological diversity. *Transactions of the North American Wildlife and Natural Resources Conference* 57:252-262.

Gilbert, L. E. 1980. Food web organization and conservation of neotropical diversity. In *Conservation biology: An evolutionary-ecological perspective,* ed. M. E. Soulé and B. A. Wilcox. Sunderland, Massachusetts: Sinauer Associates.

Glenn, S. M., S. L. Collins, and D. J. Gibson. 1992. Disturbances in tallgrass prairie: Local and regional effects on community heterogeneity. *Landscape Ecology* 7:243-251.

Harris, L. D. 1974. *The fragmented forest.* Chicago: University of Chicago Press.

Hobbs, R. J., and L. F. Huenneke. 1992. Disturbance, diversity, and invasion: Implications for conservation. *Conservation Biology* 6:324-337.

Imboden, C. 1988. Foreword. *Ecology and conservation of grassland birds,* ed. P. D. Goriup. Technical Publication 7. International Council for Bird Preservation.

Information and Education Working Group. 1992. *Partners in flight.* Fish and Wildlife Foundation, Washington, D.C.

Johnsgard, P. A. 1979. *Birds of the Great Plains.* Lincoln: University of Nebraska.

Johnson, R. G., and S. A. Temple. 1986. Assessing habitat quality for birds nesting in fragmented tallgrass prairies. In *Wildlife 2000,* ed. J. Verner, M. L. Morrison, and C. J. Ralph. Madison: University of Wisconsin Press.

Johnson, W. C. 1994. Woodland expansion in the Platte River, Nebraska: Patterns and causes. *Ecological Monographs* 64:45-84.

Keystone Center. 1991. *Final consensus of the Keystone Policy Dialogue on biological diversity on federal lands.* The Keystone Center, Keystone, Colorado.

Kirsch, L. M. 1974. Habitat management considerations for prairie chickens. *Wildlife Society Bulletin* 2:124-129.

Knopf, F. L. 1986. Changing landscapes and the cosmopolitism of the eastern Colorado avifauna. *Wildlife Society Bulletin* 14:132-142.

———. 1992a. Focusing conservation of a diverse wildlife resource. *Transactions of the North American Wildlife and Natural Resources Conference* 57:241-242.

———. 1992b. Faunal mixing, faunal integrity, and the biopolitical template for diversity conservation. *Transactions of the North American Wildlife and Natural Resources Conference* 57:330-342.

———. 1993. Avian assemblages on altered grasslands. *Studies in Avian Biology* 15:247-257.

Knopf, F. L., and M. L. Scott. 1990. Altered flows and created landscapes in the Platte River headwaters, 1840-1990. In *Management of dynamic ecosystems,* ed. J. M. Sweeney. West Lafayette, Indiana: North Central Section of The Wildlife Society.

Krueger, K. 1986. Feeding relationships among bison, pronghorn, and prairie dogs: An experimental analysis. *Ecology* 67:760-770.

Lingle, G. R. 1992. History and economic impact of crane watching in central Nebraska. *Proceedings of the Sixth North American Crane Workshop* 6:33-37.

Marsh, R. E. 1984. Ground squirrels, prairie dogs and marmots as pests on rangeland. In *Proceedings, Conference for Organization and Practice of Vertebrate Pest Control.* Fernhurst, England: ICI Plant Protection Division.

Martin, T. E. 1992. Landscape considerations for viable populations and biological diversity. *Transactions of the North American Wildlife and Natural Resources Conference* 57:283-291.

Mengel, R. M. 1970. The North American Central Plains as an isolating agent in bird speciation. In *Pleistocene and recent environments of the central Great Plains,* ed. W. Dort and J. K. Jones. Lawrence: University of Kansas Press.

Miller, B., C. Wemmer, D. Biggins, and R. Reading. 1990. A proposal to conserve black-footed ferrets and the prairie dog ecosystem. *Environmental Management* 14:763-769.

Mlot, C. 1990. Restoring the prairie. *BioScience* 40:804-809.

National Research Council. 1992. *Review of EPA's Environmental Monitoring and Assessment Program (EMAP), Interim report.* Washington, D.C.

Olson, T. E., and F. L. Knopf. 1986. Naturalization of Russian olive in the western United States. *Western Journal of Applied Forestry* 1:65-69.

O'Meilia, M. E., F. L. Knopf, and J. C. Lewis. 1982. Some consequences of competition between prairie dogs and beef cattle. *Journal of Range Management* 35:580-585.

Pulich, W. M. 1976. *The golden-cheeked warbler.* Austin: Texas Parks and Wildlife Department.

Rein, W. V. 1991. *The United States needs a national biodiversity policy.* Washington, D.C.: World Resources Institute.

Samson, F. B. 1980. Island biogeography and the conservation of prairie birds. *North American Prairie Conference* 7:293-305.

Samson, F. B., and F. L. Knopf. 1993. Managing biological diversity. *Wildlife Society Bulletin* 21:509-514.

———. 1994. Prairie conservation in North America. *BioScience* 44:418-421.

Scott, J. M., B. Csuti, J. D. Jacobi, and J. E. Estes. 1987. Species richness: A geographic approach to protecting future biological diversity. *BioScience* 37:782-788.

Skagen, S. K., and F. L. Knopf. 1993. Towards conservation of midcontinental shorebird migrations. *Conservation Biology* 7:533-541.

Swaminathan, M. S. 1992. Expanding human capacity to conserve biodiversity. In *Global biodiversity strategy.* Prepared by World Resources Institute, The World Conservation Union, and United Nations Environment Programme. Washington, D.C.: World Resources Institute.

Terborgh, J. 1986. Keystone plant resources in the tropical forest. In *Conservation biology: The science of scarcity and diversity,* ed. M. E. SoulÇ. Sunderland, Massachusetts: Sinauer Associates.

The Conservation Foundation. 1988. *Protecting America's wetlands: An action agenda.* Washington, D.C.: The Conservation Foundation.

Thompson, D. Q., R. L. Stuckey, and E. B. Thompson. 1987. Spread, impact, and control of purple loosestrife (Lythrum salicaria) in North American wetlands. *Fish and Wildlife Research* no. 2. U.S. Fish and Wildlife Service.

Udvardy, M. D. F. 1958. Ecological and distributional analysis of North American birds. *Condor* 60:50-66.

U.S. Department of Agriculture (USDA). 1990. *Land areas of the national forest system as of September 30, 1990.* U.S. Forest Service Publication FS-383.

U.S. Environmental Protection Agency (USEPA). 1990. *Environmental Monitoring and Assessment Program.* EPA/600/9-90/001.

U.S. Fish and Wildlife Service (USFWS). 1986. *North American Waterfowl Management Plan.* Washington, D.C.

Westemeier, R. L. 1988. An evaluation of methods for controlling pheasants on Illinois prairie chicken sanctuaries. In *Pheasants: Symptoms of wildlife problems on agricultural lands,* ed. D. L. Hallett, W. R. Edwards, and G. V. Burger. Bloomington, Ind.: North Central Section of The Wildlife Society.

Wiens, J. A. 1977. Climatic instability and the "ecological saturation" of bird communities in North American grasslands. *Condor* 76:385-400.

———. 1979. On competition and variable environments. *The American Scientist* 71:234-235.

Wilcove, D., M. Bean, and P. C. Lee. 1992. Fisheries management and biological diversity: Problems and opportunities. *Transactions of the North American Wildlife and Natural Resources Conference* 57:373-383.

Wilcox, B. A. 1984. In situ conservation of genetic resources: Determinants of minimum area requirements. In *National parks, conservation, and development,* ed. J. A. McNeely and K. R. Miller. Washington, D.C.: Smithsonian Institute Press.

Williams, C. D., and J. E. Williams. 1992. Bring back the natives: A new strategy for restoring aquatic biodiversity on public lands. *Transactions of the North American Wildlife and Natural Resources Conference* 57:416-423.

World Resources Institute, The World Conservation Union, and United Nations Environment Programme. 1992. *Global biodiversity strategy*. Washington, D.C.: World Resources Institute.

Zimmerman, J. L. 1992. Density dependent factors affecting the avian diversity of the tallgrass prairie community. *Wilson Bulletin* 104:85-94.

10. Climate and Vegetation in Central North America: Natural Patterns and Human Alterations

John A. Harrington, Jr.
Kansas State University
Jay R. Harman
Michigan State University

The characteristic patterns of climate and natural vegetation in central North America are strongly interrelated and result from an atmospheric circulation system that responds to both global and continental scale mechanisms. Climatic patterns arise in the region between the Rocky Mountains and the Great Lakes or Mississippi River from the interactions between two major components of the global atmospheric circulation system: the tropical Hadley Cell, and the extratropical, upper-level Westerlies. The Westerlies, particularly, are influenced and steered by thermal inequalities over the earth's surface as well as the terrain barriers such as the Rocky Mountain Cordillera. In addition, both the Hadley Cell and the Westerlies shift north and south on an annual cycle.

To some extent, natural vegetation in the midcontinent mirrors the regional climate in both species composition and plant structure. For example, there are three major biomes along a west-to-east gradient of increasing moisture availability. Grasslands associated with semiarid climates of the Great Plains give way to a transitional zone of oak savanna, sometimes called the *prairie-forest ecotone,* and finally to areas of predominantly deciduous forest in more humid climates to the east.

Both the vegetation and climate of the midcontinent vary over time. Climate system fluctuations over time scales ranging from periods of several days to several years to the millennia between glacial and interglacial periods change the patterns and frequencies of surface weather types. These climatic fluctuations affect the patterns and intensity of plant stresses and change the vegetation cover, particularly in grassland environments of the midcontinent where a range of natural vegetation, rather than a single climate type, exists in response to a number of climatic states (Küchler 1972).

This chapter summarizes the climatic controls and natural vegetation of the midcontinent region and describes natural climatic variation and the associated

Originally published in the *Great Plains Quarterly* 11 (Spring 1991):103-112 by the Center for Great Plains Studies, University of Nebraska, Lincoln.

Stanley R. Johnson and Aziz Bouzaher (ed.), *Conservation of Great Plains Ecosystems,* 135–148.

vegetative response. Because human adaptations associated with settlement and the establishment of economic activities have had significant and long-reaching impacts on native vegetation, this chapter also assesses the ways forest clearing, sod-busting, cattle ranching, and fire prevention activities have altered vegetation patterns, leaving only scarce remnants of the primeval natural cover.

Atmospheric Circulation and the Regional Climate

The circulation system of the earth's atmosphere primarily determines midcontinent climate. Latitudinal differences in energy set in motion a global wind system that partially offsets an equator-to-pole imbalance in radiation receipts and surface temperatures. This circulation system contains two major components: the Hadley Cell, a vertical and north-south oriented cell associated with tropical areas, and the mid-latitude Westerlies, an extratropical circumpolar swirl of primarily west-to-east winds (Palmén and Newton 1969). The transition area between these two circulation components shifts north and south across the midcontinent throughout the course of the year. During the summer months, the Westerlies are restricted to the northernmost sections of the midcontinent. In July and August, the Polar Front Jet Stream, the central core of the Westerlies, is commonly found between 45°N and 50°N. As the cool season progresses, the circumpolar Westerlies expand southward and influence more and more of the midcontinent, until by the middle of winter the entire area is under the influence of the upper-level Westerlies and their associated jet stream(s). The boundary zone between the Hadley Cell and the Westerlies migrates slowly northward during the spring. By late June, the annual cycle is complete and a majority of the midcontinent is again feeling the direct influence of tropical air streams and the Hadley Cell.

These annual changes in the global atmospheric circulation system produce midcontinent weather and climatic conditions that exhibit strong seasonal differences. Outbreaks of exceptionally cold Arctic air, blizzards, severe thunderstorms and tornadoes, hail, dust storms, searing heat, and torrential rains are all characteristic aspects of this annual suite of weather types. The magnitudes and frequencies of these important weather events vary from one year to the next in response to shifting patterns in the global circulation.

This discussion focuses on the warm season in order to emphasize the weather and climatic conditions that correspond with the timing of important plant processes such as growth, reproduction, and seedling survival. Warm season environmental conditions greatly influence agriculture, an activity of considerable economic and cultural importance in the midcontinent.

Throughout nearly all of the midcontinent, the predominant native vegetation is dormant during the winter with its freezing temperatures. During the warm season, however, the several aspects of global atmospheric circulation combine to produce weather conditions that stress growing plants and regulate both natural vegetation and agriculture. The Westerlies, the associated Polar Front Jet Stream, and the associated surface storm track dominate northern sections of the midcontinent. In midsummer these features are aligned with the prairie-forest ecotone from central Canada into western sections of the Great Lakes region (Harman and Braud 1985; Harrington and Harman 1985).

In contrast, the descending branch of the Hadley circulation controls climate in the southern and central midcontinent. This subsident motion produces the "Bermuda High," a large, semipermanent surface high pressure cell centered over eastern sections of the Atlantic Ocean. Surface wind flow on the west side of this system is predominantly southerly (Bryson 1966; Wendland and Bryson 1981). This pattern pumps warm, moist air northward from the western Gulf of Mexico into eastern sections of the midcontinent. Across western sections of the Great Plains, however, the northward flow originates over the warm, dry regions of West Texas and Mexico. Corcoran (1982) has suggested that the orientation of the Westerlies and resultant vertical motions in the atmosphere influence these surface air streams; drier air subsiding from middle levels of the troposphere further enhances the west-to-east contrast in atmospheric moisture. The combined influence results in a distinct gradient in atmospheric moisture, warm season rainfall totals, and associated plant moisture stresses. Changes in structure and floristics of natural vegetation parallel precipitation increases and potential evapotranspiration stress decreases from west to east.

The Regional Climate and Environmental Gradients

Scientists have for a long time recognized that moisture conditions are a primary regulator of regional vegetative cover, but our understanding of climate-vegetation relationships has improved markedly since Hare (1951) asserted that "climate is the ultimate ecological control." An important consideration is the scale of the investigation. Gleason (1923) postulated that plants react as individuals to environmental stresses and that the co-occurrence of plants in recognizable communities is a result of random processes. Site factors, such as soil, fire, regime, and biotic interactions "determine local-scale patterning" whereas "[c]limate can be considered as a first-order constraint on range limits" (Borchert 1950; Collins 1969; Neilson et al. 1989). Much of the recent advance in understanding continental-scale climate-vegetation relationships has come from investigations of climate system dynamics, specifically of the processes whereby atmospheric motions generate the surface conditions that limit vegetation distributions by stressing plants. Earlier studies relied only on implications of statistical and spatial correlations between surface climatic data and boundaries of major plant formations.

Naturalists who drew inferences from the spatial correspondence between plant formation boundaries and mapped isolines of climatic statistics, such as mean annual rainfall, provide some of the earliest insights into the influence of climate on vegetation. One particularly striking boundary is the transition zone marking the change in dominant vegetative structure from grass through savanna to forest. Currently the prairie-forest ecotone arcs southeastward from central Canada across Minnesota, Wisconsin, northern Illinois, and into Indiana. Another arcuate boundary can be traced north and eastward from the western Gulf of Mexico across eastern Texas, eastern Oklahoma, Missouri, and Illinois. The two ecotonal arcs merge in northwestern Indiana. The area between and west of these ecotones is the Prairie Peninsula, a wedge of upland prairie grasslands that seems to penetrate eastward into the deciduous forests (Transeau 1935).

Transeau (1905) recognized that patterns of annual precipitation and evaporation ratios corresponded with patterns of plant distribution. Grasslands predominated where potential evaporation was greater than precipitation, whereas forests were found where the ratio was reversed. This statistical and spatial relationship held for the Prairie Peninsula, which he believed could be accounted for by a rate of evaporation higher than precipitation. The surrounding forests occur in areas of higher precipitation. Transeau and others have demonstrated a number of other correlations between isolines of surface climatic conditions and the prairie-forest ecotone. "Midsummer relative humidity is certainly lower in the Peninsula than north or south of it," observed Transeau (1935).

In addition, Borchert (1950) identified spatial correlations between the margins of the central North American grassland region and positive temperature deviations during drought years, below normal precipitation during major drought years, the number of rain days in July and August, and the amount of winter snowfall on the northeastern boundary and the amount of winter season precipitation on the southeastern boundary. All his observations implied that more moisture was available to plants in the wooded area than in the Prairie Peninsula. Mather (1959) used Thornthwaite water budget concepts, essentially a hydrologic accounting system, to demonstrate an association between a statistically derived climatic parameter that estimates crop moisture status, the moisture index, and the major midcontinent vegetation types. Manogaran (1983) used this same water budget approach to demonstrate a good spatial relationship between soil moisture deficit and the northeastern boundary of the Prairie Peninsula and to match isolines of water surplus and the moisture index with the peninsula's western and southeastern boundaries.

As our understanding of the mechanics of the atmospheric circulation system has improved, fresh and more thorough studies of vegetation-climate relationships have been undertaken. One fruitful area of research has focused on the relationship between upper-level wind flow patterns, the associated vertical motions, and resulting climatic responses. This new orientation, made possible by advances in our understanding of upper-atmospheric meteorology, particularly when combined with computer-based models, dramatically altered the fields of both meteorology and climatology (Basu 1984), spawning what Neilson (1986) has called "the New Meteorology" in biogeographic investigations.

One aspect of this process-oriented approach to climatic studies was the development and use of the air mass concept in plant-climate studies. An air mass, a relatively homogeneous body of near-surface air with temperature and moisture referable to a particular source area, is produced by air subsiding from the controlling upper-level circulation; polar air masses are the result of subsidence from the Westerlies whereas tropical air masses result from downward vertical motions at the poleward margin of the Hadley Cell. Borchert (1950) demonstrated that maritime polar air masses, moving over the Rocky Mountain Cordillera from a Pacific Ocean source region, are most frequent over grassland areas during the warm season. These modified Pacific air masses are both warmed and dried by passage over the mountains. Bryson (1966) found similar results in his thorough study of North American air mass regions. Patterns of frequently occurring warm season air masses demonstrated spatial correlation with major plant formations. Maritime tropical air flowing northward from western sections of the Gulf of Mexico occupies areas of deciduous forests in southeastern sections of the midcontinent, cooler continental polar air is more common in the forested areas north and east of the Prairie Peninsula, and modified Pacific air influences the grassland region.

Subsequently, other researchers have examined linkages between flow patterns in the Westerlies, the associated vertical motions, and the resulting surface weather patterns. Surface weather features, including moisture, temperature, and wind, are generated and steered by the flow pattern aloft. These conditions then apparently limit the distribution of plants according to the differential response of various species to stress. Harrington (1980) suggested that the most frequent pattern of summer upper-level flow across the northern midcontinent helped determine the distribution of surface moisture and thus reinforced the location of the prairie-forest ecotone. Corcoran (1982), working on the prairie-forest ecotone in the southeastern midcontinent, found that, in grassland areas, subsidence of dry air from middle levels of the troposphere frequently diluted the warm, moist air moving northward from the Gulf of Mexico while forested areas farther to the east usually received their moist air undiluted.

By combining detailed knowledge of the dominant atmospheric motions with knowledge of how the resultant environmental conditions affect plants, these recent investigations have clarified the linkages between isolines of climatic variables and the patterns of natural vegetation distribution. A connection between the global atmospheric circulation system and gradients in surface conditions that help explain the geographic distribution of plants has been identified (Harrington and Harman 1985).

The Pattern of Natural Vegetation

Regional climatic patterns strongly affect the broad-scale distribution of vegetation in midcontinent North America, but other environmental factors, such as disturbance frequency, soil type, landform position, and local vegetational history, influence distribution as well. Thus, the vegetation at any given site reflects both climatic influences and the intricacies of the local environment. Pollen studies indicate that the current dominant species of vegetation in central North America have responded differently to the climatic changes of the Pleistocene Epoch, which began about 1.0 million to 1.5 million years ago (Davis 1976). These pollen-based studies indicate that, while some individual species may still be responding to changes in climate with minor adjustments in geographic distribution, the current climate-vegetation relationships were established around 2,000 years ago (Bernabo and Webb 1977). Thus, the following discussion underscores the role of climate, and particularly moisture availability and plant responses to moisture stress, in regulating the broad-scale patterns of the dominant species of vegetation.

Before the wide-scale disturbance associated with Euro-American settlement, three major vegetation formations occupied the midcontinent: grasslands, savanna, and forests. Subdivisions of the central North American grassland included (1) the shortgrass or bunchgrass steppes of the western High Plains, extending north into Saskatchewan; (2) an area of mixed-grass prairie that extends from central Texas northward across western Oklahoma; central Kansas, central Nebraska, eastern sections of the Dakotas, and into Canada; and (3) the tallgrass prairies in eastern sections of Oklahoma, Kansas, and Nebraska, western parts of Iowa and Minnesota, and in the Prairie Peninsula.

The eastern margin of this vast grassland region, the prairie-forest ecotone, was marked by a variety of vegetation types. Floodplain forests occupied the alluvial soils adjacent to major stream courses, whereas oak savannas grew in areas of coarse-textured soils on the intervening uplands. Farther north, aspen parklands separated grassland from boreal evergreen forest in Canada.

In the United States, oak forests are found to the northeast and southeast of the prairie grasslands and their border zone. Other deciduous trees become increasingly important farther eastward through Indiana as moisture supplies increase and environmental stresses decrease. Mixed deciduous and coniferous forests are found farther to the northeast and southeast.

The Grasslands

Variations within the grassland region of the midcontinent are a function of both environmental controls and individual species tolerances. Climate and disturbance, particularly fire and grazing, are commonly cited as the two most important environmental controls on a regional basis, while topography, soils, and seed source availability account for most local variation. Grass species apparently migrated into the midcontinent from several directions, including the western intermountain basins, the southeastern coastal plains, and the southwestern desert basins of the United States (Brown and Gersmehl 1985).

Varying migration rates, prevailing climatic gradients (including year-to-year fluctuations), and differences in individual species' ability to compete for scarce resources such as water account for considerable place-to-place variation in the plant species present across the midcontinent. North-to-south gradients in temperature, day length, and precipitation influenced the natural distribution of grass species and also influence the location of the dominant agricultural crops of today. Spring wheat, rye, oats, and barley are grown in northern areas, whereas corn dominates in central sections, and winter wheat and grain sorghum are most common southward. The north-south gradient also affects photosynthesis. The percentage of C_4 plants, species that use a photosynthesis pathway associated with warmer climates, decreases northward as average minimum temperatures during the growing season decline (Teeri and Stowe 1976). Overshadowing the north-to-south gradient and coinciding with the subdivision of the grassland region into shortgrass steppe, mixed-grass prairie, and tallgrass prairie is the increase from west to east of available moisture for plant growth.

Shortgrass Steppe

In westernmost sections of the midcontinent, the dominant natural vegetation consists of grasses that are generally 20 to 50 centimeters high. Blue grama and buffalo grass dominate, but western wheat grass, needle-and-thread grass, and wire grass are also important. These grasses are timed to grow with the spring rains. Most go dormant in summer to avoid the stresses of depleted soil moisture. Borchert (1950) has shown that "the shortgrass region is climatically distinctive in most summers." A steep gradient in warm season rainfall is a characteristic of the midcontinent; precipitation totals decrease rapidly westward from the 100th meridian.

The shortgrass steppe may have evolved in response to grazing pressure from the native ungulates including bison, pronghorn, and wapiti, or elk. Because animals generally graze grass species to the same level, the shoots of taller grasses are more susceptible. Their lowness and the focus of their growth on root production means the shortgrasses may suffer comparatively less from grazing than do taller species. Shorter grasses can also tolerate years of below normal rainfall better than tallgrasses can. Thus, both climate and a disturbance factor, grazing, helped reinforce shortgrasses as the dominant cover.

Mixed-grass Prairie

Geographically located between the tallgrass prairies to the east and the shortgrass steppes to the west, the area of mixed-grass vegetative cover has been considered by some to be a transition zone or ecotone, in part because it contains species present in both the shortgrass and tallgrass areas. Shortgrasses, such as blue grama and buffalo grass, occupy a lower layer, while the midgrasses, little bluestem, needle-and-thread grass, side-oats grama, and western wheat grass, penetrate above this level to heights of approximately 125 centimeters, imparting a distinctive structure to the mixed-grass region.

Boundaries between the mixed-grass prairie and the neighboring grassland vegetation types are indistinct. A line corresponding to the 20-inch annual precipitation isohyet has been used to define the eastern margin of the mixed-grass prairies, but other generalized boundary locations include the 97°W meridian and the 1,500-foot elevation contour (Weaver and Albertson 1956). Küchler (1972) has shown that the location of the mixed-grass prairie in Kansas shifts eastward in dry years and westward in wet. These are not species migrations, but there are changes in species dominance within the boundary areas. Effects of these changes in dominant species are most striking in the border zones between the major grassland vegetation types.

Müller and Weaver (1942) found that the seedlings of plants characteristic of uplands or western areas resisted drought better than did the tallgrass or lowland prairie species. When available moisture exceeds a threshold value, the growth of a plant above the ground is directly proportional to annual water use (Webb et al. 1978). Thus, plant populations within the mixed-grass prairie fluctuate dramatically in response to climatic variations (Coupland 1958). Comparison of climate factors to vegetation types shows that the mixed-grass prairie coincides with a region receiving 50 percent or less of normal rainfall during July of major drought years (Borchert 1950).

Tallgrass Prairie

Grasses that can grow more than two meters high characterize this region. Tallgrass prairie is not only taller but also has the greatest species diversity of the three grassland types. Dominants include big bluestem, Indian grass, and switchgrass but, on drier upland sites, little bluestem, needlegrass, June grass, prairie dropseed, and side-oats grama are common. In fact a hillside may replicate the east-to-west groupings of grasses from top to bottom, with the taller grasses in the moister lowland soils and the shorter ones in the drier soils near the hilltop (Weaver 1954). Thus, moisture availability regulates the plant populations of midcontinent grasslands on both local and regional scales.

Prior to Euro-American settlement of the midcontinent, the Prairie Peninsula extended eastward across Iowa, Illinois, Indiana, and into Ohio. The origin of this tallgrass peninsula has been the subject of considerable debate, centered primarily on the combined or separate roles of climate and fire within the region. What caused these prairies? Numerous authors have argued that the Prairie Peninsula is a fire-maintained grassland in an area of forest climate, whereas those scientists who have examined regional climatic data and presented the results of their studies have concluded that the Prairie Peninsula has a climate that differs significantly from that of adjacent forest areas (Transeau 1935; Borchert 1950; Manogaran 1983). As with most disagreements, a perspective that integrates both sides of the argument will probably emerge. Theorists are now, however, beginning to recognize fire as a natural environmental component of the tallgrass prairie as well as most other midcontinent ecosystems (Loucks 1970; Parton and Risser 1979; Vale 1982).

The Forests: The Prairie-Forest Ecotone

Two distinct vegetation types, riparian forests and oak savannas, make up the transitional zone separating the midcontinent grasslands from eastern areas of nearly continuous forest cover. Floodplain, gallery, or riparian forests were found penetrating into the grassland region along most major stream courses. Able to tap a river's abundant local groundwater supply, these forests were somewhat immune to the major midcontinent droughts. Cottonwoods, willows, elms, and ashes, their seedlings germinating during spring flooding, are the dominant trees of these riparian areas where site conditions moderate an environment otherwise inimical to trees.

The upland prairie-forest ecotone is usually characterized by park-like vegetation of oaks and prairie grasses. The best examples of this savanna vegetation were found in belts 75 to 175 kilometers wide in the "Cross Timbers" areas of Oklahoma and Texas and in "oak openings" in Minnesota and Wisconsin (Vankat 1979). Burr oak dominated the northern savannas, whereas post and blackjack oaks dominated the tree strata of the Cross Timbers. Frequent burning that destroyed other trees was necessary to maintain oak savannas. Their fire-resistant bark allowed the older, larger oaks to survive the frequent fires (Cottam 1949), and coppicing from surviving roots of burnt younger trees quickly reestablished the oak community, while less fire-tolerant trees, such as green ash in the north, succumbed to the fire. In addition, the roots of oak seedlings grow very rapidly, allowing pioneer oak species to compete successfully with grasses for limited available moisture (Holch 1931).

The Deciduous Forests

The midcontinent deciduous forests are divided into six major units according to composition and environment. Passing through increasingly cooler and wetter conditions from the grasslands toward the north and east, we find forests of the oak-hickory association, the maple-basswood association, and the hemlock-white pine-northern hardwoods association. Toward the wetter and warmer east and south, the oak-hickory association gives way to the oak-pine association and then to the southern mixed hardwood association. The sixth unit, the floodplain forests, occupies areas subject to periodic flooding. In each unit the success of the dominant species reflects environmental conditions.

Oak-Hickory Association

Oak forests with varying amounts of hickory are found on the driest sites within the westernmost sections of deciduous forests in Minnesota, Wisconsin, Illinois, Missouri, Oklahoma, and Texas, where they are commonly subjected to late summer drought. Locally such forests may be found on hilltops, ridges, slopes with south or west exposures, and areas of thin soils (Curtis 1959). Emergent oak seedlings survive in this xeric habitat because they can obtain soil moisture through rapid root extension following spring germination (Ferrell 1953). Oak seedlings are not as dependent on current photosynthetic production as other tree seedlings because they tap stored carbohydrates in the acorn during periods of limited soil moisture.

In general, the number of species within oak-hickory forests declines toward the west (Braun 1950). Dominant oaks, in addition to the species of the savannas, include white, southern red, northern red, and black oak. The hickories, bitternut, mockernut, red, black, and shagbark, are found with increasing frequency in areas to the east. Oak-hickory forests have the greatest species diversity in the Ozark Mountains of northern Arkansas and southern Missouri.

Maple-Basswood Association

Deciduous forests of sugar maple, American basswood, and American elm occur on relatively moist upland sites to the north and east of the prairie-forest ecotone. A mosaic of oak-hickory and maple-basswood forest types exists across sections of Minnesota and Wisconsin in response to variations in soil texture (Braun 1950). Grimm (1984) indicates that the probability of fire at individual sites, a reflection of soil texture and moisture conditions, is an important factor controlling the distribution of maple-basswood forests.

Both seedling growth and nutrient cycling characteristics help differentiate the maple-basswood and oak-hickory forests. Sugar maple seedlings use stored carbohydrate for early spring growth, but later require current year photosynthetic production (Kozlowski and Ward 1957), disadvantaging maples relative to oaks during a summer dry period (Curtis 1959). Nutrient cycling also helps to distinguish oak and maple forests. Nutrients that remain in sugar maple leaves when the leaves fall are transferred to the soil. Thus, maple-basswood forest soils are more fertile than oak-hickory forest soils, in part because oaks store nutrients in their trunks during the cool season period of dormancy rather than returning them to the soil (Curtis 1959).

Hemlock-White Pine-Northern Hardwoods Association

Areas of mixed coniferous-deciduous forest extend from northeastern Minnesota eastward through the Great Lakes region to the Atlantic Ocean (Nichols 1935). Within these forests, water is the single most important ecological factor and its relative plenitude determines the variations in the association (Cowles 1900). Primary subdivisions include bog or swamp forest on organic soils, the "pineries" on the drier, sandy soils, and upland mesophytic forests consisting of a mixture of coniferous and deciduous species on finer textured soils. Adding cone-bearing evergreen trees to the deciduous species helps to differentiate this forest association and also indicates a transition to cooler and moister environments.

Oak-Pine Association

In the southeastern midcontinent, particularly in southern Arkansas and parts of east Texas, southern pines add a similar coniferous element to deciduous forests. Some scientists consider the areas of southern shortleaf and loblolly pines to be a fire-produced subtype of the oak-hickory forest, but others recognize the relative dominance of the pines in areas of coarse textured, infertile soils, or south-facing slopes (Vankat 1979). Oak-pine forests occur in areas that receive an average of 40 or more inches of rainfall annually, well to the south and east of the prairie-forest ecotone. Fire is necessary for maintenance of the pine component within these forests because it removes understory competition and prepares a mineral seedbed for subsequent pine establishment, but local variations in environment also account for the existence of this forest association.

Floodplain Forests

Floodplains, bottomlands, and glacial lake plains subject to periodic flooding support forested wetlands well fertilized by silt. Hardwood swamp communities tolerate wet, poorly aerated soil. Northern floodplain forests are similar in composition to the riparian forests that extend westward into grassland areas, but in the southern floodplain forests oaks, tupelos, and bald cypress are the usual dominants of seasonally flooded sites.

Human Alterations

Few sites remain where the natural vegetative cover of the midcontinent has been preserved from human alteration. Disturbance and change are natural parts of all ecosystems, but the pronounced effects of human activity can be seen in almost every tract of land in the midcontinent. Substantial and long-lasting changes in the vegetative cover accompanied the westward expansion of European settlement across the United States. Native Indians had also altered the environment, but their impact was minor compared with Euro-American practices of plowing and clear cutting.

Different vegetation types underwent different types of environmental alteration, but each area is now adjusting to new cycles of nutrient and energy flow and possibly to new equilibrium states. The most obvious and direct environmental changes in the midcontinent are logging, plowing, introduction of new fire regimes, and alterations in grazing pressures. In addition, virtually all of this area is currently subjected to the indirect and controversial effects of acid rain and increased levels of atmospheric CO_2.

Logging, either clear cutting or selective harvesting of specific tree species, has greatly altered the forested areas of the midcontinent. In most areas, today's forests have grown back recently after human disturbance. These second growth forests represent the vegetative response to forest clearing for agriculture and to clear cutting of selective species, such as white pine in the north or loblolly pine in the south. In many cases, both the composition and structure of today's forests are markedly different from those of several hundred years ago.

Fire is a natural element in all midcontinent ecosystems. Prior to western settlement, fires set by lightning or by native Indians characterized both grassland and forest. In most areas, Euro-American settlement has reduced fire frequency, but where fire suppression efforts and unwise logging practices allowed fuel loads to build, severe slash and crown fires resulted. In the oak savannas, fire suppression has allowed the oaks to fill in the openings, destroying the park-like character of this vegetation type. Burning is also necessary to many prairie grassland ecosystems. Fire suppression has led to a woody plant invasion, particularly along the forest margins. Today, controlled burning is used to restore and maintain prairie grassland.

Plowing has also altered the midcontinent. Beginning in the 1830s, farmers equipped with steel moldboard plows turned the sod on the overwhelming majority of the tallgrass prairie. Much of this area is farmed today, but in abandoned fields near the prairie-forest ecotone, tree cover has followed the plow, encroaching into former prairie areas where fire suppression has protected it. Plowing also affected areas of original deciduous forests. Most of today's deciduous forests in the midcontinent have gone through a cycle of logging, plowing, abandonment, and old-field succession before eventually regaining a forest cover. These wooded areas resemble the prehistoric forests to varying degrees depending upon the extent and frequency of disturbance as well as the makeup of the original forest (Dodge and Harman 1985).

Grazing is both a natural component and a human-induced alteration of midcontinent ecosystems. Deer, bison, and other native animals fed on the vegetation of both forests and grasslands. Overgrazing is brought about by an increased number of grazing animals and by fencing that restricts animals to specific tracts of land. The main vegetative response to grazing pressure is a change in composition. Decreaser species become less important as increasers gain in relative dominance while invaders enter the site.

The present vegetative cover of the midcontinent provides some evidence of presettlement conditions, but it bears direct testimony to the powerful influences of human alteration. Despite the efforts of Euro-American settlers to change the landscape to suit their economic goals, the overriding imprint of natural environmental controls remains. Gross patterns of ranching, agriculture, and forest activity are clearly arranged in association with climatic gradients of temperature and available moisture.

References

Basu, J. E. 1984. Jerome Namias: Pioneering the science of forecasting. *Weatherwise* 37(4):191-194, 199-201.

Bernabo, J. C., and T. Webb III. 1977. Changing patterns in the Holocene pollen record of northeastern North America. *Quaternary Research* 8:64-96.

Borchert, J. R. 1950. The climate of the central North American grassland. *Annals of the Association of American Geographers* 40:1-39.

Braun, E. L. 1950. Deciduous forests of eastern North America. Philadelphia: Blakiston.

Brown, D. A., and P. J. Gersmehl. 1985. Migration models for grasses in the American midcontinent. *Annals of the Association of American Geographers* 75:383-394.

Bryson, R. A. 1966. Air masses, streamlines, and the boreal forest. *Geographical Bulletin* 8:228-269.

Collins, D. D. 1969. Macroclimate and the grassland system. In *The grassland ecosystem: A preliminary synthesis,* ed. R. L. Dix. Science Series no. 2. Range Science Department, Fort Collins, Colorado.

Corcoran, W. T. 1982. Moisture stress, mid-tropospheric pressure patterns, and the forest/grassland transition in the south central states. *Physical Geography* 3:148-159.

Cottam, G. 1949. The phytosociology of an oak woods in southwestern Wisconsin. *Ecology* 30: 271-287.

Coupland, R. T. 1958. The effects of fluctuations in weather upon the grasslands of the Great Plains. *Botanical Review* 24:273-317.

Cowles, H. C. 1900. The physiographic ecology of northern Michigan. *Science* 39:708-709.

Curtis, J. T. 1959. *The vegetation of Wisconsin.* Madison: University of Wisconsin Press.

Davis, M. B. 1976. Pleistocene biogeography of temperate deciduous forests. *Geoscience and Man* 13:13-26.

Dodge, S. L., and J. R. Harman. 1985. Woodlot composition and successional trends in south central lower Michigan. *Michigan Botanist* 24:43-54.

Ferrell, W. K. 1953. Effect of environmental conditions on survival and growth of forest tree seedlings under field conditions in the Piedmont region of North Carolina. *Ecology* 34:667-688.

Gleason, H. A. 1923. The vegetational history of the Middle West. *Annals of the Association of American Geographers* 12:39-85.

Grimm, E. C. 1984. Fire and other factors controlling the big woods vegetation of Minnesota in the mid-nineteenth century. *Ecological Monographs* 54:291-311.

Hare, F. K. 1951. Some climatological problems of the Arctic and Sub-Arctic. In *Compendium of meteorology,* ed. T. F. Malone. Boston: American Meteorological Society.

Harman, J. R., and D. H. Braud. 1975. Flow patterns at the 500 mb level and the prairie-forest ecotone in Canada. *Michigan Academician* 7:347-356.

Harrington, J. A., Jr. 1980. Upper atmospheric controls, surface climate, and phytogeographical implications in the western Great Lakes region. Ph.D. dissertation, Michigan State University, East Lansing.

Harrington, J. A., Jr., and J. R. Harman. 1985. A synoptic climatology of moisture stress gradients in the western Great Lakes region. *Physical Geography* 6:43-56.

Holch, A. E. 1931. Development of roots and shoots of certain deciduous tree seedlings in different forest sites. *Ecology* 12:259-298.

Kozlowski, T. T., and R. C. Ward. 1957. Seasonal height growth of deciduous trees. *Forest Science* 3:168-174.

Küchler, A. W. 1964. *Potential natural vegetation of the conterminous United States.* Special publication 36. New York: American Geographical Society.

———. 1972. Oscillations of the mixed grass prairie in Kansas. *Erkunde* 26:120-129.

Loucks, O. L. 1970. Evolution of diversity, efficiency, and community stability. *American Zoologist* 10: 17-25.

Manogaran, C. 1983. The Prairie Peninsula: A climatic perspective. *Physical Geography* 4:153-166.

Mather, J. R. 1959. The moisture balance in grassland climatology. In *Grasslands,* ed. H. B. Sprague. Publication 53. Washington, D.C.: American Association for the Advancement of Science.

Müller, I. M., and J. E. Weaver. 1942. Relative drought resistance of seedlings of dominant prairie grasses. *Ecology* 23:387-398.

Neilson, R. P. 1986. High-resolution climatic analysis and Southwest biogeography. *Science* 232:27-34.

Neilson, R. P., G. A. King, R. L. DeVelice, J. Lenihan, D. Marks, J. Dolph, B. Campbell, and G. Glick. 1989. *Sensitivity of ecological landscapes and regions to climatic change.* Corvallis, Oregon: U.S. Environmental Protection Agency, Environmental Research Laboratory.

Nichols, G. E. 1935. The hemlock-white pine-northern hardwood region of eastern North America. *Ecology* 16:403-422.

Palmén, E., and C. W. Newton. 1969. *Atmospheric circulation systems.* New York: Academic Press.

Parton, W. J., and P. G. Risser. 1979. Simulated impact of management practices upon the tallgrass prairie. In *Perspectives in grassland ecology,* ed. N. French. Ecological Studies 32. New York: Springer-Verlag.

Teeri, J. A., and L. G. Stowe. 1976. Climatic patterns and the distribution of C_4 grasses in North America. *Oecologia* 23:1-12.

Transeau, E. N. 1905. Forest centers of eastern America. *American Naturalist* 39:875-889.

———. 1935. The prairie peninsula. *Ecology* 16:423-437.

Vale, T. R. 1982. *Plants and people: Vegetation change in North America.* Resource publications in geography. Washington, D.C.: Association of American Geographers.

Vankat, J. L. 1979. *The natural vegetation of North America.* New York: John Wiley and Sons.

Weaver, J. E. 1954. *North American prairie.* Lincoln, Nebraska: Johnsen Publishing Co.

Weaver, J. E., and F. W. Albertson. 1956. *Grasslands of the Great Plains: Their nature and use.* Lincoln, Nebraska: Johnsen Publishing Co.

Webb, W., S. Szarek, W. Lauenroth, R. Kinerson, and M. Smith. 1978. Primary productivity and water use in native forest, grassland, and desert ecosystems. *Ecology* 59:1239-1247.

Wendland, W. M., and R. A. Bryson. 1981. Northern hemisphere airstream regions. *Monthly Weather Review* 109:255-270.

11. Great Plains Climates and Biota: Past, Present, and Future

Dexter Hinckley
U.S. Environmental Protection Agency

Organisms living in and around the Great Plains have evolved through long periods of changing climates. Their very existence attests to the fact that they have genetic adaptations enabling them to survive change. However, in the next century, these organisms, domesticated and wild, may be exposed to rates of change more rapid than they have experienced during their evolutionary history. This is a cause for concern because some organisms may not have the physiological, dispersal, dormancy, or behavioral adaptations needed to survive. An increase in the frequency and duration of severe droughts would be especially worrisome.

This chapter, a discussion of climate change in the Great Plains, will cover present climatic characteristics of the region, after briefly describing how past climates helped shape the topography, hydrology, soils, flora, and fauna of the region. Future climate changes likely to be caused by global warming from the atmospheric build-up of carbon dioxide, methane, and other greenhouse gases will also be discussed, as will the effects of these potential changes on crops, livestock, and natural ecosystems. Finally, strategies and techniques that could be used to cope with climate change will be presented.

Past Climates and Biota

The climatic history of the Great Plains shows many changes. During the Paleozoic Era, 450 million years ago, most of North America was under water, from the Atlantic to the Pacific Ocean, and from the Hudson Bay to the Gulf of Mexico. Today, these ancient seas are marked by distinctive fossil faunas—corals, trilobites, sharks, and primitive bony fishes (Grossman et al. 1969).

This chapter does not present policies of the U.S. Environmental Protection Agency or any other agency of the U.S. government.

Stanley R. Johnson and Aziz Bouzaher (ed.), *Conservation of Great Plains Ecosystems*, 149–159.

At the beginning of the Cenozoic Era, 60 million years ago, the whole region was uplifted and grasslands similar to those now in existence came into being. Then, as now, North American weather patterns were dominated by Westerlies, with masses of moist air coming off the Pacific, only to be virtually wrung dry by passing over three mountain ranges. Several hundred miles to the east, the drying effect of the mountains was ameliorated as masses of moist air were sucked north from the Gulf of Mexico. A huge grassland developed from the arctic to the subtropics, varying from desert to arid, and from semiarid to humid to wet grasslands, depending on the vagaries of local climate patterns (Madson 1982).

Ten million years ago, some of the mammals roaming these plains were similar to modern forms but certainly not those now living wild in the region. There were mastodons, mammoths, rhinos, hippos, camels, tapirs, giant ground sloths, giant armadillos, dire wolves, and saber-toothed tigers, and others so bizarre that they are known only by their scientific names.

During the Pleistocene Epoch, four major glaciations occurred, each of which had profound effects on the climate and topography of the region. When the glaciation was at its peak, weather was cool and rainy a thousand miles south of the ice fronts. As the glaciers retreated north, melt water created lakes; the largest, Lake Agassiz, covered 100,000 square miles in the northern Plains, leaving a legacy of rich, black soils. Similar lakes to the west did not have the benefit of good drainage and left a poor legacy of salt lakes and saline soils. Perhaps the most important legacy from the ice ages is the vast underground supply of "fossil" water in aquifers.

Sometime during the most recent glaciation (11,000 years ago, possibly earlier), humans crossed the Bering Bridge and arrived in the Great Plains. These Paleoindians were the true discoverers of the New World. Unfortunately for the resident mammals, they also were very good hunters, who were well organized and equipped with an array of primitive but lethal weapons. This was the period of "Pleistocene overkill" (Martin and Wright 1967).

About 6,000 years ago, a period of heat and drought began. Called the Altithermal, it lasted 2,000 years and turned the Great Plains into near-desert. This may have contributed to the success of Paleoindians as hunters by concentrating their prey at water holes. Most big animals disappeared from the Plains (Frazier 1989). However, some grazers survived drought and hunting. As recently as 1600, there were 45 million bison and approximately the same number of pronghorns (Shelford 1963). During the 1800s, overhunting and competition with cattle brought the bison close to extinction and greatly reduced the pronghorn populations, especially east of the Rockies. In 1889, there were only 541 bison and, in 1924, there were 30,000 pronghorns. By 1969, bison and pronghorns had rebounded to 11,000 and 350,000 (Grossman et al. 1969).

Decimation of the bison herds had profound effects on the ecosystems of the Great Plains. Migrating herds of bison had kept tallgrass out of the Kansas borderland, made the land clear for prairie dog towns, and ensured the germination of the durable shortgrasses upon which the bison fed (Grossman et al. 1969). After the hunters and bison had gone, the farmers and ranchers encountered communities of grasses and flowers determined by rainfall rather than the grazing and trampling of bison.

Another relic of the Pleistocene is making a comeback in the Great Plains. This is the continent's tallest bird, the whooping crane. Nesting in the Wood Buffalo National Park of northern Alberta and wintering on the mud flats of Aransas, Texas, the wild population had reached a low of 14 in 1938 (Grossman et al. 1969). There are now 150 in that wild flock and 100 in captivity. An attempt to establish a separate population migrating with the sandhill cranes has not been successful. Many of the "whoopers" with the sandhills have been killed by avian tuberculosis; others have died when blundering into electrified fences, until now only 11 survive. The U.S. Fish and Wildlife Service is attempting to establish a nonmigratory population in Florida.

Present Climate and Biota

Colorado and Kansas today are on rolling terrain, built up with sediment washed down from the Rocky Mountains over the millennia or deposited by retreating glaciers. Although the land seems level, it gently slopes downward toward the east over 500 miles. In fact, eastern Kansas is nearly a mile lower than the western edge of the Plains at Colorado's Front Range.

The precipitation from east to west gradually decreases from about 30 inches per year to less than 15 inches. Although mixed grasses are the natural vegetation in the Central Plains region, now it is nearly all planted in fields of grain from spring wheat in the Dakotas to winter wheat in Kansas and Oklahoma (Grossman et al. 1969).

Continental climate can be described as one with temperature extremes (large diurnal and seasonal ranges of temperatures), small annual precipitation totals, and low relative humidities. The climate now prevailing in the Great Plains certainly fits that description. In 1936, there were two regional records in North Dakota: February 15, a low of –60°F and on July 6, a high of 121°F. The frost-free season is as short as three to four months in the Canadian plains and nearly year-long in southern Texas. Annual precipitation averages 15 inches in North Dakota and the Panhandle of Oklahoma but is 56 inches near the Louisiana-Texas border and in southeastern Oklahoma. As Madson (1982) puts it, "The special quality of fine prairie weather isn't one of intrinsic merit, but of contrast with what has gone just before. Sprinkled through the prairie year are certain very fine times that are usually moderations of unbearable extremes, or which come as refreshing novelty after a long run of dull."

Winter storms can take the form of blizzards, a cyclonic storm characterized by high winds, extreme cold, and moderate-to-heavy snowfall. However, most of the precipitation, typically in the form of thunderstorms, comes during the early summer.

This may be followed by a dry season but there is great annual variation in precipitation. Droughts, typified by lack of rain during spring and summer, have occurred every 20 years or so, and most recently in 1987–89. The eastern margin of the Great Plains may average only one drought year in 20; the western margin may have drought conditions 10 years out of 20. For much of the Great Plains, the drought of the 1930s (the Dust Bowl years) began in 1929 and did not end until 1942. Overgrazing, deep plowing, and lack of irrigation compounded the damage, leading to wind erosion and "black blizzards" that were life-threatening.

After the 1930s, removal of marginal lands from cropping, soil conservation practices, and use of groundwater for irrigation reduced losses from droughts, although none of the recent droughts have been so prolonged. There was some concern that 1988 marked the beginning of another long sequence of drought years, but dust from Mount Pinatubo in the Philippines may have shortened the forecast drought.

In the eastern Great Plains, the former tallgrass country, precipitation equals or exceeds evaporation. Everywhere else, rainfall is less than evaporation, partly from effects of high temperatures and partly because of the wind sweeping across the Plains (Shelford 1963).

Naturally, vegetation has many mechanisms for surviving seasonal dry spells (long roots) or dry years (tubers or dormant seeds). Small mammals can obtain water from vegetation; large ones can travel long distances to find water. Still, there are noticeable shifts in vegetation and animals during droughts. Droughts can be especially devastating to the inhabitants of the prairie potholes, a valuable legacy of the last ice age. These potholes cover parts of a 300,000-square-mile area of the Dakotas, Iowa, Minnesota, Montana, Alberta, Saskatchewan, and Manitoba. It is believed that 50 to 75 percent of all the waterfowl produced in North America in nondrought years come from this region (Mitsch and Gosselink 1986). During drought years, waterfowl utilizing the remaining wetlands may be crowded together and killed by infectious diseases or botulin poisoning. The surviving whooping cranes are one of the species at risk.

At the other extreme, the region can suffer floods. Stalled storm systems over Texas create flash floods, turning dry gullies (arroyos) into raging torrents. However, the storm surges associated with hurricanes coming ashore over a rising tide are even more deadly. The most lethal was the Galveston flood of September 8–9, 1900, which drowned at least 8,000 people.

Also deadly are the tornados spawned by strong thunderstorms. With winds as high as 230 miles per hour, tornados can occur in any state east of the Rockies, but Oklahoma has the dubious distinction of the greatest incidence. Tornados sometimes come in swarms. During one 24-hour period over April 3–4, 1974, there were 148 in the Midwest, killing 315 people. On the average there are 850 tornados, causing 100 deaths, each year (Miller 1987).

Future Climate

Up to this point, past or present patterns of weather or characteristics of biota have been described. Now, one must gaze into a very murky crystal ball and try to determine probable changes in climate and effects on the Great Plains rangelands, crops, and natural ecosystems. Unfortunately, in this case, the past is not a good guide to the future. The warming that occurred at the end of past glaciations was quite different from that now expected. The retreat of the glaciers over several thousand years led to gradual warming and drying south to north in the Plains, allowing ample time for immigration of grasses and forbs to replace the coniferous forests and tundra along the glacial fringe. If the general circulation models used to simulate the effects of greenhouse gas build-up on climate are realistic, future warming will be far more rapid.

By 2030, less than 40 years from now, according to the Intergovernmental Panel on Climate Change (IPCC), the warming in central North America could be 3.5 to 7.2°F in winter and 3.6 to 5.4°F in summer. Precipitation could increase from 0 to 15 percent in winter and decrease 5 to 10 percent in summer. Most ominous, the increased temperature and slight decrease in precipitation could reduce soil moisture 15 to 20 percent during the summer (Houghton 1990).

Karl and Heim (1991) have analyzed regional records for the past 90 years and compared the results with the IPCC projections to determine if any changes in temperature or precipitation have already occurred. IPCC concluded that the seasonal mean minimum temperature is increasing in the central United States during all seasons but most significantly during spring and summer. However, their projections were for seasonal means and increases in mean summer temperature will not be detectable until 2005 or 2010, and summer precipitation decreases will not emerge from normal variability until after 2030.

Since the atmospheric concentrations of carbon dioxide and methane are likely to continue to increase for many decades after 2030, despite efforts to reduce emissions, the warming and drying may continue. The Great Plains could be facing not just a decade-long drought, such as that of the 1930s, but a much longer dry period of several centuries, more like the Altithermal of 4,000 to 6,000 years ago. To be sure, another ice age eventually will bring rain to the Great Plains but, meanwhile, there will be problems "back at the ranch" and "down on the farm."

Effects of Future Droughts

"The Potential Effects of Global Climate Change on the United States," a report to Congress, focused on the effects of a "doubled-CO_2" climate and did not explore the consequences of continued increases beyond that level (Smith and Tirpak 1990). The investigators assessing potential effects on crops considered not only the effects of changes in temperature and precipitation but also the effects of CO_2 fertilization and a longer growing season. Incidentally, because of differences in carbon assimilation, some crops (such as wheat, soybean, and cotton) respond more than corn and sorghum to elevated concentrations of CO_2. Unfortunately, certain weeds and pests also respond positively to higher levels of CO_2 (Strain and Cure 1986).

The report's executive summary contains two main findings about effects in the Great Plains:

- **Decline in Crop Acreage.** Studies of crop yield and economic adjustment indicate that grain crop acreage could diminish in the region. The direction of changes in wheat and corn yields depends on the direct effects of CO_2 on crop growth and the severity of climate change. If climate becomes hotter and relatively drier, yields could decrease. Whatever the climate change, relative productivity may decline. As a result, crop acreage was estimated to drop 4 to 22 percent, a reduction that would adversely affect the economy of the region. These studies did not consider use of new technologies or introduction of new crops.
- **Increased Demand for Irrigated Acreage.** The demand for irrigation on farms that continue to grow grain crops would increase. Irrigated acreage, currently at 10 percent of the total acreage and growing, could increase by 5 to 30 percent. How this demand would be satisfied was not stated, although the Ogallala aquifer could be a candidate source. Other effects of global warming could change ground and surface water supplies and, possibly, surface water quality. Changes in precipitation could affect the leaching of pesticides into groundwater and runoff to surface water in some cases, although the direction of change cannot be determined because runoff and leaching of pesticides and soils are very sensitive to rainfall variability.

Livestock raised in regions subject to heat waves may experience even greater stress, resulting in lower fertility, reduced growth, lower milk production from dairy cows, and mass mortality on chicken farms. Livestock production also may be affected indirectly through increases in the price of feed, reductions in the availability of grazing land and water, and more severe attacks by pests such as the horn fly (Smith and Tirpak 1990)

The report to Congress did not cover potential effects on wildlife and natural vegetation in the Great Plains. As indicated earlier, the plants and the animals of the Great Plains have evolved through long periods of changing climates. They have physiological, morphological, reproductive, and in the case of animals, behavioral, adaptations enabling them to survive fires and periodic droughts, as well as diurnal and seasonal fluctuations in temperature. However, these adaptations were shaped by the range of variations experienced over the evolutionary history of the organisms. For example, a whooping crane may be able to travel 500 miles, but no farther, without "refueling." Absence of wetlands at that interval, or less, along the flyway could be lethal. Similarly, a lizard may be able to cope with 120°F but dies when exposed to

130°F. The large grazers, specifically the bison and the pronghorn, will probably be able to survive, if helped by moving their herds to grasslands farther north. There are too many barbed-wire fences in their way for them to resume their ancestral migrations. At least 20 fish species of the Great Plains and Southwest face extinction from a several-degree warming because they cannot move northward to cooler waterways (Matthews and Zimmerman 1990).

Exposed to a multiyear drought, the plant community will lose forbs and grasses, those being replaced by cacti and shrubs with very deep tap roots. Even if these desert plants can expand their domain quickly enough to keep up with the changing climate, it is quite likely that dust will once again start blowing from large areas where roots no longer hold the soil. Furthermore, it has been estimated that, nationwide, 9 to 18 percent of the rarest plant species are vulnerable to extinction from a 5.4°F warming (Kutner and Morse 1992).

Helping Agriculture Cope with Climate Change

A National Academy of Sciences report, entitled "Policy Implications of Greenhouse Warming," concluded that people in the United States probably will have no more difficulty adapting to global warming than to the most severe conditions in the past, such as the Dust Bowl (NAS 1991a). A member of the Synthesis Panel (Matthews 1991) took strong exception to this conclusion, saying that "The analysis does not support the conclusion that greenhouse warming will be no more demanding than past climatic changes. If the change is unprecedented in the experience of the human species, how can it be claimed that people will have no more difficulty adapting to future changes than those of the past?" The Adaptation Panel, another group of experts participating in the study, analyzed options to help farmers cope with increasing droughts, but it did not recognize the possibility that future droughts will be even more prolonged, and perhaps more severe, than the Dust Bowl years.

Expanded Irrigation

Irrigation, of course, is the prime prospect. The Ogallala aquifer, a possible source, supplies irrigation for approximately 14 million acres in the Great Plains that are too far from surface supplies. The aquifer also provides water for municipal and industrial purposes. In some states, high withdrawal and low recharge rates of the aquifer have already resulted in "mining" of the resource (i.e., the rate of water withdrawal is greater than rate of recharge). Any action of irrigators to increase irrigation efficiency as an attempt to cope with projected water shortages, while economically beneficial, may lead to increased salinity problems if sufficient water is not applied to meet soil leaching requirements (Smith and Tirpak 1990).

In addition, serious conflicts may arise between agriculture and other users, such as power plants that need cooling water. It is even possible that demand for limited water supplies to meet other uses—including urban, industrial, fish and wildlife, recreation, and navigation—may make water too expensive for agriculture (Frederick and Kneese 1990). This seems to be happening already in California. Obviously, water managers in the Great Plains region are faced with some hard choices in the decades to come.

Other Changes

Other options available to help farmers cope with climate change include changing planting times for existing crops and substituting new varieties or crops. "Drought-evaders," plants with large root systems or the ability to cut transpiration losses, are the most promising prospects. Actually, a cultivated crop's most important ability is to produce the greatest amount of grain or forage during dry conditions. In this respect, the cereal grains do well, except under the severest drought stress (Troeh et al. 1980).

If the growing season is prolonged, earlier planting would be appropriate; even double-cropping, especially in irrigated areas, should be possible. Hard winter wheat could even be substituted for spring wheat. Winter wheat's fall planting avoids waiting for the soil to be workable in the spring, and its early harvest comes before the heat and drought of summer can cut its yield (NAS 1991b). Other substitutions are possible: sorghum for corn, and drought-resistant grasses for those not so adapted. However, all these require some cooperation on the part of the weather, and that cannot be guaranteed. Even the National Academy's study admits that rapid climate change, transient changes such as cooling prior to warming, or major changes in the variability of temperature, precipitation, storminess, and so forth, could, of course, challenge their essentially optimistic view of adaptation potential (NAS 1991b).

Helping Natural Ecosystems Cope with Climate Change

The relatively new field of conservation biology will be facing its greatest challenge in helping plants and animals in natural ecosystems on the Great Plains adapt to changes more rapid than those experienced throughout much of the history of life on earth. Such adaptation efforts will have to proceed on several parallel fronts.

Habitat Preservation

The National Academy of Sciences (1991a) recommends a twofold approach:

- Establish and manage areas encompassing full ranges of habitats
- Purchase land or easements that will allow vulnerable species to migrate to new habitats

Implementing such suggestions will require rethinking the principles of habitat protection.

Ducks Unlimited already conducts programs to help protect against the effects of climate change. Their agriculturally compatible landowner partnership program, "Prairie CARE," directly assists landowners who want to convert farm operations from those that increase surface water runoff to operations that conserve water and preserve wetlands (Clay et al. 1991).

A dynamic approach to habitat selection and protection is required if we are to cope with the effects of climate change on habitats. It has been suggested that north-south corridors should be protected to help species migrate north. In regions where farmlands create a barrier to northward migration, this would require taking large areas out of cultivation. Such corridors would not be able to bridge human-built barriers, such as cities and highways, or natural barriers, such as the Great Lakes.

Furthermore, corridors could make it all too easy for weeds and pests to move north. Such species are adapted for rapid dispersal in response to newly available habitats. There is a danger that weedy and shrubby vegetation could become so well established that more desirable species (e.g., trees) could effectively be excluded. For animals, there is the danger that such a corridor could act as a death trap in which dispersing individuals would be more likely to suffer mortality through predation, hunting, or road-kill (Hobbs et al. 1990). Nonetheless, corridors can partially compensate for the problems of habitat fragmentation by helping to ensure dispersal, recolonization, and gene flow, linking small habitats that would otherwise be isolated.

Restoration Ecology

In helping ecosystems adapt to climate change, the line between protection and restoration will be quite fuzzy. Many of the techniques developed initially to help restore strip-mined land or eutrophic lakes can be modified to help ecosystems stressed by climate change. Restoration ecology is quite distinct from wildlife ecology, silviculture, or other resource management sciences in that it seeks to create, or "re-create," entire ecosystems.

The most ambitious restoration project for the Great Plains region is the proposed "Buffalo Commons." Frank and Deborah Popper (1987) advocate returning about 139,000 square miles in ten Great Plains states to open country for use as a wildlife refuge.

This has the support of some conservation groups but has not been greeted with wild enthusiasm by the U.S. Department of Agriculture or by many Great Plains residents.

Changed management of grassland and rangeland ecosystems may help them serve as "sinks" for CO_2. The draft U.S. Action Plan in support of the Framework Convention on Climate Change specifically mentions long-term field studies to evaluate soil-cover management in the Great Plains and how it influences the flux of carbon and nitrogen trace gases. The way that plants distribute carbon above and below ground, while adapting to changing carbon concentrations in the atmosphere, is being studied to understand how these properties affect the carbon balance of terrestrial ecosystems. This knowledge will be translated into new technologies for managing ecosystems (Council on Environmental Quality 1992).

All of these adaptation techniques, whether for crops or for wildlife, require additional research before they can be implemented on a regional scale. Within each state and Indian nation, there will be many hard decisions about water allocation and land use. It is hoped that future conferences will help accelerate the research required and help define strategies for coping with short-term climate variability and long-term climate change.

Overall, the problems of the Great Plains must be approached in a complete ecological approach. Many people need to understand human ecology, plant ecology, and animal ecology. Not until the Great Plains "ekos" (the environment of its people, its plants, its animals) is understood will its people be able to deal with the Great Plains intelligently (Hoyt 1938).

Conclusions

Climate in the Great Plains has changed dramatically over the eons with profound effects on the topography, hydrology, soils, and biota we see today. The current climate is "continental," one with temperature extremes (large diurnal and seasonal ranges of temperatures), small annual precipitation totals, and low relative humidities. Global warming, induced by increased atmospheric concentrations of carbon dioxide, methane, and other greenhouse gases, is likely to make the Great Plains warmer and drier. This, in turn, will have far-reaching effects on water supplies, crops, livestock, wildlife, and the regional economy. Adaptation options are available for both domesticated and wild species but their implementation will be complicated and expensive. Now is the time to accelerate research on adaptation and to develop strategies for coping with climate change. These efforts must be based on improved understanding of human, plant, and animal ecology in the Great Plains.

References

Clay, R., T. Neraasen, and R. Coley. 1991. Management strategies in anticipation of climatic change and the resulting impact on wetlands. In *Symposium on the impacts of climate change and variability on the Great Plains,* ed. G. Wall. Department of Geography Publication Series, Occasional Paper No. 12. University of Waterloo, Ontario, Canada.

Council on Environmental Quality. 1992. U.S. National Action Plan (USNAP). *Federal Register* (Dec. 4).

Frazier, Ian. 1989. *Great Plains.* New York: Farrar, Straus, & Giroux.

Frederick, K. D., and A. V. Kneese. 1990. Reallocations by markets and prices. In *Climate change and U.S. water resources,* ed. P. E. Waggoner. New York: John Wiley & Sons.

Grossman, Mary Louise, Shelly Grossman, and John N. Hamlet. 1969. *Our vanishing wilderness.* New York: Madison Square Press, Grosset & Dunlap.

Hobbs, R. J., D. A. Saunders, and B. M. T. Hussey. 1990. Nature conservation: The role of corridors. *Ambio* 19(2):94-95.

Houghton, John. 1990. *Scientific assessment of climate change: The policymakers' summary of the report of Working Group I to the Intergovernmental Panel on Climate Change.* Geneva: World Meteorological Organization and United Nations Environment Program.

Hoyt, John C. 1938. *Drought of 1936, with discussion on the significance of drought in relation to climate.* Washington, D.C.: U.S. Government Printing Office.

Karl, T. R., and R. R. Heim, Jr. 1991. The greenhouse effect in central North America: If not now, when? In *Symposium on the impacts of climate change and variability on the Great Plains,* ed. G. Wall. Department of Geography Publication Series, Occasional Paper No. 12. University of Waterloo, Ontario, Canada.

Kutner, L., and L. Morse. 1992. Further results from climate change study. *Biodiversity Network News* 5(3):6.

Madson, John. 1982. *Where the sky began: Land of the tallgrass prairie.* Boston: Houghton Mifflin Company.

Martin, Paul S., and H. E. Wright, Jr. (ed.). 1967. *Pleistocene extinctions: The search for a cause.* New Haven: Yale University Press.

Matthews, J. 1991. Footnote on pp. 45-46. In *Policy implications of greenhouse warming.* Committee on Science, Engineering, and Public Policy. Washington, D.C.: National Academy Press.

Matthews, W. J., and E. C. Zimmerman. 1990. Potential effects of global warming on native fishes of the southern Great Plains and the Southwest. *Fisheries* 15(6):26-32.

Miller, P. 1987. Tornado! *National Geographic* 171:690-715.

Mitsch, William J., and James G. Gosselink. 1986. *Wetlands.* New York: Van Nostrand Reinhold Company.

National Academy of Sciences (NAS). 1991a. *Policy implications of greenhouse warming.* Committee on Science, Engineering, and Public Policy. Washington, D.C.: National Academy Press.

———. 1991b. *Policy implications of greenhouse warming: Report of the Adaptation Panel.* Washington, D.C.: National Academy Press.

Popper, Deborah, and Frank Popper. 1987. The Great Plains: From dust to dust. *Planning* 53:12-18.

Shelford, Victor E. 1963. *The ecology of North America.* Urbana: University of Illinois Press.

Smith, Joel B., and Dennis A. Tirpak (ed.). 1990. *The potential effects of global climate change on the United States: Report to Congress.* U.S. Environmental Protection Agency, Office of Policy, Planning, and Evaluation, Office of Research and Development. New York: Hemisphere Publishing Co.

Strain, Boyd R., and Jennifer D. Cure (ed.). 1986. *Direct effects of increasing carbon dioxide on vegetation.* DOE/ER-0238. Washington, D.C.: U.S. Department of Energy.

Troeh, Frederick, J. Arthur Hobbs, and Roy L. Donahue. 1980. *Soil and water conservation for productivity and environmental protection.* Englewood Cliffs, New Jersey: Prentice-Hall, Inc.

12. Air Quality, Climate Change, and Their Possible Impacts on the Terrestrial Ecosystems of the North American Great Plains

Sagar V. Krupa
University of Minnesota
Allan H. Legge
Biosphere Solutions

The chemical climate (air quality) of the atmosphere is as important as its physical climate (temperature, precipitation). The two components are fundamentally interrelated. The earth has evolved in an atmosphere of life-supporting chemical constituents and has a certain ability for equilibrium and resilience to absorb changes in this chemical climate (Lovelock 1987). However, problems arise when the system becomes overloaded (such as possible global warming). The concentrations of many atmospheric trace gases, gases that occur in relatively small concentrations but are very important, positively or negatively, to life on earth, have been increasing since the onset of the Industrial Revolution. For example, ambient concentrations of carbon dioxide have increased from roughly 280 parts per million at the turn of the century to the present concentrations of about 350 to 360 parts per million (EarthQuest 1990). Many trace gases, when present in high enough concentrations over sufficient duration, can adversely affect ecosystems.

Based on their physical properties, air pollutants can be classified as gases and as solid or liquid particles of varying sizes. These pollutants are transferred from the atmosphere to ecosystems and other surfaces by dry (during periods without precipitation) and wet deposition (Legge and Krupa 1990). The importance and contribution of these two processes vary in time and space. For example, dry deposition of sulfur dioxide is known to be more important closer to its sources (e.g., a coal-fired power plant), while farther away wet deposition (acidic rain) of its transformed product sulfate appears to dominate (U.S. NAPAP 1989). Depending on the meteorological conditions, air pollutants can be transported from a few kilometers to thousands of kilometers, across provincial, state, national, and international boundaries. During this process, depending on the atmospheric conditions, the initially emitted pollutants—primary pollutants—can be transformed to secondary pollutants such as nitrogen

Stanley R. Johnson and Aziz Bouzaher (ed.), *Conservation of Great Plains Ecosystems*, 161–182.

dioxide to ozone, or sulfur dioxide to sulfate particles. Thus, air pollutants may also be classified according to their spatial distribution as important at the local (sulfur dioxide), regional-continental (ozone), or global (carbon dioxide) scale. In this context, carbon dioxide, although required by plants for photosynthesis, is considered to be a pollutant since it is emitted predominantly through fossil fuel combustion and has a major role in possible global warming.

Although a particular air pollutant can be present in sufficiently high concentrations at a given time and place to exert an adverse or, in some cases, positive effect on an ecosystem component, it should be clearly understood that ambient air is always composed of pollutant mixtures. While knowledge of the joint effects of multiple air pollutants on ecosystems is limited, such exposure can result in additive, more than additive, or less than additive effects in comparison to single pollutants.

Traditionally, plant biologists have studied the effects of air pollutants through their impact on plant canopies and/or on soils. In recent years, however, the importance of human-made air pollutants in altering the global climate and their consequent impact on ecosystems is of utmost international concern. Many environmental scientists have viewed global climate change as simply alterations in temperature and moisture (see Parry 1990). While this may be true, global climate change should be viewed as a system of processes and products (Table 12.1). Because ecosystems respond to the products of climate change, they also respond to the alterations in some of the processes themselves (Krupa and Kickert 1989). Thus, the various components of air quality should be examined on an integrated basis in order to analyze their impacts on ecosystems.

Atmospheric Trace Gases and Global Change

There is increasing international concern about human influence on our global climate. The issues of concern are (1) depletion of beneficial, naturally occurring stratospheric (atmosphere between 10 and 50 km above the surface of the earth) ozone and a consequent increase in tropospheric (atmosphere in the first 10 km above the surface of the earth) ultraviolet-B radiation; (2) increases in the surface emissions of atmospheric trace gases resulting in an enhancement of the "greenhouse effect" (a term used to define the heating of the atmosphere, similar to the air inside a greenhouse) and global warming, changes in precipitation and wind patterns; and (3) possible drastic alterations in terrestrial and aquatic ecosystems.

The greenhouse effect, global warming, and climate modification are governed by the interactions between tropospheric and stratospheric processes (Wuebbles et al. 1989). A key atmospheric constituent in these interactions is ozone. Ozone concentrations vary with altitude above the earth's surface roughly as approximately: 0 to 10 kilometers

Table 12.1. Summary of atmospheric processes and products relevant to global climate change

Processes	Products
Loss in beneficial stratospheric (approximately 15-50 km above the surface) ozone	Increased ultraviolet-B radiation at the surface[a]
Increases in greenhouse (climate warming) gases Carbon dioxide Ozone Carbon monoxide Methane Chlorofluorocarbons Organo-bromines Nitrous oxide Carbonyl sulfide Sulfur dioxide (?) Nitrogen oxides	Changes in Temperature Precipitation Radiation Evaporation Wind Secondary aerosols (particles less than 2.0 μm in size)[b]
Changes in the concentrations of water molecules	
Increased ultraviolet-B penetration[a]	Decreased tropospheric (approximately 0-10 km above the surface) ozone

Source: Modified from Runeckles and Krupa (1994).

[a]Increased ultraviolet-B (280-320 nm[b] wavelength) penetration to the surface may decrease tropospheric ozone and affect other greenhouse gases.
[b]1 μm = 1/1,000 millimeter; 1 nm = 1/1,000,000 millimeter.

(troposphere), 10 percent by volume; 10 to 35 kilometers, 80 percent; and above 35 kilometers, 10 percent (Cicerone 1987). In the troposphere, ozone concentrations also vary with the latitude (Pruchniewicz 1973).

Stratospheric ozone serves as a shield against biologically harmful solar ultraviolet radiation, initiates key stratospheric chemical reactions, and transforms solar radiation into heat and the mechanical energy of winds. Also, the downward intrusions of stratospheric air supply the troposphere with the ozone necessary to initiate chemical reactions driven by the sun in the lower atmosphere. In addition to the protective effect of stratospheric ozone against ultraviolet radiation, clouds reflect a large part of the incoming solar radiation, causing the reradiation of the entire earth to be about twice what it would be in the absence of clouds (Cess 1976). Clouds cover about one-half of the earth's surface, doubling the proportion of sunlight reflected back into space to 30 percent.

Ever since the seminal papers of Johnson (1971) and Molina and Rowland (1974), human activities have been projected to substantially deplete the stratospheric ozone through increases in the global concentrations of key atmospheric trace chemicals such as methane, nitrous oxide, synthetic chlorofluorocarbons, and organo-bromines (EarthQuest 1990).

Attempts to predict the future effects of continued increases in stratospheric source gases such as chlorofluorocarbons have given rise to various mathematical models. Simulated chlorofluorocarbon releases led to decreases in the ozone layer at all latitudes (Isaksen and Stordal 1986). Larger decreases in the ozone layer were calculated for high latitudes (higher than 40°) than for low latitudes.

Although ultraviolet-B radiation (wavelength band = 280-320 nm, 1 nm = 1/1,000,000 mm) constitutes only 1.5 percent of the total solar radiation reaching the surface prior to its reduction by the earth's atmosphere, decreased amounts of stratospheric ozone will permit disproportionately large amounts of specific wavelengths of ultraviolet-B radiation to penetrate through the atmosphere (Cutchis 1974).

With or without the predicted changes, the incoming solar radiation to the surface of the earth is in the ultraviolet (short) to near infrared (long) wavelength bands. After some absorption, surfaces reradiate heat energy back to the atmosphere at long wavelength infrared. This energy is trapped by certain atmospheric chemical constituents and by clouds, leading to a warming of the atmosphere above the surface—the greenhouse effect. The critical concern at this time is whether human influence has increased and accelerated this greenhouse effect toward progressive global warming that will create disastrous ecological consequences (Houghton and Woodwell 1989).

Surface concentrations of greenhouse trace gases are increasing (EarthQuest 1990). Using climate models Wuebbles et al. (1989) predicted that global surface temperatures will increase by 1.5° to 4.5° from the nineteenth to the twenty-first centuries. This effect is attributed primarily to increasing carbon dioxide concentrations.

Many predictions of global climate change have been made through the use of one- or two- rather than three-dimensional climate models. Certainly the application of three-dimensional models is limited by the present-day availability of computer power. Many of the models have been unable to fully include the role of clouds. Studies on the role of clouds based on the Earth Radiation Budget Experiment show atmospheric cooling over North America (Ramanathan et al. 1989). Clouds appear to have a net cooling effect globally of about four times as much energy as would be trapped by doubling carbon dioxide levels. However, while in mid- and high latitudes the net cooling from clouds is large, over the tropics their cooling is nearly canceled by heating. Even within these general tendencies, a great deal of geographic patchiness and variability must be expected. Long-term average values describing the global change cannot be as valuable as defining the standard deviations of the mean values at a local level. Given the geographic patchiness, locations may vary in the future in greater or fewer number of days of high or low temperatures and wet or dry conditions that will be ecologically significant. Recent studies show that in the United States as well as in mainland China and the former Soviet Union, while daytime temperatures have remained relatively constant, nighttime temperatures have increased measurably in the last decades (Karl et al. 1991). However, this is an area in which the knowledge base and understanding are limited and, as a result, major research is required to reliably predict future changes.

Emissions of Key Primary Pollutants in the Great Plains and Air Quality

Tables 12.2, 12.3, and 12.4 provide summary statistics on the emissions of sulfur dioxide, nitrogen oxides, and volatile organic compounds in the North American Great Plains during 1985. While sulfur dioxide is known to play a critical role in the occurrence of "acidic rain," nitrogen oxides and volatile organic compounds serve as the building blocks for the production of ozone at the surface. The states within the Great Plains contributed up to 27 percent of the total sulfur dioxide emissions within the contiguous United States (Table 12.2). Similarly, the provinces within the Great Plains of Canada contributed up to 30 percent of the total emissions of that country. Texas was the largest emitter of sulfur dioxide (1,336 billion grams or gigagrams per year) within the U.S. Great Plains. However, in comparison, Ohio and Indiana emitted 2,323 and 1,758 gigagrams of sulfur dioxide during 1985. In the Canadian Great Plains, Alberta emitted the most sulfur dioxide (535 gigagrams) compared with 1,446 gigagrams for Ontario and 704 gigagrams for Quebec.

Texas was also the largest emitter (2,279 gigagrams) of nitrogen oxides in the U.S. Great Plains during 1985 and was also the largest emitter in the entire contiguous United States (Table 12.3). Similarly, Alberta was the largest emitter of nitrogen oxides (458 gigagrams) within the Canadian Great Plains, but was second to Ontario (550 gigagrams) in all of Canada. The U.S. and Canadian Great Plains as a whole contributed 35 to 36 percent of the total nitrogen oxide emissions within the contiguous United States and all of Canada.

Although the ratings for Texas and Alberta for volatile organic compound emissions during 1985 were very similar to the emission ratings for nitrogen oxides, the geographic locations within the Great Plains of the United States and Canada contributed up to only 29 percent and 25 percent of the appropriate totals for the two countries (Table 12.4).

Table 12.2. The Great Plains state- and province-level sulfur dioxide emissions for source categories in 1985

State/Province	Utilities	Industrial Sources[a]	Nonferrous Smelters	Other Sources	Total[b]
			gigagrams[c]		
U.S. Great Plains					
Colorado	61	8	0	15	84
Illinois	939	274	0	49	1,262
Iowa	177	57	0	29	262
Kansas	92	37	0	15	144
Minnesota	98	30	0	22	151
Missouri	884	74	60	51	1,069
Montana	15	32	24	12	83
Nebraska	42	5	0	11	58
New Mexico	71	58	97	20	246
North Dakota	131	49	0	20	200
Oklahoma	74	33	0	29	136
South Dakota	31	2	0	6	39
Texas	475	655	25	210	1,336
Wisconsin	344	100	0	23	467
Wyoming	125	42	0	20	187
Total (A)[b]	3,559	1,456	206	532	5,724
Canadian Great Plains					
Manitoba	3	0	459	6	468
Saskatchewan	69	9	0	7	84
Alberta	86	442	0	7	535
Total (B)[b]	158	451	459	20	1,087
			percent		
(A) as a percentage of the entire contiguous United States	24	39	29	27	27
(B) as a percentage of all of Canada	21	52	27	56	30

Source: Modified from USEPA (1989).

[a]Excluding smelters.
[b]Values may not sum to totals due to independent rounding.
[c]1 gigagram = 1 billion grams.

Table 12.3. The Great Plains state- and province-level nitrogen oxide emissions for source categories in 1985

State/Province	Transportation	Utilities	Industrial Sources	Other Sources	Total[a]
			gigagrams[b]		
U. S. Great Plains					
Colorado	116	103	2	45	266
Illinois	310	391	41	138	879
Iowa	129	72	6	34	242
Kansas	135	125	13	111	384
Minnesota	163	114	9	35	322
Missouri	194	231	27	34	486
Montana	83	23	9	26	141
Nebraska	103	32	1	14	149
New Mexico	64	97	4	92	257
North Dakota	49	96	1	21	168
Oklahoma	148	74	19	119	361
South Dakota	42	14	1	8	66
Texas	715	468	236	860	2,279
Wisconsin	155	101	5	56	316
Wyoming	46	84	8	65	203
Total (A)[b]	2,452	2,025	382	1,658	6,519
Canadian Great Plains					
Manitoba	74	2	1	7	84
Saskatchewan	92	36	4	14	145
Alberta	192	67	175	24	458
Total (B)[b]	358	105	180	45	687
			percent		
(A) as a percentage of the entire contiguous United States	31	34	46	44	35
(B) as a percentage of all of Canada	30	45	73	22	36

Source: Modified from USEPA (1989).

[a]Values may not sum to totals due to independent rounding.
[b]1 gigagram = 1 billion grams.

Table 12.4. The Great Plains state- and province-level volatile organic compound emissions for source categories in 1985

State/Province	Transportation	Evaporation[a]	Combustion	Process Sources	Total[b]
			gigagrams[c]		
U.S. Great Plains					
Colorado	162	71	26	14	273
Illinois	317	374	76	93	859
Iowa	107	67	14	6	194
Kansas	99	73	15	35	221
Minnesota	169	139	52	23	384
Missouri	178	161	60	49	448
Montana	52	17	68	11	148
Nebraska	68	37	7	2	115
New Mexico	87	28	26	7	149
North Dakota	32	14	5	5	56
Oklahoma	138	80	31	61	311
South Dakota	37	20	21	1	79
Texas	685	446	61	921	2,113
Wisconsin	159	160	53	9	381
Wyoming	40	10	9	21	80
Total (A)[c]	2,330	1,697	524	1,258	5,811
Canadian Great Plains					
Manitoba	56	15	7	2	80
Saskatchewan	67	14	36	33	149
Alberta	136	35	15	149	335
Total (B)[c]	259	64	58	184	564
			percent		
(A) as a percentage of the entire contiguous United States	29	26	22	38	29
(B) as a percentage of all of Canada	25	16	16	41	25

Source: Modified from USEPA (1989).

[a]Includes solvent evaporation and petroleum marketing and storage emissions.
[b]Values may not sum to totals due to independent rounding.
[c]1 gigagram = 1 billion grams.

Figure 12.1 shows the United States divided into the U.S. Environmental Protection Agency's (EPA) administrative regions. Figures 12.2, 12.3, and 12.4 show summary statistics of the ambient concentrations of the three key pollutants (sulfur dioxide, nitrogen dioxide, and ozone) in the EPA regions. In these figures, the states within the U.S. Great Plains are in EPA Regions V, VI, VII, and VIII. During 1991 the highest ambient, annual average concentration of sulfur dioxide was recorded in Region V (Figure 12.2). Conversely the lowest concentration was recorded in Region VIII. This was also true for nitrogen dioxide (Figure 12.3). In comparison, Region VI exhibited the highest composite average of the second highest daily one-hour ozone concentrations (Figure 12.4). Region VI includes Texas, the state with the highest emissions of nitrogen oxides and volatile organic compound (precursors for ozone) within the U.S. Great Plains (Tables 12.3 and 12.4).

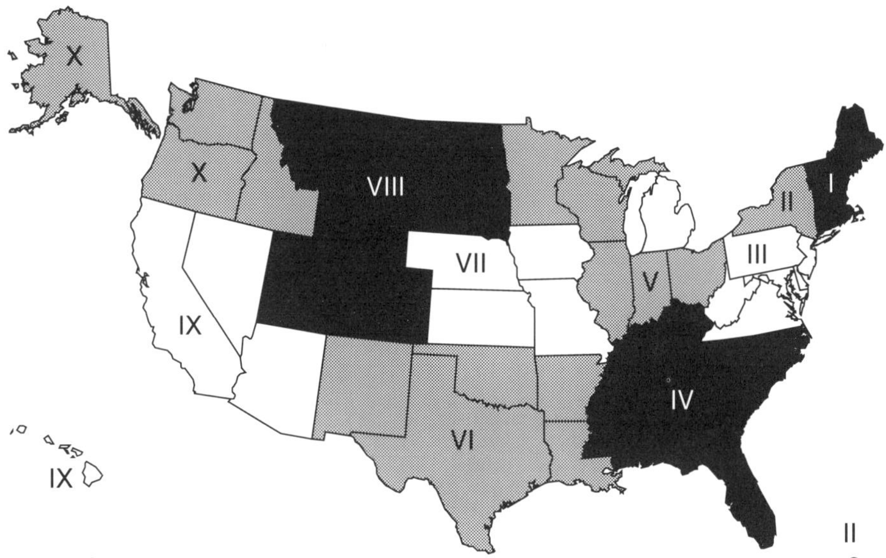

Source: USEPA 1992.

Note: All or parts of Regions V, VI, VII, and VIII constitute the Great Plains of the United States.

Figure 12.1. Ten administrative regions of the U.S. Environmental Protection Agency

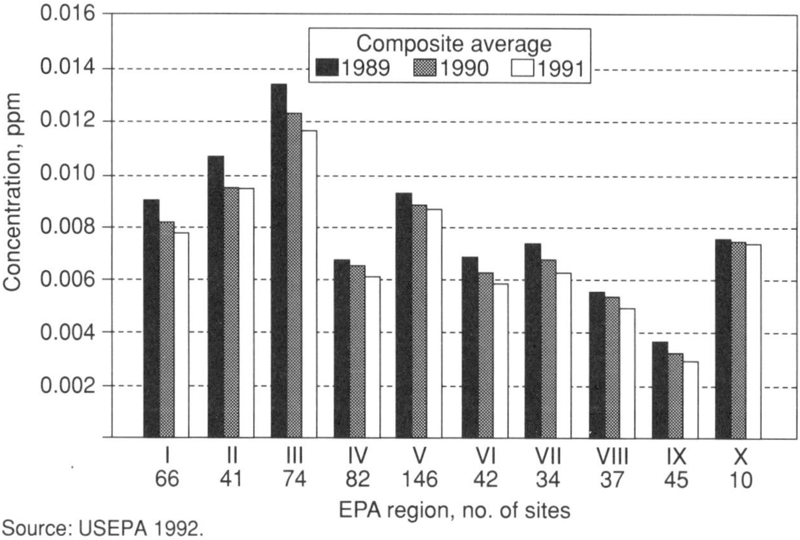

Source: USEPA 1992.

Note: All or parts of Regions V, VI, VII, and VIII constitute the Great Plains of the United States.

Figure 12.2. Regional comparisons of the 1989, 1990, 1991 composite averages of the annual average sulfur dioxide concentrations

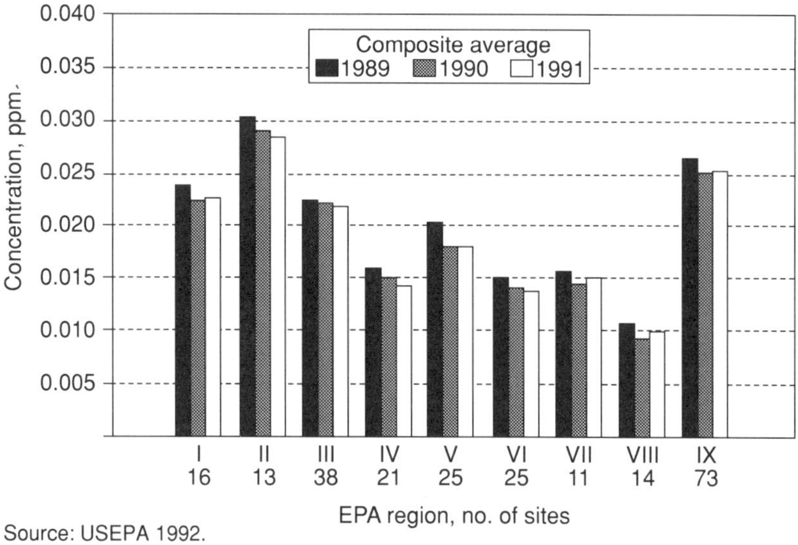

Source: USEPA 1992.

Note: All or parts of the Regions V, VI, VII, and VIII constitute the Great Plains of the United States.

Figure 12.3. Regional comparisons of 1989, 1990, 1991 composite averages of the annual mean nitrogen dioxide concentrations

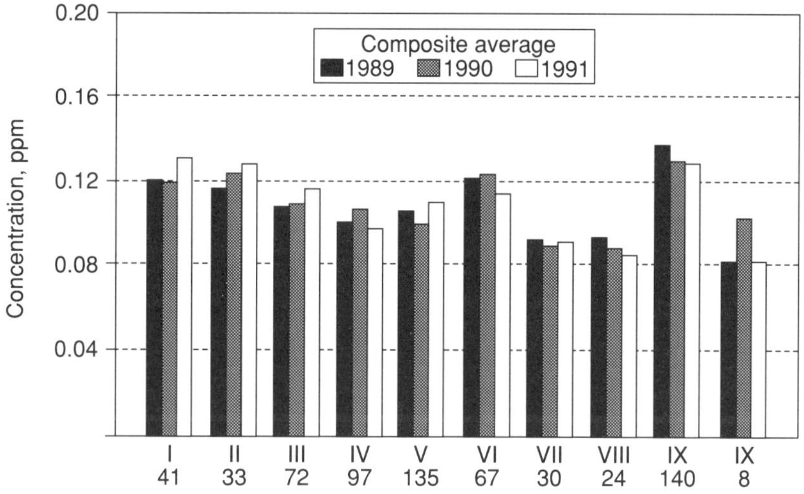

Source: USEPA 1992.

Note: All or parts of the Regions V, VI, VII, and VIII constitute the Great Plains of the United States.

Figure 12.4. Regional comparisons of the 1989, 1990, 1991 composite averages of the second highest daily one-hour ozone concentrations

Table 12.5 summarizes the relationships among various characteristics of the North American Great Plains region and the corresponding features of the contiguous United States and all of Canada. Within the Great Plains states of the United States, transportation constitutes the largest source of nitrogen oxides and volatile organic compounds (Tables 12.3 and 12.4). The population among these states accounts for 25 percent of the population of the 48 states within the contiguous United States and reflects the prevalence of a number of large urban centers. In contrast, in the Canadian Great Plains, the population accounts for only about 17.5 percent of the entire population of Canada. Yet sulfur dioxide, nitrogen oxides, and volatile organic compound emissions constitute 30, 36, and 25 percent of the entire country (Table 12.5). As in the United States, while transportation constitutes the largest source of nitrogen oxides and volatile organic compound within the three provinces (Tables 12.3 and 12.4), sulfur dioxide is contributed primarily by industrial sources and smelters (Table 12.2). This reflects the prevalence of rich natural resources in those areas and the relatively sparse population.

Because air pollutants exist as gases and particles and are deposited onto plant canopies and soil through dry deposition (in the absence of precipitation), wet deposition and acidic rain are of much concern. Table 12.6 provides an example of annual wet deposition of chemical constituents. It will suffice to state that the annual depo-

Table 12.5. Emissions of sulfur dioxide, nitrogen oxides, and volatile organic compounds

Characteristics	Observation[a]
United States	
Number of states	33
Approximate population	25
Emissions of sulfur dioxide	27
Emissions of nitrogen oxides	35
Emissions of volatile organic compounds	29
Canada	
Number of provinces	25
Approximate population	17.5
Emissions of sulfur dioxide	30
Emissions of nitrogen oxides	36
Emissions of volatile organic compounds	25

[a]All values are cumulative from appropriate Great Plains U.S. states or Canadian provinces, as a percentage of the corresponding total for contiguous United States or entire Canada.

sition of most of the major acid-forming components by precipitation in the U.S. Great Plains is roughly 30 to 50 percent less than their corresponding deposition values for Ohio, Pennsylvania, and western New York. Furthermore, there is a curvilinear relationship between the annual deposition of the major acid-forming substances of hydrogen, sulfate, and nitrate, starting with a minimum in Minnesota, attaining a maximum in Ohio and Pennsylvania, and again reaching a minimum toward Vermont. There is a very strong relationship among emission regions, meteorology at the regional and interregional scale, and the composition of individual or types of precipitation events. This type of resolution or analysis is lost in the calculation of annual average values. It is important to note that climatic characteristics vary within the day and between days, and so does plant biology. Therefore, any establishment of cause-effect relationships between the two must consider the variability of both phenomena. At this time, this issue has not been satisfactorily addressed by scientists.

In the preceding narrative the emphasis has been on three criteria pollutants (pollutants for which ambient air quality regulations exist both in the United States and in Canada) relevant to terrestrial ecosystems. At the present time there are no such regulations for carbon dioxide, although fossil fuel combustion is a dominant source

Table 12.6. Annual wet deposition of selected chemical constituents during 1990 in the Great Plains states of the United States

	Wet Deposition Range				
Location	Calcium	Ammonium	Nitrate	Sulfate	Acidity
	(kilograms per hectare)				(pH)[a]
Colorado	0.64-3.32	0.59-2.65	1.96-9.67	2.25-6.87	5.70-4.87
Illinois	1.75-3.13	3.33-5.03	13.5-17.4	20.1-30.8	4.62-4.41
Iowa[b]	2.79-2.87	4.44-4.72	10.2-11.9	11.5-14.2	5.11-5.0
Kansas	1.74-3.91	2.91-4.36	7.39-14.4	5.36-18.6	5.62-4.90
Minnesota	0.94-1.42	1.98-3.83	5.82-7.39	5.50-6.04	5.71-5.04
Missouri[b]	1.90-2.37	3.74-3.80	14.3-15.8	18.1-24.3	4.72-4.62
Montana	0.29-0.66	0.50-1.04	1.51-4.03	1.00-3.87	5.53-5.15
Nebraska[b]	1.05-1.70	2.50-3.83	4.97-8.06	3.91-7.43	5.73-5.43
New Mexico	0.82-0.99	0.66-1.31	2.95-3.99	3.15-3.77	5.37-4.95
North Dakota	0.55-1.11	1.15-1.82	2.71-3.63	2.73-3.54	5.75-5.28
Oklahoma	1.33-1.94	1.96-2.34	4.75-8.61	4.44-9.94	5.80-5.03
South Dakota[b]	0.51-1.05	1.18-2.77	3.40-5.48	2.24-3.72	5.74-5.35
Texas	1.04-3.63	0.97-4.65	2.69-11.8	3.76-22.3	5.34-4.70
Wisconsin	1.24-2.36	3.26-4.36	8.59-13.3	9.84-18.1	5.06-4.59
Wyoming	0.43-1.87	0.42-1.15	0.59-3.85	1.61-12.3	5.57-5.10

Source: NADP 1990.

[a]pH is measured on a scale of 0 → 7 → 14
(Acid) (Alkali)
(The pH value of distilled water at 25°C and 1 atmospheric pressure is 5.68.)
[b]Only two monitoring sites with acceptable data.

of current ambient concentrations of carbon dioxide (EarthQuest 1990). Nevertheless, ambient concentrations of carbon dioxide have not been monitored routinely. Up to now, carbon dioxide measurements have been part of the efforts to monitor global climate change and special research projects, such as the Alberta Government-Industry Acidic Deposition Research Program (Legge and Krupa 1990). Based on these available data, it can be concluded that the annual average carbon dioxide concentrations at the present time are in the range of 350 to 360 parts per million, with hourly maxima at certain locations being as high as approximately 490 parts per million (Legge and Krupa 1990).

Responses of Crop and Tree Species to Elevated Carbon Dioxide and Ozone Concentrations and Climate Change

The key aspects of air quality and climate change and their impacts on crops and forests relate to the following variables: increasing ambient concentrations of carbon dioxide and ozone, as well as possible alterations in surface ultraviolet-B radiation, temperature, and precipitation patterns. Although a small number of studies have examined the combined effects of carbon dioxide and ozone, as well as ultraviolet-B radiation and others, no current studies have examined the joint effects of all of the aforementioned variables (Krupa and Kickert 1993). Tables 12.7 and 12.8 summarize the relative biomass responses of some important crop and tree species of the North American Great Plains to either elevated carbon dioxide or ozone concentrations.

Table 12.7. Some important crop species of the North American Great Plains and their relative sensitivity to elevated surface concentrations of carbon dioxide or ozone

Species[a]	Relative Positive Response—Carbon Dioxide[b]	Relative Negative Response—Ozone
Alfalfa	1.57	Sensitive
Barley	1.25-1.70	Tolerant
Beans	1.61-1.82	Sensitive-Intermediate
Corn	1.29-1.55	Sensitive
Cotton	1.95-3.09	Sensitive
Forage grasses	Unknown	Sensitive
Oats	1.42	Sensitive
Potato	1.44-1.64	Sensitive
Sorghum	2.98	Intermediate
Soybean	1.20-1.55	Sensitive-Tolerant
Wheat	1.26-1.37	Sensitive-Intermediate

Source: Krupa and Kickert (1989).

[a]Biomass responses to elevated carbon dioxide or ozone are unknown for these important crops grown in parts of the Great Plains: canola, sugar beet, and sunflower.
[b]Mean relative yield increase in comparison to control (ambient carbon dioxide concentration).

Table 12.8. Some important tree species of the North American Great Plains and their relative response to elevated surface concentrations of carbon dioxide or ozone

Species[a]	Relative Positive Response—Carbon Dioxide	Relative Negative Response—Ozone
Aspen	Unknown	Sensitive-Resistant
Birch	Nonresponsive	Intermediate
Black cherry	Unknown	Sensitive-Resistant
Blue spruce	Sensitive	Unknown
Cottonwood (eastern)	Sensitive	Sensitive
Green ash	Nonresponsive	Sensitive-Resistant
Hickory	Nonresponsive	Unknown
Jack pine	Intermediate	Resistant
Loblolly pine	Intermediate	Sensitive
Lodgepole pine	Nonresponsive	Resistant
Pin oak	Unknown	Sensitive
Ponderosa pine	Sensitive	Intermediate
Red maple	Unknown	Intermediate
Sycamore	Unknown	Sensitive
Walnut	Sensitive	Resistant
White ash	Unknown	Sensitive
White oak	Unknown	Sensitive
White pine (eastern)	Sensitive	Sensitive-Resistant
White spruce	Sensitive	Unknown

Source: Modified from Krupa and Kickert (1989).

[a]Biomass responses to elevated carbon dioxide or ozone are unknown for these species found in parts of the Great Plains: balsam fir, black oak, black spruce, burr oak, larch, live oak, red pine, and willow.

Although carbon dioxide is a prime factor in projected global warming, increased carbon dioxide levels in the troposphere will influence crop production directly because of their effects on photosynthesis (Table 12.7) (Rogers and Dahlman 1990). Although C-4 plants (plants in which photosynthetic carbon dioxide is fixed in a molecule containing four carbon atoms) such as corn, sorghum, millet, and sugar cane are more efficient users of carbon dioxide than C-3 plants (plants in which photosynthetic carbon dioxide is fixed in a molecule containing three carbon atoms) such as wheat, rice, and almost all other crops, C-3 species benefit most from increased carbon dioxide levels. Because the C-4 plants have inherently better photosynthetic efficiency, C-4 plants use water more efficiently, but as Rogers and Dahlman (1990) point out, increased water use efficiency is an important consequence of increased carbon dioxide levels for both C-3 and C-4 plants.

Increased carbon dioxide levels cause shifts in the partitioning of assimilated carbon among different plant parts like roots, leaves, stems, and pods. Rogers and Dahlman (1990) suggest that a doubling of average carbon dioxide concentrations could lead to greater than 50 percent increases in potato and alfalfa yields and almost 30 percent increases in corn and soybean yields, if all other environmental factors remain unchanged. However, they caution that better computer simulation models are needed that can deal with the interactive effects of carbon dioxide with other environmental factors.

Marie and Ormrod (1990) believe that the available evidence indicates that current surface-level ozone concentrations can cause substantial crop losses in many parts of the world (several ozone-sensitive crop species are listed in Table 12.7). Other environmental stresses, particularly drought, can dramatically modify the response to ozone; where drought is severe, it tends to minimize the adverse effects of ozone. There is also evidence indicating that warmer air temperatures decrease ozone sensitivity (Marie and Ormrod 1990). However, much more needs to be known about these and other interactions before it will be possible to determine the relative importance to crop production of changes in carbon dioxide, ultraviolet-B, and ozone levels, with or without the influence of temperature and water availability.

If we consider alterations in temperature and precipitation patterns only, according to Dekker and Achununi (1990) all sectors of agriculture are sensitive to weather and climate, and any substantial change will alter total productivity. However, this does not necessarily imply reduced productivity. In developed areas, agriculture is supported by a complex system of research, education, information, finance, and farm supply, all overlaying the agricultural potential of the available soils. Global warming will decouple soil from climate since, although soil characteristics are partly the result of past climates, soil properties change slowly relative to climate change. Public policy and agricultural management will attempt to develop strategies for maintaining production in areas with the best present-day soils, in spite of shifts in climate. As an example, Dekker and Achununi (1990) cite the shift from wheat to barley in the northern Canadian wheat belt, which they attribute to shortened growing seasons. In contrast, the growth in U.S. soybean acreage has been almost entirely driven by the prospects for increased profitability.

Of equal or greater importance than direct effects of rising temperatures are the indirect effects on the hydrologic cycle, leading to shifts in the dependence on irrigation. Projected drier summers in parts of the U.S. corn belt will probably create a shift from the production of corn to grain sorghum (Dekker and Achununi 1990).

If the climate modification is gradual, plant breeders are likely to be able to develop new cultivars that are better suited to the changed conditions. However, as cropping practices are altered, parallel changes in pest control and management must be made in anticipation of the upsurge of pests and diseases well suited to the changed conditions.

In contrast to the previous discussion of crops, Shinn (1990) summarized the forest responses to increased atmospheric carbon dioxide. Responses of crop plants to elevated carbon dioxide have been studied more than similar responses for trees. One of the difficulties in extrapolating information from crops to trees is the acclimation of response to carbon dioxide. There have been few studies on trees with periods of exposure sufficient to reach acclimation. There is evidence that photosynthetic stimulation to elevated carbon dioxide declines with long-term exposure (Shinn 1990).

Increased leaf surface temperature (by as much as 4°C in one study), as a secondary response to elevated carbon dioxide, increases the water vapor pressure gradient, which acts as a feedback to increase the plant transpiration rate. However, the most important feedback is the possible increase in infrared radiation into space, a key factor in possible global warming (Shinn 1990).

Much of the early evidence of the direct effects of carbon dioxide on forest species has been gathered from studies on growth response (Table 12.8). While these studies are integrative in nature, they are not diagnostic in the context of physiological ecology. Elevated carbon dioxide levels have been shown to increase the growth response of several conifer species. In the present context, however, short-term studies could be misleading and there is no consensus on generalized tree responses.

Tree longevity is the most significant determinant of species distributions in the successional transition during stress. Hence, the forest stands that are harvested should change most drastically. While old-growth forests are rapidly disappearing under the pressure of lumber economics, those forests that replace them will be decidedly different and will require a different management strategy. The direct effects of carbon dioxide will be part of the management plan, as will be the characteristics of the new physical environment. The increased frequency of drought and temperature extremes, altered seasonal precipitation and runoff, and the resulting actual evapotranspiration compared with potential evapotranspiration will be the important determinants of the indirect effects of carbon dioxide.

Because of tree decline and other similar problems, many studies have been conducted to determine the effects of ozone on tree species (Kickert and Krupa 1990). However, much of the knowledge is based on the exposure of tree seedlings in artificial systems, often as monocultures. Researchers do not know how ozone exposure patterns relate to changes in competition between individuals of the same species or between two or more species. However, they do know that there are differences in sensitivity or response to ozone at and within species genotype levels.

Increases in tropospheric ozone can reduce photosynthesis and dry matter in some sensitive tree species. According to Kickert and Krupa (1990), this amounts to a decrease in the rate of carbon cycling in a forest ecosystem and has two implications: (1) it is uncertain whether concurrent exposures to elevated levels of ozone and carbon dioxide would have any adverse effect on the rate of carbon cycling in ozone-sensitive tree species; and (2) aside from changes in the ambient carbon dioxide levels, a decrease in the rates of growth and carbon cycling in ozone-sensitive tree species could decrease the rate of cycling of other nutrients.

Watershed hydrologic responses due to chronic ozone effects on vegetation are probably more significant in conifer forests and in humid temperate zones. Results from studies on the spruce trees in the Ore Mountains along the heavily polluted (from sulfur dioxide) border of the Czech Republic and eastern Germany showed a decrease in canopy interception with defoliation, increases in the rate of snowmelt, soil erosion, soil moisture in the root zone, and runoff through streamflow. Similarly, tree injury attributed to high acidic deposition in an uninhabited, uncut Norway spruce watershed of the northern Black Forest in western Germany may be due to significant increases in hydrological water yield between 1973 and 1986 (Caspary 1989).

While the previous discussion has been directed to the components of specific ecosystems, landscape ecology focuses on spatial variation of surface cover types and ecological system structures and processes above the level of a single ecosystem. Patchiness, landscape fragmentation, and dynamic interaction between spatial elements of mosaics in the landscape are key concepts. The dynamics of the potential role of ozone relate to the processes of its horizontal and vertical diffusion or mixing and deposition of the pollutant to various surfaces in the landscape mosaic of multiple ecosystems and abiotic surface covers. Defined in this way, researchers know almost nothing of the effects of ozone on forested landscapes. We can speculate that, in landscapes where spatial forest fragmentation is greatest and in regions where ambient ozone levels are relatively high, ozone exposures would have the greatest effect because trees along the edges of many patches would be exposed more than trees in the interior of the patches. Over time, the most ozone-sensitive species might be eliminated in proportion to their proximal distance from the edge.

In North America, computer models of deciduous forest systems have been used. Kickert and Krupa (1990) raise concerns regarding such models, since: (1) they are simply regional, without reference to explicit landscape elements, (2) they refer to "air pollution" as a cause of change in forest species composition over time but do not explicitly use a time series or the dynamics of ambient pollutant exposure data as input, and (3) only radial growth of trees is modeled as an effect while height growth effects are ignored. The only published model of the effects of ozone on forests at the landscape level is from Austria. However, this model has not been validated with real evidence (Kickert and Krupa 1991). Presently, no experimental data are available to show the joint effects of ozone × ultraviolet-B radiation × carbon dioxide × temperature × moisture. In the final analysis, Kickert and Krupa (1990) make a case that air pollution effects on forest ecosystems should be viewed as a chain of events, rather than as single cause-effect relationships.

Two other issues that require some discussion are the impacts of possible increases in surface ultraviolet-B radiation and the occurrence of acidic precipitation. Although at the present time there is no definitive evidence that ultraviolet-B radiation levels have increased across the United States (Scotto et al. 1988, Frederick et al. 1989), numerous studies have been conducted on the effects of elevated ultraviolet-B levels on crops (Krupa and Kickert 1989). However, only one field study reports reduced crop (soybean) yield (Teramura and Sullivan 1988). In contrast to crops, there have been only six studies on the effects of ultraviolet-B radiation on a total of 31 tree species (Biggs 1990). Problems with experimental techniques limit the use of these data to predict tree responses under field conditions, including trends as related to changes in natural ultraviolet-B radiation at the surface (Biggs 1990).

At present, while sulfur dioxide is considered to be a problem at the local scale in certain very specific cases, its contribution to acidic precipitation is of greater concern. However, there are no published reports of adverse effects of ambient acidic precipitation on crops, although there are reports of such effects using artificial, simulated rain of extreme composition and exposure (Legge and Krupa 1990). In contrast, wet deposition of nitrate in high-elevation forests appears to provide a fertilization effect leading to prolongation of tree growth into the fall and thus the trees' increased vulnerability to ozone and/or frost damage (U.S. NAPAP 1990).

Conclusions

Since the beginning of the industrial age, numerous studies have been conducted on the impact of air pollutants on terrestrial ecosystems (crops and forests). Although ambient air is always composed of pollutant mixtures, in determining the relative air quality and its ecosystem effects at a given geographic location and time, a predominant number of the aforementioned studies have been univariate (single cause and effect) in nature. Based on these studies, surface-level ozone is now generally accepted as the most important air pollutant in the context of adverse vegetation effects.

The building blocks for surface-level ozone are mainly emissions of human-made nitrogen oxides and volatile organic compounds. Within the North American Great Plains, Texas and Alberta are the top regions for such emissions in the United States and Canada. This appears to be mainly due to the prevalence of natural gas and/or oil industries in the two regions and the consequent urbanization. Nevertheless, the total emissions of nitrogen oxides and volatile organic compounds within the North American Great Plains represent roughly 25 to 36 percent of the corresponding total emissions within the contiguous United States and all of Canada.

Because of concerns about the sulfur dioxide emissions (direct effects on human health and welfare and a role in acidic precipitation) and nitrogen oxides (precursor for ozone and a role in acidic precipitation) and because of the increases in surface ozone (effects on human health and welfare), ambient air quality standards (U.S.) or objectives (Canada) have been established for the three atmospheric constituents.

While in the United States there are primary (to protect human health) and secondary (to protect human environment such as ecosystems and materials) ambient air quality standards, in Canada there are three air quality objectives: desirable, acceptable, and tolerable. There are many similarities between the values of the air quality standards in the United States and the corresponding values for the air quality objectives in Canada. Nevertheless, the derivation of these standards or objectives is from the results of univariate experiments.

In the context of global climate change, in almost all of the studies, elevated carbon dioxide concentrations have produced increases in plant biomass (both crop and tree species). The question remains whether this stimulation will offset any adverse effects of elevated surface ozone concentrations. Researchers do not know this, since multivariate experiments are few and rare.

Global climate change includes: (1) increases in ambient carbon dioxide levels, (2) changes in ambient ozone concentrations, (3) possible changes in surface ultraviolet-B radiation, and (4) frequencies in the number of days that may be warmer or cooler, or drier or wetter during the growth season. Computer models are continuously being developed and improved to examine the joint effects of all these variables on crops and forests. But computer model outputs are only as good as the data used to run those models. Since appropriate experimental data virtually do not exist within this context of joint effects of multiple variables and global change, present-day computer model outputs have a significant degree of uncertainty associated with them. Nevertheless, such efforts can educate; scientists created the concept of Occam's Razor, which is the principle that forces many researchers to look for the simplest explanation. But it does not mean that nature necessarily must oblige.

Acknowledgments

We wish to convey our appreciation to Leslie Johnson (word processing) and Sid Simms (graphics production) for their valuable help in the preparation of this manuscript. This work was supported by the University of Minnesota Agricultural Experiment Station through its senior author.

References

Biggs, R. H. 1990. Effects of increased UV-B radiation on forest responses. Presented at the 83rd Annual Meeting of the Air and Waste Management Association, Paper no. 90-152.3, Pittsburgh, Pennsylvania (June 24-29).

Caspary, H. J. 1989. Water yield changes in catchments as an ecohydrological indicator of forest decline. In *Proceedings of the International Congress on Forest Decline Research: State of knowledge and perspectives,* Lake Constance, Federal Republic of Germany (Oct. 2-6).

Cess, R. D. 1976. Climate change—Appraisal of atmospheric feedback mechanisms employing zonal climatology. *Journal of Atmospheric Science* 33:1831-1843.

Cicerone, R. J. 1987. Changes in stratospheric ozone. *Science* 237:35-42.

Cutchis, P. 1974. Stratospheric ozone depletion and solar ultraviolet radiation on earth. *Science* 184: 13-19.

Dekker, W. L., and V. R. Achununi. 1990. Greenhouse warming and agriculture. Presented at the 83rd Annual Meeting of the Air and Waste Management Association, Paper no. 90-151.2, Pittsburgh, Pennsylvania (June 24-29).

EarthQuest. 1990. Vol. 4(2):1-20, Boulder, Colorado: University Corporation for Atmospheric Research.

Frederick, J. E., H. E. Snell, and E. K. Haywood. 1989. Solar ultraviolet radiation at the earth's surface. *Photochemistry and Photobiology* 50(8):443-450.

Houghton, R. A., and G. M. Woodwell. 1989. Global climatic change. *Scientific American* 4:36-44.

Isaksen, I. S. A., and F. Stordal. 1986. Ozone perturbations by enhanced levels of chlorofluorocarbons, nitrous oxide (N_2O) and methane: A two-dimensional diabatic circulation study including uncertainty estimates. *Journal of Geophysical Research* 91(D4):5249-5263.

Johnston, H. S. 1971. Reduction of stratospheric ozone by nitrogen oxide catalysts from supersonic transport exhaust. *Science* 173:517-522.

Karl, T. R., G. Kukla, V. N. Razuavayev, N. Vyacheslav, M. Changery, R. G. Quayle, R. R. Heim, D. R. Easterling, and B. F. Cong. 1991. Global warming: Evidence for symmetric diurnal temperature change. *Geophysical Research Letter* 18:2252-2256.

Kickert, R. N., and S. V. Krupa. 1990. Forest responses to tropospheric ozone and global climate change: An analysis. *Environmental Pollution* 68:29-65.

———. 1991. Modeling plant response to tropospheric ozone: A critical review. *Environmental Pollution* 70(2):271-383.

Krupa, S. V., and R. N. Kickert. 1989. The greenhouse effect: Impacts of ultraviolet-B (UV-B) radiation, carbon dioxide (CO_2), and ozone (O_3) on vegetation. *Environmental Pollution* 61(4):263-393.

———. 1993. The greenhouse effect: The impacts of carbon dioxide (CO_2), ultraviolet-B (UV-B) radiation, and ozone (O_3) on vegetation (crops). *Vegetatio* 104/105:321-328.

Legge, A. H., and S. V. Krupa (ed.). 1990. *Acidic deposition: Sulphur and nitrogen oxides.* Chelsea, Michigan: Lewis Publishers.

Lovelock, J. E. 1987. *Gaia: A new look at life on Earth.* New York: Oxford University Press.

Marie, B. H., and D. P. Ormrod. 1990. Effects of tropospheric ozone on plants in the context of climate change. Presented at the 83rd Annual Meeting of the Air and Waste Management Association, Paper no. 90-151.3, Pittsburgh, Pennsylvania (June 24-29).

Molina, M. J., and F. S. Rowland. 1974. Stratospheric sink for chlorofluoromethanes: Chlorine atom catalyzed destruction of ozone. *Nature* 249:810-812.

National Atmospheric Deposition Program (NADP). 1990. *National Trends Network Annual Data Summary.* Fort Collins, Colorado: NADP.

Parry, M. L. 1990. *Climate change and world agriculture.* London: Earthscan, in association with the International Institute for Applied Systems Analysis and United Nations Environment Programme.

Pruchniewicz, P. G. 1973. The average tropospheric ozone content and its variation with season and latitude as a result of the global ozone circulation. *Pure and Applied Geophysics* 106:1058-1073.

Ramanathan, V., R. D. Cess, E. F. Harrison, P. Minnis, B. R. Barkstrom, E. Ahmad, and D. Hartmann. 1989. Cloud-radiative forcing and climate: Results from the earth radiation budget experiment. *Science* 243:57-63.

Rogers, H. H., and R. C. Dahlman. 1990. Influence of more CO_2 on crops. Presented at the 83rd Annual Meeting of the Air and Waste Management Association, Paper no. 90-151.1, Pittsburgh, Pennsylvania (June 24-29).

Runeckles, V. C., and S. V. Krupa. 1994. The impact of UV-B radiation and ozone on terrestrial vegetation. *Environmental Pollution* 83:191-213.

Scotto, J., G. Cotton, F. Urbach, D. Berger, and T. Fears. 1988. Biologically effective ultraviolet radiation: Surface measurements in the United States, 1974 to 1985. *Science* 239:762-764.

Shinn, J. H. 1990. Effects of increased CO_2 on forest response. Presented at the 83rd Annual Meeting of the Air and Waste Management Association, Paper no. 90-152.2, Pittsburgh, Pennsylvania (June 24-29).

Teramura, A. H., and J. H. Sullivan. 1988. Effects of ultraviolet-B radiation on soybean yield and seed quality: A six-year field study. *Environmental Pollution* 53(1-4):466-468.

U.S. Environmental Protection Agency. 1989. *The 1985 NAPAP emissions inventory (Version 2): Development of the annual data and modeler's tapes.* Document no. EPA-600/7-89-012a. Research Triangle Park, North Carolina: USEPA, Office of Research and Development.

———. 1992. *National air quality and emissions trends report, 1991.* Document no. EPA-450/R-92-001. Research Triangle Park, North Carolina: USEPA, Office of Research and Development.

U.S. NAPAP (National Acid Precipitation Assessment Program). 1989. *Models planned for use in the NAPAP integrated assessment.* Washington, D.C.: U.S. NAPAP.

———. 1990. *Draft of final assessment report,* Vols. 1-3. Washington, D.C.: U.S. NAPAP.

Wuebbles, D. J., K. E. Grant, P. S. Connell, and J. E. Penner. 1989. The role of atmospheric chemistry in climate change. *Journal of the Air Pollution Control Association* 39:22-28.

Land Resources

13. Conservation, Restoration, and Management of Great Plains Landscapes

Brent J. Danielson and Erwin E. Klaas
Iowa State University

The settlement and domestication of the Great Plains is essentially complete. The most spectacular large mammal concentrations on the continent have been almost entirely destroyed. In the span of about four human generations, the Plains's former inhabitants have been conquered and the landscape greatly changed. Relative to the western mountains and deserts, the Great Plains has almost no large wilderness areas and relatively little publicly owned land exists. Of these, the largest tracts are in designated national grasslands that are managed chiefly for livestock grazing. The tallgrass prairie has been largely replaced with row crop agriculture, transportation corridors, and urban development. Wetlands in the Prairie Pothole region have been drained and put into agricultural production. Much of the shortgrass prairie in the western half of the region is still intact, but the native mammals have been replaced with cattle and other livestock. But, even on the High Plains where rainfall is low, irrigation has allowed large areas of grassland to be plowed and converted to grain crop production.

Only recently have states begun to inventory what is left of their natural areas. In states where inventories are relatively complete, the proportion of land that is protected in some way by private or public ownership is extremely small, on the order of 2 percent (Table 13.1). The majority of these protected areas are relatively small and usually are used for hunting, fishing, or other forms of recreation. Virtually none of them is fully protected from human influences. The result has been a more-or-less random pattern of small and often very isolated reserves scattered across the landscape. Furthermore, these reserves are often managed by a number of different groups in an uncoordinated manner. Clearly, the future of ecosystem conservation and landscape management in the Great Plains will depend on extensive and coordinated ecological restoration.

This chapter should not be cited without permission of the authors. The views expressed are those solely of the authors and may not represent the official views of the National Biological Service or Iowa State University.

Stanley R. Johnson and Aziz Bouzaher (ed.), *Conservation of Great Plains Ecosystems*, 185–198.

Table 13.1. Approximate area of open space protected by federal, state, and private ownership in Great Plains states

State	State Area	Type of Ownership					
		Federal	State	Private	Total	Total	Areas
	acres					percent	number
Iowa	36,016,000	200,000	343,000	64,000	607,000	1.7	1,955
Kansas	52,657,280	127,448	356,111	15,300	498,859	0.9	201
Nebraska	49,507,200	625,718	301,023	75,931	1,002,672	2.0	445
Oklahoma	44,737,920	1,892,310	810,064	39,070	2,741,444	6.1	116
South Dakota	49,354,240	3,163,000	1,266,000	8,100	4,437,100	9.0	NA

Sources: (Iowa) John Fleckenstein, Iowa Land Heritage Program, Department of Natural Resources, Des Moines.
(Kansas) Kent Montei, Kansas Department of Wildlife and Parks, Pratt.
(Nebraska) Gerry Steinauer, Nebraska Natural Heritage database, Game and Parks Commission, Lincoln.
(Oklahoma) Ian Butler, Oklahoma Natural Heritage Inventory, Norman.
(South Dakota) Doug Backlund, South Dakota Natural Heritage Program, Game, Fish & Parks, Pierre.

Notes: Municipal parks and transportation rights-of-way are not included. The statistics may include publicly owned areas that are not primarily managed as natural areas, such as prison farms and military bases.

The entire Great Plains region is showing signs of ecological and environmental stress. Not only is the integrity of natural ecosystems at great risk, but also the economic and social viability of rural areas is declining. As biodiversity conservation has emerged recently as a globally significant environmental issue, the idea of sustainable development has gained in acceptance as a workable concept. Because neither of these concepts is well understood by everyone, there is a serious need to develop common ground for discussion and action. We hope that the ideas we present concerning ecosystem conservation and landscape management will help to accomplish this end.

Landscape and restoration ecology are young sciences and still suffer from the lack of a firm theoretical base of their own. However, they are founded in ecological theory and can draw on a wealth of information that provides some general principles about how land might be restored and managed to maintain biodiversity and ecological and evolutionary processes (Noss 1992).

Setting Goals

In a recent comprehensive and detailed plan for land conservation in North America, Noss (1992) states that the ultimate goal for land conservation should be the perpetual maintenance of a region's native biodiversity. He suggests that primary goals for ecosystem management should be scientifically based as well as comprehensive and idealistic so that conservation programs can maintain a long-term vision.

Landscape management on an area as large as the Great Plains will probably have to be employed on a more convenient scale, such as the ecoregions described by Omernick (1987). Ecoregions are relatively homogeneous and unique combinations of biological communities and abiotic environmental factors (Cowley 1967). For example, the Flint Hills area of eastern Kansas is an ecoregion defined by a unique combination of rolling terrain, shallow rocky soils, and tallgrass prairies.

Ecoregions contain one or more complete and functional ecosystems that are self-sustaining, self-regulating, and capable of self-recovery following natural stochastic perturbations (Noss 1990; Karr 1990). The term *complete* refers to the full complement of native species in their natural abundance and distribution.

For the most part, it is likely that any functional system is also likely to be a complete system. However, since ultimate goals are rarely achieved, adding "completeness" ensures that nearly functional systems do not totally lack the ecological redundancy (Walker 1992) that is a major factor in sustaining a community in the face of environmental uncertainty.

Approaches to a Landscape Management Strategy

In developing a conservation strategy, the size and configuration of the landscape are overriding influences in determining the long-term and large-scale dynamics of the system to be conserved. The concept of conserving land in a single large preserve or several smaller preserves (acronym SLOSS) is an attempt to address the advantages and disadvantages that arise from large versus small reserves (Wilson 1992). A large reserve with its larger populations and greater diversity will be relatively more stable than a single small reserve but may be much more susceptible to a single catastrophic event. Many small reserves may collectively contain the same, or even greater diversity than the single large reserve, and are less prone to catastrophe. However, each of the small reserves is more likely to suffer from the stochastic variability inherent in small populations.

The paradox of the SLOSS concept is the subject of much controversy among conservation biologists, but it is unlikely that there is any one solution that is applicable to all ecoregions. It is not controversial, however, that to save the greatest diversity of species, we should strive to conserve the greatest total area possible. Wilson (1992) has suggested that each ecosystem must be studied in turn to decide the best design.

Two major features of any landscape influence a species' ability to sustain its population over time: composition and the physiognomic arrangements of the components (Dunning et al. 1992). Because we cannot empirically predict a species' ability to persist, we will use a simplified computer model to demonstrate the effects of these features on a hypothetical species' persistence.

The first feature to consider is the landscape's composition. In general, the more land restored to natural conditions the better (but see Howe et al. 1991; Pulliam and Danielson 1991; and Danielson 1992a,b for exceptional cases). However, the total amount of land to be managed is usually fixed by some economic, political, or social reality. Thus, we will keep landscape composition constant for the moment and manipulate only the physiognomic feature of these components to determine the effects of landscape physiognomy on the efficiency with which the species can exploit the available habitats. Assume that this hypothetical landscape is composed of three types of habitat distributed in discrete patches: sources, sinks, and unusable habitat (Pulliam 1988, Pulliam and Danielson 1991; Danielson 1992b). A source patch is a high-quality patch that is very productive for our hypothetical target species. A sink patch is a marginal quality patch that is just barely usable. Sink patches, on average, do not produce enough offspring to compensate for mortality. Unusable habitat patches are all the remaining habitats that the species cannot use.

For the sake of simplicity, we will make all of these patches the same size. It is easiest to think of each patch as a single individual's yearly home range. However, thinking of these patches as micropopulations does not change the qualitative results, since each has a degree of productivity that is dependent upon the type of habitat.

Randomly Scattered Patches

In our first example, the patches are distributed at random over the entire landscape (Figure 13.1a). Using computer simulation techniques, we can seed this landscape with a substantial number of individuals of our hypothetical species (200 breeding adults) and then monitor the species' population dynamics.

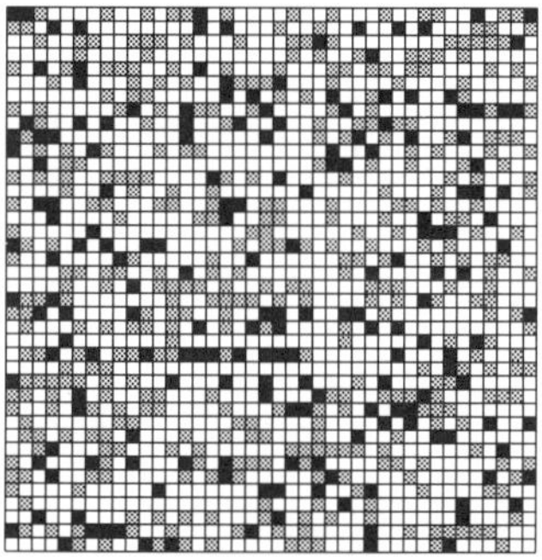

(a) Random dispersion results in some clusters, composed of various numbers of patches. As an arbitrary example, a population simulated in this landscape lasted about 164 years before becoming extinct (Figure 13.2).

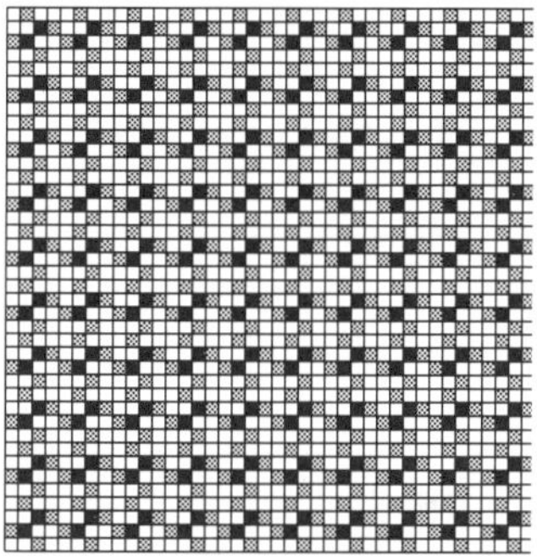

(b) A hypothetical landscape with evenly dispersed patches and no clusters of patches. As an arbitrary example, a population simulated in this landscape persisted only 46 years (Figure 13.2).

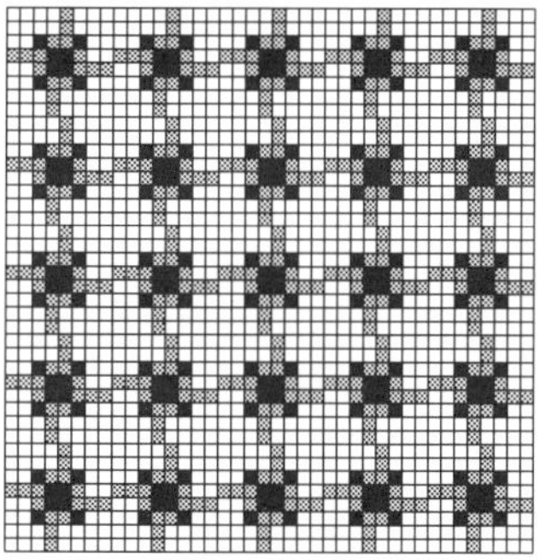

(c) A hypothetical landscape with evenly dispersed clusters of patches. This landscape contains patches arranged to provide numerous local subpopulations in clusters of usable habitat patches that are well connected to other clusters. As an arbitrary example, a population simulated in this landscape persisted for more than 480 years (Figure 13.2).

Figure 13.1. A set of hypothetical landscapes, each composed of the same numbers with three habitat types

Notes: Each solid square represents a source habitat patch that, on average, produces a surplus number of individuals. The stippled squares represent sink habitat patches that, on average, produce a slight deficit. The open squares represent habitat patches that are unusable by the species in question.

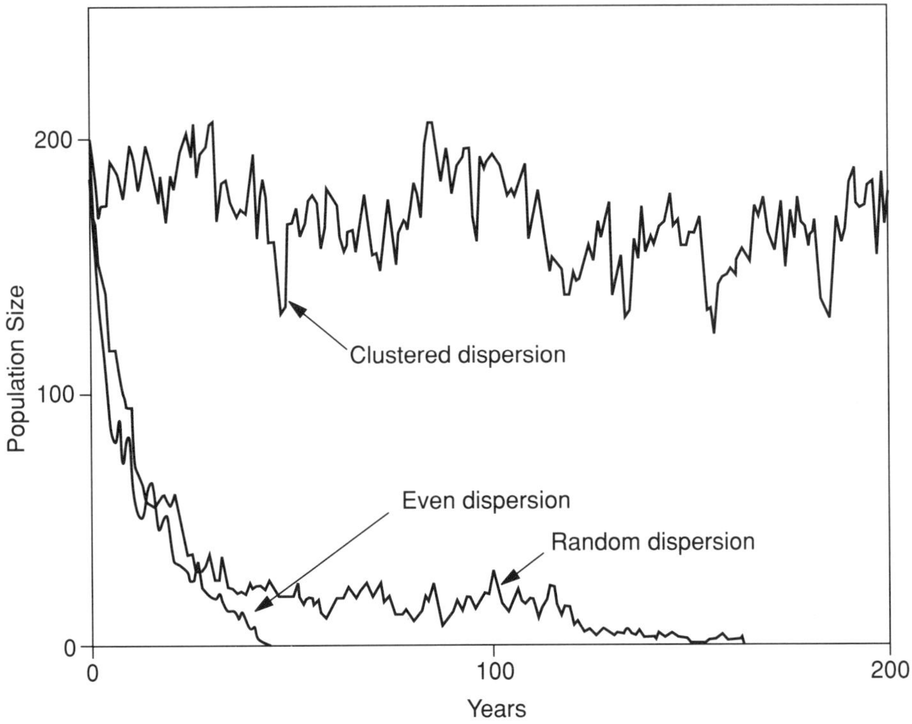

Figure 13.2. The long-term dynamics of a species that exists in three different landscapes

Notes: Each landscape has exactly the same amounts of each habitat type. The population in the clustered landscape was still going strong when the simulation was terminated after 480 years.

To some extent, the outcome of this example depends on the details of our simulated species' life-history characteristics. If the species is a rather poor disperser, that is, if it cannot easily move between patches to find the available higher quality patches, then the population will eventually become extinct (Figure 13.2). A few individuals may survive in the small clusters of source patches for a while, but they eventually die out because in years when one of the clusters has a surplus, few of those surplus individuals will succeed in finding another patch. In years when the cluster does poorly due to stochastic variation in life and death rates, there is little likelihood that immigration from outside the cluster will be sufficient to maintain the species' presence in that cluster. Thus, one by one, these subpopulations disappear. Sometimes the clusters are recolonized, but more often than not, they are permanently lost and eventually the entire population is lost.

Evenly Dispersed Patches

In our second example, we arrange the habitats to minimize isolation but keep the landscape's composition the same. The amount of resources is roughly the same as in the first example. But, by changing the physiognomic arrangement of patches, we intend to increase the efficiency with which the target species exploits the habitat. To do this, we distribute the same amount of each type of habitat in a coherent and coordinated fashion that attempts to spread the conserved habitat out as evenly as possible (Figure 13.1b). This landscape plan attempts to keep all the patches well connected to all other patches and prevents isolated patches from being underutilized. (Note that diagonal dispersal is as likely as vertical or horizontal dispersal in this simulation.) However, in this instance, the landscape is highly fragmented and the habitats are spread too thin. Although there are good connections between all patches, there are no clusters of source patches that will reliably produce a large enough surplus to quickly colonize and exploit patches that have been vacated due to the death of the resident. In fact, the habitats are spread so thinly that the population becomes extinct much more quickly than in the previous random dispersion pattern (Figure 13.2).

Clustered Patches

Taken together, the first two examples suggest that the best result will be achieved by larger clusters of usable habitat connected to each other by additional habitat that facilitates movements among the clusters. Again, using the same amounts of each type of habitat, we construct a landscape with a highly clustered pattern in which the clusters are evenly dispersed and connected to other clusters via corridors (Figure 13.1c).

In this new landscape, the species does much better, averaging between 180 and 200 individuals. After 400 breeding cycles, the population is still going strong. At a more local scale, from time to time individual subpopulations wink in and out of existence as they become temporarily extinct, only to be recolonized from nearby clusters. Thus, given a constraining amount of land that can be committed to conserving biodiversity, the physical arrangement of the habitats appears to have a strong effect on the ability of a species to persist in a landscape.

The point of this modeling exercise is to demonstrate the need for large-scale coordination and land-use planning when creating conservation reserves that require little or no artificial manipulation of species or communities. Uncoordinated conservation efforts in fragmented landscapes will require extensive assistance in the form of stocking and transplanting programs to maintain the population of each species. Such an intensive management approach requires constant monitoring of local abundances to know when and where restocking should occur. Obviously, this will be much more expensive in time, resources, and knowledge than a self-sustaining population resulting from a carefully planned and coordinated large-scale master plan.

Conserving Ecological Communities

Computer models are, by necessity, simplistic. In a real-world ecological community, there are many directly or indirectly interacting species, species with different dispersal abilities, habitat requirements, and so on. Each species will use any given habitat quite differently. What constitutes a patch of source habitat for one species could constitute a sink or even unusable habitat for another. In such circumstances, conserving one type of habitat at the expense of another type is likely to result in positive effects for some species but negative effects for others. There are, however, circumstances when two or more species may respond similarly to changes in the availability of a habitat even though they have different habitat preferences (e.g., Danielson 1992a,b).

Somehow, a sufficiently diverse set of interconnected habitats is needed. But that brings about another problem: What should the corridors that connect these patches look like? A suitable corridor for one species could easily be impassable to another species, since some species follow ridge tops while others follow drainages. Some prefer to move through open habitat, some through forests. To satisfy all species, as many types of corridors are needed as there are types of habitats, and that, in the end, is likely to be strongly influenced by the number and kinds of species that are to be maintained.

Adding to the ecological complexity of the real world is the fact that, through time, local communities are usually quite dynamic as some species become locally extinct while others colonize the local area from other parts of the larger landscape. These temporary local communities undergo extinctions and recolonizations as the result of interspecific interactions, stochastic variation in population abundances, or the cumulative biotic and abiotic changes that occur in every environment. This begs the question then: Is there a single, optimal set of patches and corridors that ensures the continued existence of the landscape's entire biological diversity? The answer is not an easy one and has generated a good deal of debate (e.g., Simberloff and Abele 1982; Quinn and Hastings 1987; Gilpin 1988).

There are several problems with the reserve-corridor approach besides the types of habitats that should be preserved, but perhaps the most difficult problem that must be overcome is that the protection and restoration of discrete patches and corridors severely limit the range of spatial scales for the patches, clusters, and corridors. Certainly, the home range of a single large carnivore such as a wolf or grizzly bear is much larger than the space required for an entire population of voles or similar small vertebrates. A corridor that is short enough to allow some species to traverse it may be much longer than the distance that an individual of another species can travel in its entire lifetime Thus, a system of connected reserves can be constructed at the appropriate spatial scale for only a few species. Clearly, a functional ecological system with a diversity of species requires a diversity of habitats arranged in a diversity of spatial scales.

Large Reserve Approach

An alternative approach to the conservation of functional and complete ecological systems is to attempt to conserve and/or restore not just natural habitats, but also natural landscapes. It is reasonable to assume that a natural landscape in a large reserve will contain the necessary diversity of landscape components and physiognomy to support a diversity of species. This approach requires large contiguous reserves, where each species is free to function at its own spatial scale yet continues to interact with species that are functioning at their own spatial scales. In a large region, with its many semi-independent local communities, a turnover of species at the local level will not endanger the diversity of the entire system. Because the local communities exist in a larger ecological context—each species is distributed according to its habitat preferences, dispersal abilities, and interactions with other species—each species will be maintained somewhere within the larger landscape.

In the Great Plains, the preservation and restoration of truly large reserves will depend on the strength of public support and long-term commitment for biological conservation. Even then, given the small amount of open space in the Great Plains protected by public or private ownership (Table 13.1), it may take a century or more to achieve the ultimate goal of preserving biodiversity.

As a working guideline, we suggest that the percentage of open space land designated for ecosystem conservation be increased to at least 10 percent of each ecoregion. Ecosystem reserves should be as large as necessary for the intended native biodiversity to be conserved in a self-sustaining fashion. This may result in reserves ranging from hundreds of thousands to millions of acres. The concept of largeness will undoubtedly vary across regions (in some instances, smaller areas with unique land forms or biotic communities should also be protected). A 100,000-acre reserve in Iowa, where land is highly valued for row crop production, may be costlier to set aside than 1,000,000 acres of grassland in Texas. Much of this initial cost is unavoidable, but some existing small areas that are not critical to the survival of a species or community could be eliminated to partially compensate for these expenses.

Expensive as this conversion may be initially, we anticipate that in the long run the eventual cost will be far smaller because these large regions will require less management than current reserves. Large reserves will be much more ecologically self-sufficient, and there will be less need for intensive management of individual species, monitoring, or the continual crisis management of declining diversity.

Example of Current Large-scale Management

The North American Waterfowl Management Plan (NAWMP), in conjunction with federal farm policy, is an example of how coordinated efforts between private and public sectors can achieve landscape-level effects. The 15-year plan was initiated in 1986 with the signing of an international agreement between Canada and the United States; Mexico signed onto the plan in 1988. The purpose of NAWMP is to increase continental waterfowl populations to mid-1970s levels by restoring and protecting about six million acres of wetland and upland habitat at a project cost of $1.5 billion. To help fund the plan, in December 1989 the U.S. Congress enacted the North American Wetlands Conservation Act, which encourages the government and private groups to form working partnerships for protecting, restoring, and enhancing wetlands across the entire North American continent. As of the end of 1992, more than $70 million have been appropriated by Congress and matched with more than $140 million from the states and the private sector. A total of 331,889 wetland acres in the United States and 632,420 acres in Canada have been directly affected (USFWS 1993).

The plan includes 12 regionally coordinated joint ventures, four of which are in the Great Plains: Gulf Coast, Playa Lakes, Rainwater Basin, and Prairie Pothole. The Prairie Pothole Joint Venture (PPJV) is closely linked with the Prairie Habitat Joint Venture in Alberta, Saskatchewan, and Manitoba, one of the first joint ventures to be organized. Within the PPJV, 398,109 acres of existing wetlands and adjoining upland nesting cover have been protected. An additional 184,446 acres have been restored and 343,818 acres have been enhanced.

The federal farm bills of 1985 and 1990 have also significantly changed the landscape of the Great Plains. The so-called "swampbuster" and "sodbuster" provisions enacted penalties for converting wetlands and grasslands and other marginal lands into row crops. The Conservation Reserve Program provided incentives for farmers to plant highly erodible cultivated land with permanent cover for ten years, and the Wetland Reserve Program provided subsidies to farmers for restoring wetlands. Farmers were also required to develop soil conservation plans for their land or face the loss of crop subsidies. In 1987, the U.S. Fish and Wildlife Service and cooperating states initiated programs to supplement the benefits of the farm bills and provide farmers additional incentives to restore wetlands.

The number of wetland projects, dollars spent, and acres affected have been carefully tracked by the agencies involved with the NAWMP and the various joint ventures, but relatively little has been done to measure the biological effectiveness of the plan and the affiliated farm programs. In terms of dollars, $220.8 million of the projected goal of $1.5 billion has been spent. Although the plan has been ambitious compared with most conservation programs, the annual budget is minuscule compared with our willingness to fund national defense and other programs.

The NAWMP and the new farm programs are a good beginning toward managing landscapes. However, many of these restored, protected, or enhanced acres are in small parcels and may not be very productive in conserving biodiversity at the ecosystem level. In fact, many of these small parcels are scattered rather haphazardly across the landscape, not unlike the example illustrated in Figure 13.1. We suggest greater emphasis on large-scale planning so that ecological integrity can be achieved in the collection of wetlands. In fact, a more effective approach might be to concentrate restorations in large, but dense, clusters of wetlands that would be close enough together to have a significant flow of plant and animal species between potholes. For instance, in northwest Iowa, pothole restoration is focusing on a four-county area. Similar projects in southwestern Minnesota, eastern South Dakota, central North Dakota and southwestern Manitoba, and southeastern Saskatchewan are being undertaken. A more coordinated effort to concentrate wetland landscapes might produce a greater degree of ecological integrity (Noss 1990; Karr 1990). Once these regional reserve centers are established, a comprehensive interstate rural development policy could develop a lower density corridor system on private lands that would eventually connect these centers.

Another set of candidates for restoration includes the Loess Hills of western Iowa, the Flint Hills of northeastern Kansas, and the Sand Hills of central Nebraska. These areas presently support unique and diverse ecological communities. (The Sand Hills and Flint Hills correspond to ecoregions defined by Omernick 1987.) These areas are not so entirely degraded that they require complete reconstruction. For the most part, a successful program will include the acquisition of new land and a strong rural development policy encouraging the restoration of these systems to something close to their original conditions. In the design stage, areas that are critical to maintaining the spatial coherence of the system could be targeted for special attention, while areas that are not critical might be sacrificed for economic purposes. Again, once these regions have functional ecological systems, a carefully designed interstate program could conceivably link these regions through encouraging more ecologically benign use of the intervening private lands.

In all of these regions, the long-term restoration and maintenance of functional ecological systems will require a significant initial commitment. However, the economic commitment can be minimized to some extent by energetic cooperation between the many local, state, and federal agencies that have a vested interest in the success of these undertakings. And, while these systems will not be restored overnight, we envision that once restoration is complete, the need for continual human intervention will decline.

Evolutionary Perspective

Long-term evolutionary advantages exist in the restoring of large, spatially complex, and diverse ecological systems. It is only in systems with ecological integrity that evolution can produce the new adaptations and even new species that will eventually inhabit the new ecological landscapes that are continually arising. When a species exists in many slightly different ecological settings (as they would in our proposed large-scale and well-connected system), the species is subject to different selective pressures in each of these subpopulations or communities. Most of the microevolution that occurs in these situations will have no appreciable effect on the evolutionary future of a species, but some will, and in which communities this will occur we cannot even make an informed guess. Of course, this localized evolution may indeed occur within small isolated reserves, but without omniscient foresight we can rarely recognize the adaptation from the maladaptation. So, how would we know which types of new variants we should actively spread to other reserves, and which to keep isolated? In a large reserve we do not have to recognize anything. Large reserves that retain their ecological integrity will also retain their evolutionary integrity, and natural processes, operating between and within individual subpopulations and communities, will make the necessary selection.

Furthermore, some evolutionary adaptations may not be possible in small isolated reserves. A species that is strongly affected by anthropogenic climatic change might adapt to utilize other habitats—but only if such habitats are available. Under changing conditions, the strength of the associations between species can change as well. If there is a sufficient pool of other species available in nearby communities, new interactions between species and between species and their new environments may evolve into novel biotic communities. Paleo-ecologists have shown that fairly widespread communities are not indefinitely stable entities unto themselves (Graham 1986; Hunter et al. 1988). Even over shorter ecological time scales, communities can be quite dynamic (see the collection of papers in Gee and Giller 1987; Ricklefs 1987; Cornell and Lawton 1992) because of a number of natural phenomena such as interspecific interactions, stochastic variation in population abundances, or the cumulative biotic and abiotic environmental change. How likely is this type of evolutionary change in small reserves? Are ecologists both scientifically and economically capable of intervening in ways that will compensate for this isolation? If a long-term solution to the conservation of natural systems in an ever-changing environment is to be successful, these sorts of evolutionary changes must be allowed to take place.

Human settlement of the Great Plains has reached, or perhaps exceeded, a level of carrying capacity that is compatible with the survival of its natural ecosystems. For any landscape management plan to succeed, the people of this region will have to radically change their attitudes toward land stewardship and regional planning. This will not be easy because public ownership of land and government regulation have never been popular. Although it may easily be argued that it is impractical to create the large reserves required for ecological and evolutionary integrity, it is our opinion that the seemingly more practical small reserves (no matter how numerous) may, in the end, prove much more costly, less likely to succeed, and wholly dependent on knowledge and foresight that are far beyond the capabilities of current conservation science. Thus, we suggest that the best and only truly practical approach is to aim high, to strive for solutions with the greatest potential for long-term ecological and evolutionary self-sufficiency.

Acknowledgment

J. Pease reviewed an early version of this manuscript. This is journal paper No. J-15311 of the Iowa Agriculture and Home Economics Experiment Station, Ames, Iowa. Project No. 102-40-17-17-3136.

References

Cornell, H. V., and J. H. Lawton. 1992. Species interactions, local and regional processes, and limits to the richness of ecological communities: A theoretical perspective. *Journal of Animal Ecology* 61:1-12.

Cowley, J. M. 1967. Biogeography. *Canadian Geographer* 11:312-326.

Danielson, B. J. 1992a. Communities in a landscape: The influence of habitat heterogeneity on the interactions between species. *American Naturalist* 138:1105-1120.

———. 1992b. Habitat selection, interspecific interactions, and landscape composition. *Evolutionary Ecology* 6:399-411.

Dunning, J. B., B. J. Danielson, and H. R. Pulliam. 1992. Ecological processes that affect populations in complex landscapes. *Oikos* 65:169-175.

Gee, J. H. R., and P. S. Giller (eds.). 1987. *Organization of communities past and present.* Oxford: Blackwell Scientific Publications.

Gilpin, M. E. 1988. A comment on Quinn and Hastings: Extinctions in subdivided habitats. *Conservation Biology* 2:290-291.

Graham, R. W. 1986. Response of mammalian communities to environmental changes during the late Quaternary. In *Community ecology.* Ed. J. Diamond and T. J. Case. New York: Harper and Row.

Howe, R. W., G. J. Davis, and V. Mosca. 1991. The demographic significance of "sink" populations. *Biological Conservation* 57:239-255.

Hunter, M. L., G. L. Jacobson, and T. Webb. 1988. Paleoecology and the course-filter approach to maintaining biological diversity. *Conservation Biology* 2:375-385.

Karr, J. R. 1990. Biological integrity and the goal of environmental legislation: Lessons for conservation biology. *Conservation Biology* 4:244-250.

Noss, R. F. 1990. Can we maintain biological and ecological integrity? *Conservation Biology* 4:241-243.

———. 1992. The wildlands project. *Wild Earth* (special issue):10-25.

Omernick, J. M. 1987. Ecoregions of the conterminous United States. *Annals of the Association of American Geographers* 77:118-125.

Pulliam, H. R. 1988. Sources, sinks, and population regulation. *American Naturalist* 132:652-661.

Pulliam, H. R., and B. J. Danielson. 1991. Sources, sinks, and habitat selection: A landscape perspective on population dynamics. *American Naturalist* 137S:s50-s66.

Quinn, J. F., and A. Hastings. 1987. Extinction in subdivided habitats. *Conservation Biology* 1:198-208.
Ricklefs, R. E. 1987. Community diversity: Relative roles of local and regional processes. *Science* 235: 167-171.
Simberloff, D., and L. G. Abele. 1982. Refuge design and island biogeographic theory: Effects of fragmentation. *American Naturalist* 120:41-50.
U.S. Fish and Wildlife Service (USFWS). 1993. *Waterfowl 2000: News from the North American Waterfowl Management Plan* 5(1).
Walker, B. H. 1992. Biodiversity and ecological redundancy. *Conservation Biology* 6:18-23.
Wilson, E. O. 1992. *The diversity of life.* Cambridge, Massachusetts: Belknap Press.

14. Rangeland Ecosystems in the Great Plains: Status and Management

M. M. Kothmann
Texas A&M University

Rangeland is defined as land on which the indigenous vegetation (climax or natural potential) is predominantly grasses, grass-like plants, forbs, or shrubs and is managed as a natural ecosystem. If plants are introduced, they are managed as indigenous species. Rangelands include natural grasslands, savannas, shrublands, many deserts, tundras, alpine communities, marshes, and meadows (Forage and Grazing Terminology Committee 1991).

The rangelands of the Great Plains contain great diversity of soils, climate, vegetation, and fauna. Annual precipitation averages 40-plus inches in the east, decreasing to 10 inches in the west. Temperature increases from an annual average of 32° in the north to 70° in the south. Great Plains rangelands are characterized by warm summers and cold, dry winters. Highly variable precipitation patterns are the norm, with peaks during spring and summer. This results in predominantly spring and summer growth of native vegetation. Grasslands are the dominant vegetation type, ranging from tallgrasses in the east to shortgrasses in the west. Shrub cover has increased markedly in the southern Great Plains during the twentieth century.

Forage produced on rangelands is not harvested or used directly by humans. Vegetation management on rangelands is primarily through control of grazing, infrequent use of burning, and brush or weed control. Grazing by livestock and game animals constitutes the primary agricultural economic use of rangeland forage. Rangelands have other extremely important resources, including water, wildlife habitat, genetic diversity of plant and animal species (West 1993), and recreation, but landowners seldom receive direct economic benefits from them.

Land use in the Great Plains changed significantly from the nineteenth to the twentieth century. While there are 310 million acres of rangeland in the Great Plains today, much of the current 246 million acres of cropland and 64 million acres of pasture was originally rangeland. The conversion of rangeland to cropland, and in some instances back to rangeland, has been a dominant issue during the twentieth century. Rangelands remaining are unsuited for intensifying agricultural use by conversion to cropland, since most land with capability for cultivation has already been converted. Livestock have replaced many of the native, large herbivores that previously grazed the Plains (Joyce

Stanley R. Johnson and Aziz Bouzaher (ed.), *Conservation of Great Plains Ecosystems*, 199–209.

and Skold 1988). Burning, which was prevalent on the Great Plains before the region was settled, has been largely suppressed. These shifts in land use have caused instabilities in the Plains ecosystems that need to be addressed.

This chapter addresses the three greatest problems affecting rangeland ecosystems in the Great Plains: inappropriate land use (i.e., conversion to cropland), poor grazing management, and suppression of burning.

Land Use: Conversion to Cropland

Extensive conversion of rangeland to cropland during the early part of the twentieth century was followed by dramatic increases in soil erosion (Laycock 1988), primarily by water (Table 14.1) in the east and by wind (Table 14.2) in the west. The principal on-site impact of accelerated erosion is reduced potential productivity. Principal off-site effects include siltation of other lands and waterways, increased flooding, and reduced water and air quality. Off-site costs of wind and water erosion can be sizable (Ervin and Johnson 1992; Laycock 1983).

Table 14.1. Estimated average annual sheet and rill erosion on nonfederal rural land for 1987, by state

State	Cropland	Pastureland	Rangeland
		tons/acre/year	
Eastern Great Plains			
Minnesota	2.5	0.3	0.1
Iowa	6.5	1.5	—
Missouri	7.0	1.9	0.6
Arkansas	3.7	1.6	2.4
Central Great Plains			
North Dakota	1.9	0.5	0.9
South Dakota	2.3	0.3	0.9
Nebraska	4.7	1.3	1.1
Kansas	2.6	1.0	1.4
Oklahoma	2.3	0.7	1.7
Texas	2.5	0.8	1.1
Western Great Plains			
Montana	1.9	0.2	0.8
Wyoming	0.7	0.3	1.3
Colorado	2.2	0.5	2.2
New Mexico	1.0	0.2	1.0

Source: SCS 1989.

Table 14.2. Estimated average annual wind erosion on cropland for 1987, by state

State	Cultivated Cropland	Noncultivated Cropland
	tons/acre/year	
Eastern Great Plains		
Minnesota	4.9	0.3
Iowa	2.0	0.2
Missouri	0.9	0.2
Arkansas	0.0	0.0
Central Great Plains		
North Dakota	4.2	0.2
South Dakota	2.7	0.2
Nebraska	2.0	0.5
Kansas	3.8	0.4
Oklahoma	4.4	1.0
Texas	11.8	2.6
Western Great Plains		
Montana	8.9	0.4
Wyoming	12.7	2.3
Colorado	11.3	0.6
New Mexico	8.6	5.0

Source: SCS 1989.

Spurred by drought and accelerated erosion of Great Plains croplands during the 1930s, the federal government purchased 5.7 million acres of highly erodible cropland and converted them back to rangeland. These are now managed by the USDA Forest Service as the national grasslands, except for 1.9 million acres in Montana that are managed by the Bureau of Land Management under the U.S. Department of the Interior (Laycock 1988). During the drought of the 1950s, the government acted through the Soil Bank Program to convert 28.7 million acres of cropland to perennial vegetation. Total cost of the program was almost $2.5 billion (Laycock 1988). After this program expired, most of this land was returned to cultivation.

Conversion of rangeland to cropland again accelerated during the 1970s and 1980s (Laycock 1983). Government policies (homestead and farm commodity laws) and economic incentives (lending policies and real estate value) have been responsible for stimulating these land-use conversions (Laycock 1991). Again the government reacted in the 1985 Food Security Act, which established the Conservation Reserve Program (CRP). Under CRP contracts, almost 34 million acres (nationally) have been converted from cultivated cropland to perennial vegetation; however, CRP contracts will begin to expire in 1995 so these lands (most of which are highly erodible) will again become eligible for cultivation (Ervin and Johnson 1992). Reentry of these lands into crop production will significantly increase air pollution from blowing dust, deplete underground water aquifers, increase water erosion, and increase ground and surface water pollution. These problems affect urban as well as rural citizens.

The effects of both the Soil Bank Program and CRP on levels of air-borne dust have been documented at Lubbock, Texas (Ervin and Johnson 1992). Graphing annual dust data and acres enrolled in Soil Bank and CRP for 1947 through 1990 clearly showed that removing these highly erodible acres from cultivation markedly reduced wind erosion. However, Ervin (1993) noted that even with this improvement, air quality in Lubbock does not meet national air quality standards.

Conversion of rangeland to cropland continues. For the period 1982-87, rangeland in the Great Plains decreased by over 5 million acres with similar trends for most Plains states. During the same period, cropland in the Plains increased (2.3 million acres), mostly in the northern states. The greatest changes in pastureland acreage were a decrease in Iowa (–672,900 acres) and increases in Texas (672,200 acres) and Oklahoma (450,400 acres), but there was little net change in total acres of pastureland in the Great Plains. Examining the acres of cropland, pastureland, and rangeland by land capability class and subclass provides an indication of the potential for shifts in land use. Nationally there are 23 million acres of cropland that are in capability classes V to VIII (Table 14.3). This land should be returned to perennial vegetation to prevent further deterioration from wind and water erosion.

Table 14.3. Land cover and use of nonfederal rural land, by land capability class for the United States in 1987

Class	Cropland	Pastureland	Rangeland
		thousands of acres	
I	28,618.0	2,021.5	430.2
II	191,026.5	31,650.3	16,267.9
III	134,135.1	39,525.0	46,366.2
IV	45,971.3	25,441.6	53,430.4
V	2,856.1	4,592.6	4,913.3
VI	16,211.4	16,907.1	131,805.9
VII	3,936.1	9,670.7	144,309.0
VIII	84.9	165.6	4,161.7
V-VIII	23,090.5	31,336.0	285,189.9
Total	422,841.4	129,974.4	401,684.6

Source: SCS 1989.

Government policies, at all levels from local to federal, should encourage sustainable land uses—uses that are within the resource's capability. Often there is conflicting government policy. One branch of the U.S. Department of Agriculture is spending billions of dollars to remove highly erodible land from cultivation while another (Economic Research Service) has identified 57.7 million acres of range and pasture lands in the Great Plains with a supposedly high or medium potential for conversion to cropland (Laycock 1988). Not only is the suitability of these lands for conversion questionable, but also this appears to be a serious policy conflict. Policies should encourage landowners to choose sustainable land uses, and policies that promote poor land use should be eliminated. Overall, the direct and indirect costs (on-site and off-site) of these policies need to be considered.

Range Condition

Range condition is a generic term applied to various techniques for evaluating the current status of rangelands relative to a specified potential. Although the Soil Conservation Service (SCS) has been the lead agency for assessing range conditions for the Great Plains, there has not been agreement among range professionals on the best methods for evaluation. The SCS defines range condition as the present status of vegetation of a range site in relation to the climax (natural potential) plant community for that site. It is an expression of the relative degree to which the kinds, proportions, and amounts of plants in a plant community resemble that of the climax plant community for the site (Jacoby 1989).

In 1987, the National Resources Inventory indicated that Great Plains rangelands were 3 percent excellent, 34 percent good, 49 percent fair, and 12 percent poor (SCS 1989). The condition of these lands is indicated in the inventory of conservation treatment needs, which indicates that 215 million acres of our nation's rangelands need management for forage improvement. Additional conservation needs were weed control or brush management (67 million acres), forage reestablishment with brush management (21 million acres), plant reestablishment for forage (10 million acres), mechanical soil treatment for forage improvement (8 million acres), and erosion control (65 million acres). It should be noted that more than 99 percent of the conservation treatment need on cropland is for erosion control, whereas on rangeland only 17 percent of the acreage need treatment for erosion control. On rangelands, the primary need is for forage and brush management.

The primary tool for forage management is the control of livestock grazing. This includes the kind of animals (species), number of animals (stocking rates), seasons of grazing, and the grazing schedule (grazing systems). Rangelands that have not been invaded by long-lived perennial shrubs or weeds can generally be improved through grazing management. Of these management factors, perhaps the most difficult for ranchers to manage is the adjustment of animal numbers to balance forage demand with the fluctuating forage supply. The failure to stock ranges properly, especially during the frequent droughts, is a major cause of range deterioration.

When shrubs and perennial weeds have invaded, plant control measures are generally needed to change vegetation. Mechanical and chemical control methods have been developed for use today (Scifres 1980), whereas fire was the primary environmental factor limiting the abundance of many shrubs on Great Plains rangelands before the area was settled in the late nineteenth century. The reintroduction of fire as a vegetation management tool appears to be a promising "new" technology.

Grazing Management

Grazing by large herbivores has been an integral, natural component of Great Plains rangeland ecosystems since the close of the Pleistocene Epoch. During the latter part of the nineteenth century, bison were replaced by cattle and sheep, and goats replaced many antelope. Deer numbers have increased significantly in portions of the Plains during the twentieth century. The control of prairie dogs over millions of acres of the Plains represented a significant change in grazing effects.

The effects on rangeland ecosystems of shifting from totally unmanaged wildlife grazing to managed livestock and wildlife grazing cannot be accurately evaluated because there were no quantitative evaluations during the nineteenth century (Lauenroth et al. 1993). While some early travelers reported grass stirrup high, others reported inadequate forage for their horses because of heavy grazing by bison. Some reports indicated clear water, while others reported water polluted by bison. Both types of reports probably are correct, depending upon the time and location.

In natural ecosystems, wildlife populations respond to changes in forage supply and other natural controls. The result is generally cyclical changes in populations with booms and busts. Periodic droughts and severe winters probably resulted in extensive die-offs of wildlife, similar to the livestock losses that occurred during severe storms of the late nineteenth century (Laycock 1988). A return to the unmanaged rangeland ecosystems of presettlement is unrealistic. Rather, the task is to develop an understanding of the structure and function of rangeland ecosystems and to develop management programs that will achieve the goals of landowners and society for productive, healthy rangelands.

The development of fencing, wells, and supplemental feed combined with set stocking rates during the twentieth century have allowed ranchers to maintain stable, though sometimes excessive, livestock numbers on rangelands (Joyce and Skold 1988). Overstocking was widespread during the early years of the ranching industry because the concepts and principles of proper stocking and grazing management were only beginning to be identified at the turn of the century (Allred 1993). During the 1930s, proper land use became a national issue (Graham 1944), and following World War II, land use moved into an era of increased scientific study and emphasis on conservation and improved management of rangelands. The American Society for Range Management (now Society for Range Management) was organized in 1948 and several western universities began offering a curriculum in range management. While much remains to be done, teaching, research, and extension programs have developed a significant body of knowledge related to sustained use and restoration of rangeland ecosystems. Great strides have been made in applying this information to improve rangeland management.

Since grazing is a natural component of Great Plains rangelands, it is not surprising that rangeland vegetation and soils are stable and productive under moderate grazing (Lauenroth et al. 1993). In fact, exclusion of grazing can be considered an ecological disturbance. Comparison of vegetation composition of rangelands that were grazed with areas excluded from grazing and other areas that were mechanically disturbed indicated that vegetation on the ungrazed range was more similar to the disturbed range than to the grazed range (Lauenroth et al. 1993).

Periods of deferment from grazing (3 to 6 months) are beneficial because they promote secondary plant succession and improve rangelands that have been overgrazed. Specialized grazing systems have been developed to facilitate the scheduling of grazing and deferment (Kothmann 1980). While grazing systems may enhance the rate of secondary succession on rangelands, grazing is beneficial only if the stocking rates are correct. Grazing systems are not successful in improving rangelands if lands are overstocked.

Range management research has evaluated different grazing intensities and management systems for restoring degraded rangelands and maintaining their productivity. Long-term studies indicate that the intensity of grazing is the most important management factor affecting vegetation (Lauenroth et al. 1993; Kothmann et al. 1978). Comparisons of infiltration and runoff in Texas found little difference between ungrazed and light or moderately grazed rangelands; however, heavy, continuous stocking reduced infiltration and increased sediment production (Blackburn et al. 1982; McCalla et al. 1984). Sustained heavy grazing reduces plant vigor and cover, but healthy, productive rangelands can be sustained indefinitely under moderate grazing intensity.

The climate of the Great Plains is characterized by high annual and seasonal variability in precipitation. Such variations result in highly variable forage production, which presents a difficult management problem for ranchers. Ranchers like to maintain relatively stable livestock herds, but the forage supply fluctuates greatly among years. If stocking rates are based on the average year's forage production, in nearly 50 percent of the years the land will be overstocked. If stocking rates are set very conservatively in order to not overgraze the range during frequent drought years, then much usable forage is left in the average and above average years. Flexible stocking rates have been recommended, but until recently ranchers have not been provided with suitable technology for assisting them in making these difficult and timely management decisions (Kothmann and Hinnant 1992a,b).

Computer programs are being developed to assist ranchers on-site with the difficult tasks of budgeting for forage and deciding on stocking rates (Kothmann and Hinnant 1990). One such program, which is being developed and tested in cooperation with ranchers in Texas, is proving to be effective and efficient. Development of new software continues. Educational programs are needed to make ranchers aware of the technology and to assist them in mastering the basic concepts involved in forage budgeting and grazing management.

Although range management technology is available to ranchers, its adoption is limited by several factors: low or negative economic returns from application of the technology, difficulty in integrating new technology into existing management systems, poor understanding of the response of nontargeted ecosystem components, and inadequate knowledge of the availability and application of the technology. Solutions to these problems lie in developing technology that is more cost effective and easier to apply and in improving the systems for transferring new technology to the rancher.

Fire Management

Fire was a dominant influence that created the vegetation type found on the Great Plains. Fires, set accidently by lightening and purposefully by Native Americans, burned native vegetation at frequent intervals. Most of the plant species that were dominant at the time of European settlement of the Plains are very tolerant to burning. Following settlement, extensive fires were reduced by direct action of settlers and by reducing available fuel by allowing heavy grazing. The control of fire throughout the Great Plains, along with heavy livestock grazing, has allowed woody vegetation to increase significantly. The most noticeable increase in woody vegetation has been in the Southern Plains in New Mexico, Texas, and Oklahoma.

Negative experiences with wildfires gave rise to the collective notion that all fires were destructive and should be suppressed. This attitude toward fire is changing in management agencies, as well as among ranchers, as fire technology is developing. The knowledge base required to use prescribed burning as a vegetation management tool is becoming more sophisticated. As with grazing, the shift in attitude is from fire as an unmanaged menace to a productively managed tool.

Prescribed burning is a planned application of fire to an area to meet specific management applications. Prescribed fire considers the influence of weather, vegetation, topography, and other variables. Objectives for prescribed burning might include suppressing the woody plants that invade rangeland, improving grazing distribution of livestock, increasing the palatability of forage, and enhancing wildlife habitat. Prescribed burning requires a carefully designed fire plan. The critical elements are knowing why (objective setting), what (target area and species), when (timing, season), and how (selection of techniques and procedures) (Scifres and Hamilton 1993).

The conservation treatment needs for nonfederal rural lands indicate that weed or brush management is needed on 67 million acres for enhanced forage production, and forage reestablishment with brush management is needed on more than 20 million acres. Much of the brush encroachment on these 87 million acres is the result of fire suppression. Extended periods of no fires result in major shifts in the balance between woody and herbaceous species, and probably a change in the composition of herbaceous vegetation. Many of the brush species that are spreading most rapidly (e.g., Juniper spp.) are readily controlled by burning. If not controlled, juniper will form closed canopies that eliminate most other woody species and nearly all herbaceous forage production. This reduced forage production reduces the site's productivity for livestock and many wildlife species and, when combined with heavy stocking, may reduce water infiltration and increase runoff and erosion.

In the past, ranchers have used chemical and mechanical methods in an attempt to control the invasion of the woody species. However, without fire as a component of the management system, the invasion is a continuing problem. Woody vegetation must be controlled at 15- to 20-year intervals. The cost of the chemical and mechanical control practices is becoming economically prohibitive for ranchers. Fire has been shown to be an effective management tool (Wright and Bailey 1982; Scifres and Hamilton 1993), but additional fire technology research is needed so that vegetation responses to prescribed burning can be evaluated and the public can be educated about the importance of burning.

Conservation Ethics

> Conservation is more than wise use and preservation of natural resources—mineral, soil and water, plant and animal, human. It may well be the foundation of a new social philosophy. Our public land policies were developed around the exploitation of natural resources and they helped materially to promote our present economy. Is it not reasonable to suppose, then, that laws and regulations which promote the conservation of resources will help to build a new attitude toward the environment in which we live? (Graham 1944)

There is an important relationship between the land and personal human welfare. Certainly, humankind influences the environment. Learning to manage land can provide for present needs and preserve options for future generations. Human civilization depends on productive and sustainable use of our resources. Between humans and their environment is interjected the power to reason. While there is nothing utilitarian about an idea in itself, a great deal of physical and material well-being depends upon ingenuity of thought. By losing sight of this fact, one of the most potent influences in human progress slips from grasp. In a world of expert knowledge and specialized techniques, a purpose for knowledge and an amazing amount of synthesis is needed. Facts can obscure the big picture that is needed to link knowledge with purpose (Graham 1944).

Paul Weiss (1942), in discussing the education of biologists, claimed that, in the world ahead, their fitness would depend only partly on biological competence: "Unless they are also made conscious of their obligation to society, they will . . . be at a loss to justify their subsidized existence to a society asking uncomfortable questions. . . . And the biologist will have to be convincing in showing cause why what he is doing should not be discontinued as a publicly supported enterprise."

What Graham and Weiss wrote 50 years ago is still relevant today. Society is struggling to develop land-use policies that encourage management of natural resources for production to meet current needs of society and conservation for future diversity and productivity. On one extreme are individuals who have little regard for current human needs, especially economic. On the other extreme are individuals seeking to exploit resources only for short-term economic gains. We must adopt rational public land-use policies that allow resources to be used to meet a diverse array of human needs (economic and noneconomic) for both the short and long terms. Conservation of rangelands does not imply nonuse. Grazing and burning are natural components of rangeland ecosystems and can be used in rational management plans to meet a diverse array of human needs.

References

Allred, K. W. 1993. The trail of E. O. Wooton. *New Mexico Resources IX* (spring). College of Agriculture and Home Economics, New Mexico State University, Las Cruces.

Blackburn, W. H., R. W. Knight, and M. K. Wood. 1982. *Impact of grazing on watersheds: A state of knowledge.* TAES Publication MP 1490. College Station, Texas.

Ervin, R. T. 1993. Personal communication.

Ervin, R. T., and P. N. Johnson. 1992. *Economic evaluation of the Conservation Reserve Program in the southern high plains of Texas.* Department of Agricultural Economics, Texas Tech University, Lubbock.

Forage and Grazing Terminology Committee. 1991. *Terminology for grazing lands and grazing animals.* Blacksburg, Virginia: Pocahontas Press, Inc.

Graham, E. H. 1944. *Natural principles of land use.* New York: Oxford University Press.

Jacoby, P. W. (chair). 1989. *A glossary of terms used in range management.* Denver: Society for Range Management.

Joyce, L. A., and M. D. Skold. 1988. *Implications of changes in the regional ecology of the Great Plains.* In *Impacts of the Conservation Reserve Program in the Great Plains: Symposium proceedings, Denver, Colorado, Sept. 16-18, 1987,* ed. J. E. Mitchell. USDA-FS General Technical Report RM-158. Fort Collins, Colorado: Rocky Mountain Forest and Range Experiment Station.

Kothmann, M. M. 1980. Evaluation of livestock needs in designing grazing systems for rangeland. In *Digestive physiology and nutrition of ruminants, Vol. 3: Practical nutrition* (2nd ed.), ed. D. C. Church. Corvallis, Oregon: O&B Books, Inc.

Kothmann, M. M., and R. T. Hinnant. 1990. Grazing management stock adjustments templates. *TAEX software catalog* (July). Texas Agricultural Extension Service, Extension Computer Technology Group, College Station.

———. 1992a. APSAT: An operational level grazing management model. *Proceedings, Western Section, American Society of Animal Science* 43:354-356.

———. 1992b. Forecasting carrying capacity for drought management. In *Beef cattle research in Texas.* Consolidated Progress Report 4950-4976. Texas Agricultural Experiment Station.

Kothmann, M. M., W. J. Waldrip, and G. W. Mathis. 1978. Rangeland vegetation of the Texas rolling plains: Response to grazing management and weather. In *Proceedings, First International Rangeland Congress,* ed. Harold Heady. Denver: Society for Range Management.

Lauenroth, W. K., D. G. Milchunas, J. L. Dodd, R. H. Hart, R. K. Heitschmidt, and L. R. Rittenhouse. 1993. Grazing in the Great Plains of the United States. In *Ecological implications of livestock herbivory in the West,* ed. M. Vavra, W. A. Laycock, and R. D. Pieper. Denver: Society for Range Management.

Laycock, W. A. 1983. Plowing of fragile grasslands in the northern and central Great Plains. In *Proceedings, The Range Beef Cow Symposium VIII* (Dec. 5-7), Sterling, Colorado.

———. 1988. History of grassland plowing and grass planting on the Great Plains. In *Impacts of the Conservation Reserve Program in the Great Plains: Symposium proceedings, Denver, Colorado, Sept. 16-18, 1987,* ed. J. E. Mitchell. USDA-FS General Technical Report RM-158. Fort Collins: Rocky Mountain Forest and Range Experiment Station.

———. 1991. The Conservation Reserve Program—How did we get where we are and where do we go from here? In *The Conservation Reserve—Yesterday, today, and tomorrow: Symposium proceedings, Washington, D.C., Jan. 14, 1991,* ed. L. A. Joyce, J. E. Mitchell, and M. D. Skold. USDA-FS General Technical Report RM-203. Fort Collins, Colorado: Rocky Mountain Forest and Range Experiment Station.

McCalla II, G. R., W. H. Blackburn, and L. B. Merrill. 1984. Effects of livestock grazing on sediment production, Edwards Plateau of Texas. *Journal of Range Management* 37:291-294.

Scifres, C. J. 1980. *Brush management principles and practices for Texas and the Southwest.* College Station: Texas A&M University Press.

U.S. Department of Agriculture, Soil Conservation Service (SCS). 1989. *Summary report, 1987 National Resources Inventory.* USDA-SCS, Statistical Bulletin no. 790. Iowa State University Statistical Laboratory, Ames.

Weiss, Paul. 1942. The training of biologists. *Science* 95:32-34.

West, N. E. 1993. Biodiversity of rangelands. *Journal of Range Management* 46:2-13.

Wright, H. A., and A. W. Bailey. 1982. *Fire ecology.* New York: John Wiley and Sons.

15. Conservation Forestry for Sustainable Great Plains Ecosystems

Gerald F. Bratton
U.S. Department of Agriculture
Peter R. Schaefer
South Dakota State University
James R. Brandle
University of Nebraska—Lincoln

Two hundred years ago, the American Great Plains was a region dominated by grassland yet enriched by patches of woody vegetation, a region of harsh climatic extremes characterized by spring flooding and summer drought, and a region devastated by wildfire yet renewed by regrowth. Forests moved onto the Plains during periods of abundant moisture only to be pushed back by drought and wildfire.

With the advent of agricultural development, the nature of vegetation has changed. Wildfire, floods, and roaming buffalo herds found little tolerance among a society whose existence was, and continues to be, dependent upon the successful husbandry of crops and livestock.

With development came dams, suppression of wildfire, extensive areas of cropland, and replacement of roaming buffalo herds with intensive grazing systems dominated by cattle. A system with great diversity became one of monocultures.

Today the Great Plains ecosystem is very different than it was 200 years ago. The Plains are intensively managed and designed to provide an abundance of food and fiber. It is in this managed ecosystem that conservation forestry has a vital role to play.

Woodlands and trees are some of the most valuable, least recognized, and vastly underestimated resources available in the region. Occupying less than 5 percent of the land area, woodlands provide a variety of wood products and enhance economic and environmental sustainability. Trees protect soil from erosion; control blowing snow; protect crops, livestock, residents, and homesites; clean the air; protect water quality and quantity; add economic and biological diversity; provide critical habitat for migratory and local wildlife; and improve aesthetic values. Trees diversify farm income by providing fuel wood; specialty crops such as Christmas trees, nuts, and fruits; and traditional products like posts, lumber, and veneer. Although trees are not presently

Stanley R. Johnson and Aziz Bouzaher (ed.), *Conservation of Great Plains Ecosystems*, 211–227.

considered a major component of traditional agriculture, when combined with various conservation practices, trees can contribute to a more sustainable farm operation. For example, field windbreaks reduce residue requirements under minimum tillage systems. Consequently, low residue crops such as soybeans may be added to the rotation without significantly increasing the danger of wind erosion. By using several conservation practices, the producer gains the flexibility to choose the best management practices for a variety of crops.

Environmental Variability: Climate

The Great Plains is a region of climatic extremes and vegetation contrasts. Average temperatures range from below 4°C in the Canadian Prairies to above 21°C in southern Texas. In the northern portions of the region, temperatures in January can easily reach –40°C. By contrast, in southwestern Texas, temperatures often exceed 40°C in July or August. Overall, there is a gradual gradient from warm to cool from the eastern border of the region to the base of the Rocky Mountains (NOAA 1983; USDA 1935). However, of greater significance to the woody plants are the rapid changes in temperature that can sometimes fluctuate over 27°C within a 12-hour period.

Precipitation shows a similar pattern. The driest areas are in southwestern Texas where precipitation averages 250 millimeters per year. In contrast, rainfall in southeastern Texas may exceed 1,000 millimeters annually. Across the midsection of the region from central Iowa to eastern Colorado, annual precipitation decreases from 830 millimeters to 300 millimeters. Along the Canadian border, northern Minnesota receives approximately 520 millimeters while northwestern North Dakota receives less than 400 millimeters (NOAA 1983; USDA 1935). In the southwest, the combination of high temperature and low precipitation results in high levels of evaporation, which severely limits tree growth.

More important than precipitation averages are precipitation extremes. Most of the Great Plains is subjected to periodic droughts of varying lengths. For example, tree ring data from archaeological sites in Ash Hollow, Nebraska, indicate that over the last 700 years there have been a total of 21 drought periods of five years or more, the longest of which lasted 38 years (Weakly 1943). Obviously, these types of droughts have a major impact on the distribution and survival of woody vegetation.

Present concerns about global climate change are relevant to discussion of woodland resources. Most general circulation models project that the North American Great Plains will become warmer and drier as atmospheric carbon dioxide increases (IPCC 1990). By the year 2030, temperatures in the central part of the region are projected to be 2° to 4°C higher, with winter precipitation increasing 0 to 15 percent and summer precipitation decreasing 10 to 15 percent. Soil moisture in summer is projected to be 15 to 20 percent lower than current averages. If correct, these changes would have dire consequences for plant growth in a semiarid environment.

History of Tree Planting

The Great Plains has a history of tree planting, beginning with the earliest settlers. Both state and federal laws encouraged tree planting. The Timber Culture Act of 1873 encouraged settlement of the region and led to more than 10 million acres of tree claims but resulted in only limited tree planting success. Repealed in 1891, the Timber Culture Act was replaced by the Kincaid Act in 1904, the Clark-McNary program in 1924, and subsequently the Prairie States Forestry Project in the 1930s. During this latter effort, nearly 212 million trees were planted in more than 18,000 miles of multirow windbreaks (Droze 1977).

A 1954 survey of these windbreaks by Read (1958) indicated that only 42 percent were in good condition and 27 percent were in poor condition or had been removed. A 1975 report by the General Accounting Office (1975) indicated that in Kansas, Nebraska, and Oklahoma, removal rates exceeded 10 percent per year, primarily because of installation of center-pivot irrigation systems. A 1980 survey by the U.S. Department of Agriculture, Soil Conservation Service, was more optimistic, indicating that windbreak plantings in Kansas, Nebraska, North Dakota, Oklahoma, and South Dakota exceeded removals by 1.8 percent (USDA 1980). Unfortunately, the optimism was short lived. From 1982 to 1987, the Great Plains region lost 24,149 windbreaks (approximately 3.2 percent).

The most recent efforts to encourage tree planting were under the Conservation Reserve Program. But in the Great Plains, less than 5 percent of the eligible acres were planted to trees; the rest were planted to grass (Deneke and Bratton 1989). This is particularly unfortunate since a survey by the Forestry Committee of the Great Plains Agricultural Council indicated that grass was chosen because it could easily be removed upon expiration of the 10-year contracts (Bratton 1987).

Description of Woodland Resources

The Great Plains region consists of a variety of intermixed vegetation types. This complicated mix of forest and grass vegetation existed in relative equilibrium until the introduction of contemporary agricultural systems and is best understood when divided into three subunits: a western transition zone, the interior grasslands, and an eastern transition zone (Figure 15.1).

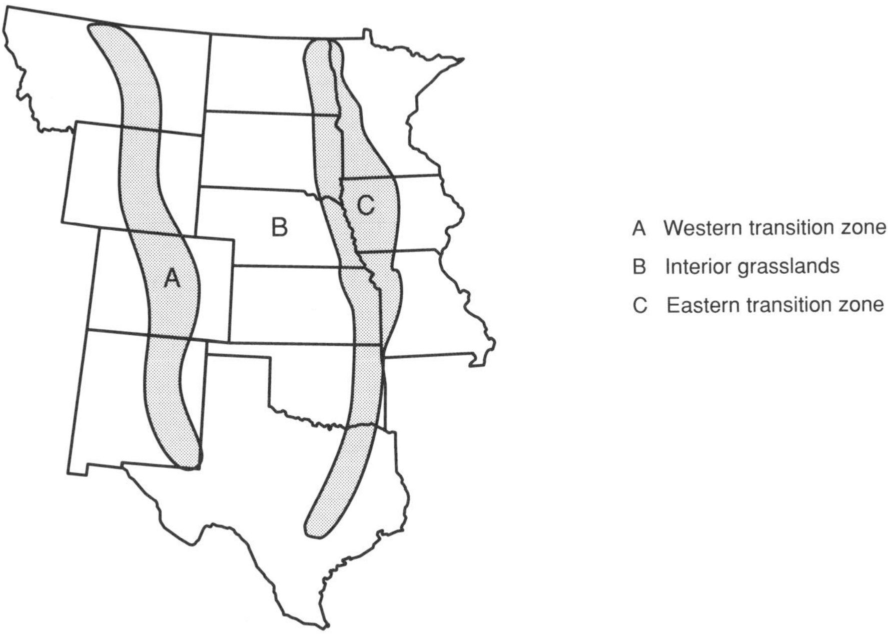

Figure 15.1. Major woodland zones in the Great Plains

Eastern and Western Transition Zones

Along the east and west boundaries, different woodland, grassland, and cropland systems intermingle in intricate landscape. Woodlands within these zones are composed of riparian forests, woody draws, and savannas, as well as orchards and plantations.

The eastern edge of the Great Plains lies along the western extent of three eastern forest types: the northern hardwood forest, the central hardwood forest, and the southern pine region. These woodlands flourish on sites that are too steep, frequently flooded, or otherwise unacceptable for agriculture, but which may be extremely profitable for timber production. As a result, many acres of high-quality timber stands extend like fingers into the Great Plains grasslands.

The western edge of the Great Plains abuts the ponderosa pine and pinyon-juniper woodlands of the Rocky Mountains. These western transition woodlands lie in the rain shadow of the Rocky Mountains, are relatively dry, and are generally less productive in traditional forest products. These woodlands are of immense beauty and provide valuable homesites, watershed protection, wildlife habitat, livestock grazing, and recreational use.

Interior Grasslands

The grassland ecosystems of the interior Great Plains were historically maintained by drought and fire. Once nearly free of trees except for narrow belts of riparian woodland or isolated woody islands, the grasslands existed because of protection afforded by wet sites or topography (Wells 1965; see also Figure 15.2). They are now fragmented and altered by farming and intensive grazing.

Now that wildfire is controlled, trees are commonplace. Some are present as a result of tree planting efforts of prairie residents. Others are native or have naturalized and exist as riparian woodlands or woody draws. Many are spreading into adjacent grassland areas.

Native Woodland Resources

While there are more than 20 million acres of native woodland in the Great Plains, they represent less than 1 percent of the total land area of the region. Comprehensive inventories of all the states in the region are incomplete. Several states have no detailed inventory, while others have inventories of only major forested areas. State foresters report that while there is a desperate need for in-depth woodland inventories to assist with proper resource planning and management, funds for complete inventories are meager.

For sake of simplicity, both transition zones and interior woodland resource data are combined and categorized into two major groups: native woodland resources (Table 15.1) and planted woodland resources (Table 15.2). The native woodland resources will be further delineated into commercial and noncommercial forestland where individual state inventories allow.

The wooded riparian areas are of special importance for several reasons:

- Improving water quality by buffering runoff from adjacent lands
- Stabilizing stream banks
- Shading streams to maintain cooler temperatures for quality fisheries
- Storing water and helping to reduce flood severity
- Providing wildlife habitat and recreation areas
- Supporting productive forests for the economic development of forest products
- Increasing biodiversity and stability in agricultural ecosystems

The health and vigor of many native woodland resources are declining. Natural resource practitioners report that expanded farming practices, uncontrolled grazing, and indiscriminate logging have seriously depleted and degraded much of the resource, which result in accelerated stream bank erosion, increased cropland erosion, less wildlife habitat, lower quality forest products, increased water pollution, and a generally less attractive environment.

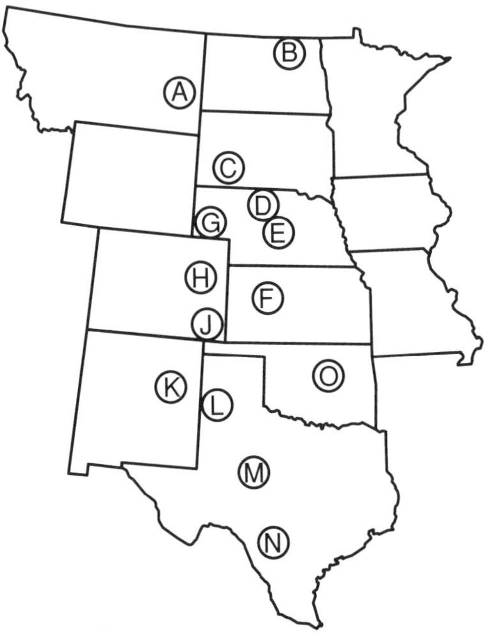

Source: Adapted from Wells 1965.

Figure 15.2. Nonriparian woodlands in the Great Plains

A Montana, eastern Wyoming, and western Dakotas. Woodland: *Pinus ponderosa, Juniperus scopulorum.*

B Pembina escarpment and Turtle Hills, North Dakota; and Erskine and Big Stone Moraines, Minnesota. Woodland: *Populus tremuloides, Quercus macrocarpa.*

C Pine Ridge escarpment, South Dakota, Nebraska, and Wyoming. Vegetation similar to A.

D Niobrara escarpments, Nebraska. Woodland: *P. ponderosa, J. scopulorum, J. virginiana, Q. marocarpa.*

E Sand Hills area, Nebraska. Woodland: isolated stands of *P. ponderosa, J. virginiana, Celtis occidentalis.*

F Western Kansas. Woodland: isolated stands of *J. virginiana.*

G Wildcat Hills, Nebraska. Vegetation similar to A.

H Cedar Point, Colorado. Woodland: *P. ponderosa, J. scopulorum.*

J Black Mesa-Mesa de Maya area, Oklahoma, New Mexico, and Colorado. Woodland: *Pinus edulis, P. ponderosa, Juniperus monosperma, J. scopulorum, Quercus undulata.*

K Canadian escarpment, New Mexico. Vegetation similar to J.

L Llano Estacodo, New Mexico and Texas. Woodland: on northwest, *P. edulis, J. monosperma, Q. undulata;* on east (break of the Plains), *Juniperus pinchotii, Quercus mohriana.*

M Callahan Divide, Texas. Woodland: *Quercus virginiana, Q. shumardii, Q. mohriana, Juniperus ashei, J. pinchotii.*

N Edwards Plateau, Texas. Woodland: Similar to M.

O Oklahoma and eastern Kansas. Woodland: *Quercus stellata, Q. marilandica, Q. muehlenbergii, Q. shumardii, J. virginiana.*

Table 15.1. Native woodland resources in the Great Plains

State	Commercial	Noncommercial	Total	Nonforest	Forest
		1,000 acres			percent
Colorado			13	13,667	<1.0
Iowa	2,050	522	2,572	33,766	7.0
Kansas	1,207	150	1,358	50,979	2.5
Minnesota[a]	660	196	856	18,509	4.4
Missouri[a]	2,532	742	3,275	17,008	16.2
Montana[a]				98,003	<1.0
Nebraska	537	108	718	48,333	1.5
New Mexico[a]			5	50,445	<1.0
North Dakota	343	148	491	43,821	1.1
Oklahoma[a]	9,400	900	10,300	40,715	25.3
South Dakota			714	47,149	1.5
Texas[a]			51	19,228	<1.0
Wyoming[a]				16,074	<1.0
Total			>20,353		

Sources: Individual state data for Iowa, Kansas, Minnesota, Missouri, Nebraska, and South Dakota were taken from USDA Forest Service inventories. Data for Colorado, Montana, New Mexico, Texas, and Wyoming were estimated using USDA, SCS National Resources Inventory information. The data for Oklahoma were obtained from the Oklahoma Division of Forestry.

[a]Reflects only the forest resources contained in the prairie unit or Great Plains area of this state.

Table 15.2. Windbreak resources in the Great Plains

State	Number	Acres	Miles
Colorado	12,999	21,938	2,345
Iowa	64,603	58,550	6,036
Kansas	77,943	113,947	20,241
Minnesota	101,597	164,281	14,853
Missouri	32,235	34,199	8,700
Montana	26,270	26,522	10,469
Nebraska	117,347	224,573	18,646
New Mexico	3,237	1,414	388
North Dakota	168,238	325,602	52,289
Oklahoma	16,955	30,539	4,159
South Dakota	94,060	262,548	18,813
Texas	7,542	15,788	1,668
Wyoming	4,806	5,607	765
Total	727,872	1,285,508	159,372

Source: USDA 1982 and 1987.
Note: This information originally appeared in an unpublished special report prepared by the Iowa State University Statistical Laboratory.

Planted Resources

There are more than 1.3 million acres of planted forest resources in the Great Plains, the majority of which are windbreaks. Windbreaks are especially important because they help to make agriculture sustainable by protecting crops, livestock, homes, roads, wildlife, and humans from wind, cold, and snow. Windbreaks help conserve energy by buffering wind on homes and livestock and conserve water (especially irrigation water) by reducing the evaporation rate caused by hot, dry summer wind. These "working trees" in agriculture protect cropland and soil areas downwind from ten to twenty times their height. They also protect areas upwind from two to three times their height.

Approximately 730,000 field and farmstead windbreaks in the Great Plains grow on more than 1,200,000 acres and stretch for nearly 160,000 miles (Table 15.2; USDA 1980). The windbreaks protect approximately 7,650,000 acres of cropland and 430,000 farmsteads. However, while approximately 55 percent of farmsteads are protected, only 3 percent of the total cropland in the region is protected, leaving 234 million acres without tree protection (Brandle et al. 1992).

The health and vigor of many windbreaks are declining. Many were planted in the 1930s and 1940s and the majority of the others are more than 30 years old. Many of these older windbreaks are in poor condition. In South Dakota, 83 percent of the windbreaks are heavily infested with grass and weeds, many are grazed and severely damaged, and most are in a state of decline (Schaefer et al. 1987). Similar conditions are seen over much of the region.

Public Incentive Programs for Woodland Resources

Five major federal cost-sharing programs are available to Great Plains landowners to improve established woodlands and to create new woodlands or special use plantings.

Stewardship Incentives Program

The SIP is designed to improve forest resources of individual states and is administered by each state forestry agency. Specific practices include: developing forest management plans; encouraging reforestation and afforestation; improving forest stands; establishing, maintaining, and renovating windbreaks and hedgerows; protecting and improving soils and water; protecting and improving riparian and wetland areas; improving fisheries habitat; improving wildlife habitat; and enhancing forest recreation.

Conservation Reserve Program

Authorized by the Food Security Act of 1985, the CRP emphasizes removing highly erodible cropland from production and curbing surplus commodities. Cost-sharing was available for various tree planting practices including: erosion control, permanent wildlife habitat, windbreaks, vegetative filter strips, and timber production.

Forestry Incentives Program

The FIP consists of four practices whose major emphasis is on production of wood products: tree planting for production and environmental improvement, forest stand improvement to increase timber growth and quality on suitable sites, site preparation for natural regeneration to encourage the production of forest products and environmental improvement, and special forestry practices significant and unique to local conditions.

Agricultural Conservation Program

The ACP, administered by the Agricultural Stabilization and Conservation Service (ASCS), encourages practices that have been the mainstay of the ASCS cost-sharing program for conservation improvements on farms and ranches. The basic forestry practices include forest tree and shrub establishment as part of a conservation effort and tree stand improvement.

Great Plains Conservation Program

Congress authorized the GPCP in 1956 to help stabilize the economy and improve agriculture. The program, administered by the Soil Conservation Service, is authorized to share with landowners 50 to 80 percent of the cost to establish various conservation measures. Covered practices include windbreaks, wildlife tree and shrub establishment, and block tree planting.

Some states have programs that supplement federal incentives and often include cost-sharing programs to address local issues. In some states, tax incentives are used to encourage various practices.

A number of private sources fund tree planting efforts. Two widely recognized sources in the Great Plains are Pheasants Forever and Quail Unlimited, both of which furnish funds to assist with installation of wildlife habitat or food plots.

Technical Assistance for Landowners

Most landowners lack the knowledge or technical expertise to develop their own woodland management programs. Thus, assistance is available through several state and federal agencies: state forestry agencies, the Cooperative Extension Service, the Soil Conservation Service, Soil Conservation or Natural Resource districts, as well as through private consultants.

The 1990 Farm Bill authorized the Center for Semiarid Agroforestry to develop and transfer conservation forestry technologies and information to natural resource professionals who, in turn, furnish technical assistance to landowners.

Woodland Ecosystems Issues

In preparing this chapter, the authors surveyed state forestry organizations in the Great Plains region to determine issues relevant throughout the region.

Resource Condition

The woodland resources of the Great Plains are declining in both health and land area. Age, improper management, farming practices and herbicides, and development of water resources are the most important underlying causes of these declines.

Age

Trees have a finite life span that varies with species and growing conditions. Tree mortality in naturally forested areas is part of the succession process, contributes to changes in forest composition over time, and generally does not result in loss of the forest. Tree mortality in planted woodlands is more severe since trees do not replace themselves through natural regeneration. As trees age or are damaged by insect, disease, or climatological events, they must be renovated or replaced with new plantings. Field observations throughout the Plains suggest that more than 50 percent of the region's windbreaks need renovation.

Significant tracts of natural forest are declining because of factors that inhibit successful regeneration. To the detriment of riparian forest species, such forces as dams, channelization, irrigation, and agricultural expansion have restricted the meandering nature of water courses, reduced the incidence of periodic flooding, and increased the fluctuations in water tables. Urban development has led to the loss of native woodland areas and they have been replaced with highly managed and often less diverse urban forests.

Management

The condition of present woodland resources is a direct reflection of past and present management. By and large, most woodland areas and windbreaks have been poorly managed. Poor species selection and lack of diversity, competition from weeds and grasses, and lack of protective fencing have resulted in widespread decline and loss of windbreaks.

The extensive use of only a few species prone to insect and disease problems has had a significant impact on the health of windbreaks in the Great Plains. For example, Siberian elm was planted extensively throughout the Plains because of its drought and cold tolerance and rapid growth rate. Unfortunately, it is also susceptible to several serious insect and disease pests, which has resulted in the loss of thousands of windbreaks. Similar problems are occurring with American elm, tatarian honeysuckle, and poplars. In each case, certain favorable traits led to heavy reliance on these species but each is also susceptible to disease or insect pests.

The lack of species diversity in windbreak plantings is serious and difficult to overcome because of the limited number of adapted species. The current overuse of green ash and eastern red cedar in windbreaks may pose similar problems in the future.

Farming Practices

During this century, agricultural production has grown tremendously with little regard for the region's natural resource base of grasses, water, wildlife, and woodlands.

Productive soils along river systems have been cleared of woody vegetation for the production of row crops leading to large-scale elimination of riparian forests. Unrestricted livestock grazing along water courses and in other naturally wooded areas has resulted in a significant loss of forest cover, high levels of soil erosion, and the degradation of associated plant, animal, soil, and water resources.

Domestic farm animals are detrimental to windbreaks and woodlands. They destroy natural regeneration in the understory and compact the soil, reducing root growth. Serious declines in forest health have occurred on western rangelands where woody draws have been heavily grazed, which results in significant losses of wildlife habitat. Although most windbreaks are not grazed, those that have been are less effective and show distinct signs of declining health.

Trees growing in agricultural settings are often subject to spray drift from the application of agricultural chemicals to adjacent fields. Herbicides used for broadleaf weed control in fields are injurious to broad-leaved trees. For example, Siberian elm, a common species for field windbreaks in the northern Plains, is very sensitive to 2,4-D. Many foresters believe that the widespread decline of Siberian elm is the result of combined effects of exposure to 2,4-D, canker diseases, and defoliating insects. It is not known to what extent herbicides are affecting woodland resources; however, the incidence of herbicide damage claims in the region has increased at least 40 percent since 1987 (Bergmeier and Smith 1992).

Undoubtedly, exposure to broadleaf weed herbicides has resulted in a decline in vigor of field windbreaks throughout the Great Plains, greatly reducing their effectiveness, predisposing them to insect and disease problems, and in many cases resulting in their death.

Water Development

Engineering along river systems for irrigation, recreation, and flood control has affected associated woodland resources in several ways. The creation of reservoirs along rivers of the region has reduced the variation in downstream flows and significantly reduced major flooding. As a consequence, the water regime has been drastically changed, altering site conditions and species composition of riparian forests and modifying wildlife populations associated with the river systems.

In cases where rainfall is adequate, the pioneer forests of cottonwood and willow have matured and are being replaced with later successional stages dominated by green ash and associated species (Johnson 1992, 1994). Along the Platte River in Nebraska, this change has led to the loss of open wetland areas used heavily by sandhill cranes, least terns, and piping plovers. On the other hand, the development of riparian forests has increased wooded habitat for neotropical migrants, another group of species threatened by loss of native habitat. In other areas, low water tables and a lack of adequate rainfall have led to the loss of the cottonwood and willow. The shrub-grassland vegetation that has replaced them is more useful to grassland wildlife species. It is difficult to say which is better, but the fact that it has changed is certain.

Naturalization of Woodland Species

The virtual elimination of grassland fires has significantly changed site conditions of much of the rangeland, especially in the southern and central Plains (Wright and Bailey 1982). As a result, woody species (such as mesquite, eastern red cedar, and Osage-orange, formerly suppressed by fire) are aggressively invading rangelands. Russian olive and Siberian elm, both introduced species, have spread throughout the region. Again, the control of wildfire, an essential component of the grassland ecosystem, has resulted in environmental changes. While some may view these changes as beneficial, there are significant negative impacts on the range resources and associated wildlife.

Genetic and Resource Diversity

Fragmentation and loss of native woodlands, coupled with limited species selection and loss of windbreaks, act to reduce genetic and resource diversity of Great Plains woodland resources. This loss in diversity limits the ability of plant species to adapt to changing conditions and leaves a species vulnerable to damaging agents. Maintaining or improving genetic and species diversity should be a basic underlying principle when managing Great Plains woodlands. The heavy reliance on a few favored species in conservation planting continues and the potential of serious harm to these resources is very high (Schaefer 1993).

Climate Change

If the predictions of climate change and time scale are accurate, significant change will occur within the next 50 years. For most tree species, this provides insufficient time for natural selection and genetic adaptation to occur. Consequently, present species may not be able to survive significantly altered environmental conditions. The potential effects on forest resources bear careful attention by state and federal governments and resource agencies.

Resource Management

Great Plains woodlands, whether native or planted, are almost entirely privately owned and are managed at the landowner's discretion. Unfortunately, the majority of woodlands on the Great Plains are poorly managed. Why is this? First and foremost, it is because both the general public and public officials appear to lack an understanding of the value of woodland resources and the role of trees and shrubs in stabilizing our managed ecosystems. The net effect is that woodland resources have a low priority nationally, regionally, and locally. This not only affects the landowners' interests in managing the resource, but also their ability to do so.

Assistance to landowners for woodland resource management is provided by several public agencies. State forestry agencies throughout the region are acutely understaffed and many have had budget and staff cut in recent years. The number of extension forestry specialists in the region is very limited. Neither Nebraska, Wyoming, nor South Dakota has an extension forestry specialist, and most states have no trained foresters at the county level. The Soil Conservation Service employs foresters at both the state and county level; however, several Great Plains states do not have a state-level SCS forester. District conservationists who have forestry training are often delegated workloads that do not allow for a forestry program. Most county ASCS and conservation district offices also lack personnel trained in woodland resource management.

Generating support necessary to address the vast tree planting and woodland management needs in the Great Plains is a major challenge. Agency politics and "turf protection" stand in the way of meeting this challenge. Closer working relationships between public agencies and land-grant institutions need to be developed.

Solutions to Woodland Ecosystems Issues

The situation just described is distressing in light of the present condition of native woodlands and the 1.3 million acres of windbreaks. However, it is much more alarming when we consider that 97 percent (234 million acres) of cropland in the Great Plains is without tree protection. The need to stabilize agricultural production and economies is real. Although not a cure-all, the increased incorporation of trees and shrubs into agricultural management systems will help to protect and stabilize agricultural resources and the integrity of production programs. Increasing the diversity of agricultural ecosystems, reducing nonpoint source pollution, reducing soil loss, protecting crops and livestock, and providing harborage for wildlife and beneficial insects all lead to healthier, more productive agricultural environments.

All of the woodland resource programs in place today work to a degree, but none of them, separately or combined, has been able to develop widespread acceptance of woody vegetation in agricultural landscapes. Present programs are based on incentives that make it easier for landowners to plant trees and shrubs or maintain existing woodlands on their property. They do nothing to persuade landowners that trees and shrubs are desirable working components of their agricultural operations. Incentives are good for people already inclined to plant trees, but they do little to encourage those who do not understand the value of woodland resources to the agricultural environment. This is the basic issue underlying most woodland problems in the Great Plains and poses the greatest challenge.

Increased information, education, and demonstration efforts are critical to increasing people's receptiveness to tree planting in the Great Plains and should focus on training and involving people (Sampson and Raschke 1990). These efforts must be directed to resource management agencies and their staffs, schools, FFA and 4-H programs, college agriculture programs, and local, state, and federal politicians. Outreach efforts to landowners must include on-site technical assistance from properly trained and knowledgeable agency staff. These programs should help people understand why and how to get involved, as well as the benefits of being involved (Sampson and Raschke 1990).

Programs must provide opportunities for landowner input so that individuals feel an ownership of the programs. Significant improvements in woodland resource management will be very difficult to achieve without a more knowledgeable public.

Increased Emphasis on Research and Development

Factual and locally relevant information is needed to fuel successful outreach programs. Increased research funding is needed in the areas of woodland-based economic development, management of woodland systems, landscape ecology, economic analysis of woodland benefits, the biology and control of insects and diseases, and the selection of adapted tree and shrub species. Windbreak design and ways to better integrate tree culture with crop and livestock production (including economic analysis) are also important research areas.

The marketing and utilization of wood products should be addressed. Ways should be sought to develop traditional timber markets. Nontimber markets should be developed for products such as energy (firewood, chips, pellets, liquid fuels), extractives (medicinals, essential oils, flavorings), fruit, nuts, seed, horticultural and landscape products, and other specialty areas.

Expanded Incentives Programs

If outreach programs are increased as suggested, incentive programs will become even more important in assisting tree planting and management efforts. Existing federal programs should be continued and funded at higher levels as the demand for tree planting and care increases.

A consistent natural resource policy must be developed across local, state, and federal agencies. Multiagency support and cooperation and systems development must be sought. Detailed resource inventories must be completed for all states. Legislation to discourage windbreak removals and encourage windbreak renovation and establishment should be promoted.

Conclusions

Although woodland resources cover only a small fraction of Great Plains land area, they are of tremendous importance to the region. Overall, the resource is considered to be in a general state of decline as a result of the combined effects of advanced age, improper management, farming practices and herbicides, and water development activities. Efforts to manage the resource effectively are hindered by a lack of understanding of the value of trees and shrubs in agricultural environments by both the general public and decision makers. A shortage of trained field personnel to provide technical assistance and limited interagency cooperation also contribute to poor forest management opportunities. Recommended approaches to address these issues include expanded information, education, and demonstration programs; increased emphasis on research and development; expanded federal, state, and local incentive programs; development of consistent natural resource policy; multiagency cooperation; a systems approach to resource management; completion of resource inventories; and legislation promoting resource management.

References

Bergmeier, R., and D. Smith. 1992. Herbicide drift—A survey in the Great Plains states (unpublished draft). Data available from R. Bergmeier, Montana State Division of Forestry, 2705 Spurgin Road, Missoula, Montana 59801.

Brandle, James R., Tom D. Wardle, and Gerald F. Bratton. 1992. Opportunities to increase tree planting in shelterbelts and the potential impacts on carbon storage and conservation. In *Forests and global change, Vol. 1: Opportunities for increasing forest cover,* ed. R. Neil Sampson and Dwight Hair. Washington, D.C.: American Forests.

Bratton, Gerald F. 1987. Survey of Conservation Reserve Program. In *Proceedings, Great Plains Agricultural Council Forestry Committee,* 39th Annual Meeting (June 22-25), Cheyenne, Wyoming. Great Plains Agricultural Council Publication 122.

Deneke, Frederick J., and Gerald F. Bratton. 1989. The Conservation Reserve Program in the Great Plains: Past, present, future. In *Proceedings, Great Plains Agricultural Council Forestry Committee,* 41st Annual Meeting (June 12-15), Santa Fe, New Mexico. Great Plains Agricultural Council Publication 130.

Droze, Wilmon H. 1977. *Trees, prairies, and people—A history of tree planting in the plains states.* Denton: Texas Woman's University.

General Accounting Office (GAO). 1975. Action needed to discourage removal of trees that shelter cropland in the Great Plains. RED-75-375. Washington, D.C.

Intergovernmental Panel on Climate Change (IPCC). 1990. Climate change. In *Climate change: The IPCC scientific assessment,* ed. J. T. Houghton, G. J. Jenkins, and J. J. Ephraums. New York: Cambridge University Press.

Johnson, W. C. 1992. Dams and riparian forests: Case study from the upper Missouri River. *Rivers* 3: 229-242.

———. 1994. Woodland expansion in the Platte River, Nebraska: Patterns and causes. *Ecological Monographs* 64:45-84.

National Oceanic and Atmospheric Administration (NOAA). 1983. Climate data. Asheville, N.C. (Originally printed by U.S. Department of Commerce, June 1968).

Read, Ralph A. 1958. *The Great Plains shelterbelt in 1954.* Great Plains Agricultural Council Publication 16.

Sampson, N., and B. Raschke. 1990. The opportunity to expand the use of trees for conservation. In *Trees are the answer: Proceedings, Great Plains Agricultural Council Forestry Committee,* 42nd Annual Meeting (June), Colorado Springs, Colorado. Ed. P. Hoefer, G. Fechner, and S. McDonald. Great Plains Agricultural Council Publication 132.

Schaefer, P. R., S. Dronen, and D. Erickson. 1987. Windbreaks: A plains legacy in decline. *Journal of Soil and Water Conservation* 42(4):237-238.

Schaefer, P. R. 1993. Genetic diversity and stress resistance: Some concerns regarding artificially established trees in the Great Plains. *Stress Mechanisms Symposium* (April 6-7), College of Agriculture and Biological Sciences, South Dakota State University, Brookings.

U.S. Department of Agriculture, Forest Service (USDA). 1935. *Possibilities of shelterbelt planting in the plains region.* U.S. Forest Service, Lake States Forest Experiment Station.

U.S. Department of Agriculture, Soil Conservation Service (USDA). 1980. *Field windbreak removals in five Great Plains states.* Publication 626-893/2247. Washington, D.C.

———. 1982 and 1987. *National resources inventory* (Basic statistics, unpublished special report prepared by Iowa State University Statistical Laboratory).

———. 1989. *Summary report, 1987 national resources inventory.* USDA/SCS Statistical Bulletin 790. Iowa State University Statistical Laboratory, Ames.

Weakly, Harry E. 1943. A tree-ring record of precipitation in western Nebraska. *Journal of Forestry* 41: 816-819.

Wells, Philip V. 1965. Scarp woodlands, transported grassland soils, and concept of grassland climate in the Great Plains region. *Science* 148:246-249.

Wright, Henry A., and Arthur W. Bailey. 1982. *Fire ecology—United States and southern Canada.* New York: John Wiley and Sons.

Water Resources

16. Total Resource Management Plan for Addressing Groundwater Concerns

A. J. Dornbusch, Jr., B. M. Vining, and J. L. Kearney
U.S. Department of Agriculture

The aquifers underlying the Great Plains provide the groundwater resources that are so important as sustenance for agriculture and as sources of drinking water. Several systems are included in this region: the High Plains aquifer, the northern Great Plains regional aquifer system, and the northern and central Midwest regional aquifer systems.

These groundwater sources, as important as they are, are increasingly endangered. Declining water levels due to overdrafts from irrigation have led to regulatory action designed to maintain or rebuild groundwater levels. Nitrates and pesticides have been found in groundwater in sections of the Great Plains. Detection of atrazine, alachlor, metolachlor, cyanazine, and simazine in these aquifers is associated with the half-life of the pesticides, the likelihood of use, the intensity of chemical use, the percentage of land used for row cropping, and wells in shallow, unconfined aquifers. Nonpoint source contamination is not believed to be the sole contributor to groundwater contamination. Point source contamination is believed to be caused by improperly constructed older wells, mixing and loading operations, spills, and pesticide dealer sites. Where nitrate levels near the 10-milligram-per-liter maximum contaminant level (MCL), both point and nonpoint source contamination are suspected.

This chapter explores the current status of the groundwater resource in the Great Plains, looks at current approaches, and proposes alternatives for consideration.

Overview of Groundwater Resources in the Expanded Great Plains

Aquifer systems in the Great Plains include the High Plains aquifer of the southern Great Plains, the northern Great Plains regional aquifer system, and the northern and central Midwest regional aquifer systems (Figure 16.1). While aquifers and the rocks forming them may be grouped or separated by various parameters, for purposes of discussion, it must be remembered that such divisions in nature are not always so distinct. When dealing with the geologic environment, variations from the expected

Stanley R. Johnson and Aziz Bouzaher (ed.), *Conservation of Great Plains Ecosystems*, 231-251.

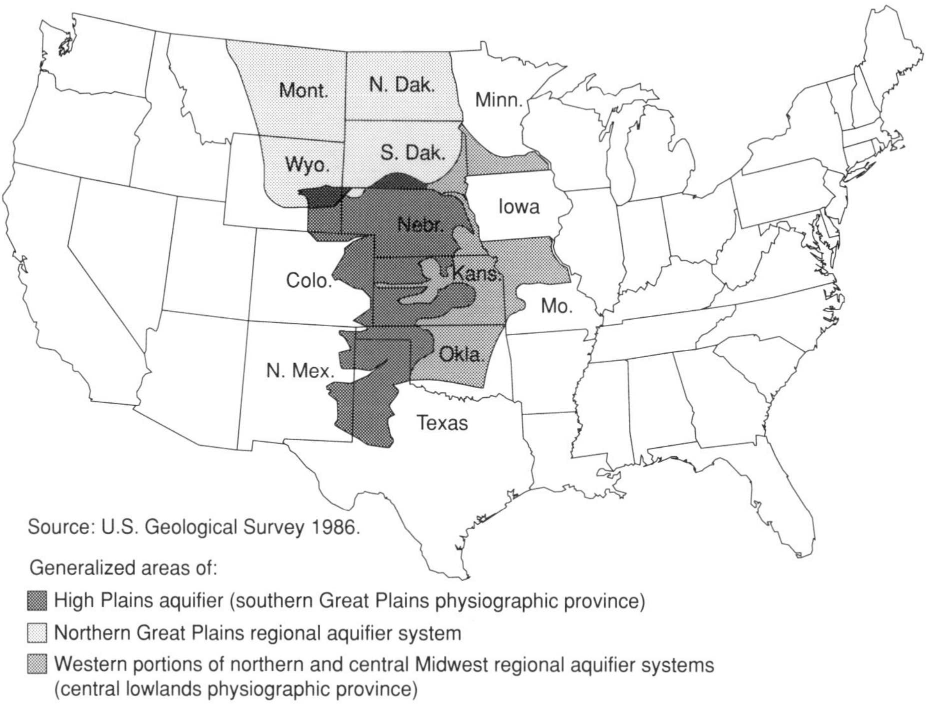

Figure 16.1. Location of regional aquifers

norm should not be considered a surprise, but a discovery that requires description, analysis, and evaluation in relationship to the environmental concerns and the conservation practices that can positively affect those concerns.

High Plains Aquifer of the Great Plains

The High Plains aquifer underlies about 175,000 square miles in parts of eight states—South Dakota, Wyoming, Nebraska, Colorado, Kansas, Oklahoma, New Mexico, and Texas. It is irregularly shaped (Figure 16.1) with a length of 800 miles north to south and a width of 500 miles in Nebraska to 100 miles in the Texas Panhandle. The High Plains aquifer is an unconfined (water table) aquifer in which the water table is near the ground surface and forms the upper boundary of the aquifer.

Dune sand deposits of very fine to medium wind-blown sand cover about 19 percent of the High Plains aquifer's area. The most extensive deposit of dune sand in west-central Nebraska covers 20,000 square miles with a thickness of about 300 feet. Large areas of southern Kansas into the Oklahoma Panhandle are also covered with dune sand. Alluvial deposits of unconsolidated gravel, sand, silt, and clay occupy most valleys within the aquifer area. Throughout the High Plains, dune sands are important recharge areas for the aquifer. Likewise, alluvial deposits and associated valley streams form important surficial links to the High Plains aquifer, particularly along the major river valleys.

Recharge to the High Plains aquifer is from precipitation and seepage from streams. Because in the High Plains area evapotranspiration greatly exceeds precipitation, areas of sandy soils that have large infiltration rates, high permeability, and low moisture storage capacity become important areas for recharge to the aquifer.

Losing streams in the High Plains area also become important locations for recharge to the aquifer during times of runoff. During times of intense rains, the streams may, in some places, transport large quantities of water. During these flows, large quantities of water infiltrate into channel and floodplain deposits, and a portion of the water percolates downward to recharge the aquifer. Although ephemeral streams are intermittent sources of recharge, it is possible that a few high flows may be a principal source of recharge to the aquifer. Management of recharge areas and the drainage areas contributing to them is an important consideration in the water quality and quantity of the High Plains aquifer.

The concentration of dissolved solids in water from the High Plains aquifer near stream valleys is affected by irrigation practices. Where the water table is near the surface, the salts that accumulate in the soil due to evapotranspiration are dissolved, flushed with irrigation water, and transported to the aquifer.

The quality of water in the High Plains aquifer generally is suitable for irrigation use, but in many places the water does not satisfy U.S. Environmental Protection Agency (EPA) drinking water regulations because of excessive concentration of dissolved solids, fluoride, chloride, sulfate, selenium, and nitrate.

The High Plains aquifer is the principal source of water for irrigation in the High Plains. An estimated 170,000 wells pump almost 18 million acre-feet of water from the aquifer to irrigate nearly 14 million acres. About 95 percent of all water pumped from the aquifer is used for irrigation. Groundwater irrigation began in the late 1800s, but development was sparse until the 1940s and 1950s when it was spurred by drought, technological advances in well drilling and pumping plants, and development of natural gas fields as an inexpensive source of energy. By 1949, 2 million acres were irrigated with most of them in the southern High Plains. During the 1980s, 14 million acres were irrigated with more than half in the northern High Plains. Annual pumpage of groundwater for irrigation was an estimated 4 million acre-feet in 1949 and increased to about 18 million acre-feet in 1980. The growth rate has decreased in some areas (such as Texas) primarily because of declining water availability and increased in other areas (such as Nebraska) with relatively large quantities of water and recharge.

Widespread changes in water levels have occurred within the High Plains aquifer (Figure 16.2), changes that are caused by variable precipitation, differing land uses, and groundwater withdrawals. Irrigation water is being withdrawn from the aquifer much faster than it is being recharged, leading to water levels declining from the beginning of development until 1980 and slowing since 1980. Despite overall declines, water levels have risen in a few scattered areas—the largest of these south of the Platte River, where the rise is caused by recharge from surface water irrigation and leakage from canals and reservoirs.

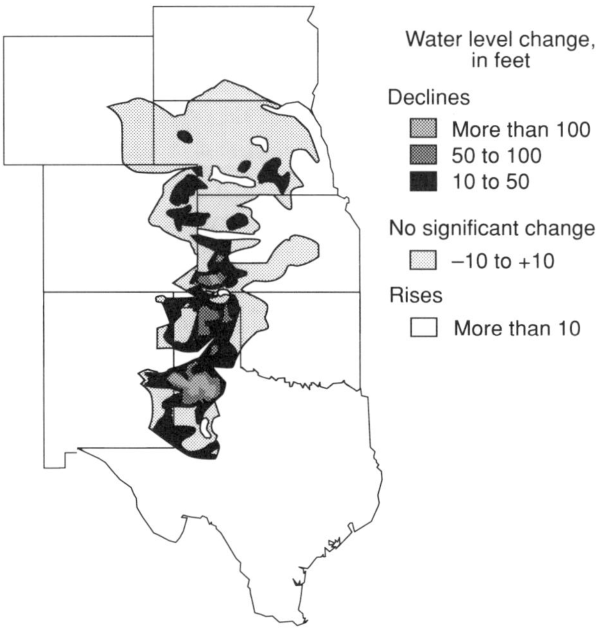

Source: Modified from Weeks et al. 1988.

Figure 16.2. Water level changes in the High Plains aquifer, predevelopment to 1980

Since 1980, water level changes have been more variable (Figure 16.3). Most states have variably sized, scattered areas of both declines and rises, as well as areas of no significant change. Factors contributing to pattern changes are above normal precipitation, a decrease of irrigated acres, improved irrigation scheduling, and more efficient irrigation technology.

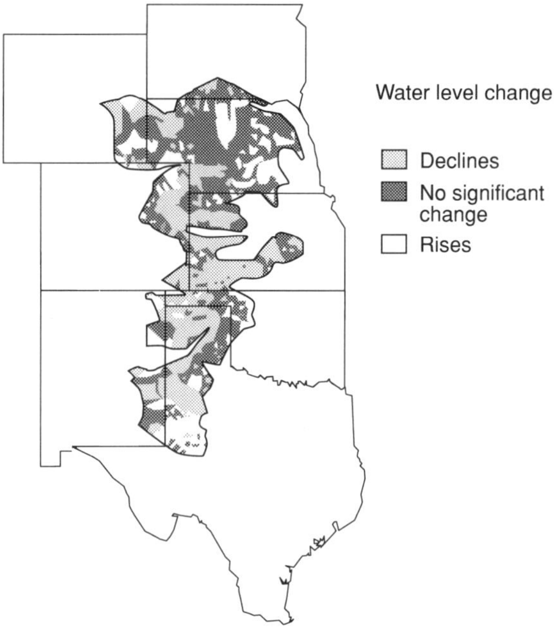

Source: Adapted from Dugan and Schild 1992.

Figure 16.3. Water level changes in the High Plains aquifer, 1980–90

Water-level changes affect changes in saturated thickness and volume of water in storage. Such changes are directly related to changes in well yield; decreased well yields and increased pumping lifts significantly increase operating expenses.

Northern Great Plains Regional Aquifer System

In the generally semiarid northern Great Plains, where average annual precipitation ranges from 12 inches in the northwest to 24 inches in the southeast, groundwater becomes the major source of water. Quantity is often a first concern with quality close behind.

The groundwater of the northern Great Plains aquifer system occurs in a 300,000-square-mile area covering North Dakota, and parts of Montana, South Dakota, Wyoming, and Nebraska. The flow system, associated with both consolidated bedrock and unconsolidated sediments of glacial and alluvial processes, is extremely complex. The bedrock aquifers are confined (artesian) aquifers occurring throughout most of the area. The unconsolidated materials that are most productive are bodies of sand and gravel from glacial outwash and alluvial channel deposition. These sand and gravel deposits, which are irregular in shape and size, as well as scattered, contain unconfined (water table) aquifers. Shallow groundwater is scarce in the unglaciated western part of the region, but important in the glaciated areas east of the Missouri River.

The geologic framework within which the confined (bedrock) groundwater flows is one of recharge in and near the highland areas of the mountains to the west and southwest, as well as around the Black Hills area, and of flow in a generally eastward and northeastward direction for more than 600 miles toward discharge areas in the eastern Dakotas and the Canadian province of Manitoba.

Virtually all eastward-flowing streams draining the recharge areas in the highlands lose a part of their flow as they cross the aquifer. Recharge also results from infiltration of precipitation falling directly on the exposed rocks in lowland areas. Not all the water entering aquifers along outcrop areas recharges the deep regional aquifer system and moves to the eastern discharge area. A large part of the recharge water discharges in a short distance through springs and seeps along the flanks of the highland areas. The fraction of the total recharge that remains in the deeper aquifer systems becomes the regional flow.

Upward leakage through confining beds appears to be a discharge mechanism for all the aquifer systems underlying the northern Great Plains. (Vertical leakage between aquifers may be detected by geochemical methods.) Leakage occurring through the matrix of confining beds, along fractures, and where confining beds are missing is an additional source of recharge from lower to upper systems. In eastern discharge areas these sand and gravel deposits function as groundwater drains for the underlying aquifer systems.

Much of the water in the deeper aquifers is not very desirable for household purposes and cannot be used for irrigation. In the past, this water was used in western areas where flowing wells could be constructed, but as the pressure head was depleted and good quality water was more needed, use of this water has decreased substantially.

Any plans to further develop the deep aquifer systems in the northern Great Plains region need to consider leakage among beds, as well as storage in the associated confining beds, because the water from adjacent beds may have an entirely different chemical quality than water from the developed aquifer system. Also, the management of groundwater resources needs to consider agricultural practices and management in areas of recharge to the aquifer system.

In much of the eastern area of the northern Great Plains, the unconsolidated sand and gravel aquifers associated with glacial drift are the most productive aquifers for drinking water, as well as other uses. The aquifers occur as broad sheet-like deposits, or as long and relatively narrow deposits associated with buried river valleys. These aquifers are usually unconfined and found at or near the surface. Those aquifers associated with the buried river valleys occur at greater depths and can have unconfined and confined components.

Outside of these major aquifers, lesser water amounts can be obtained from small isolated deposits of sand and gravel. The small aquifers are randomly distributed throughout the glacial drift and may range from a few acres to a few hundred acres in size. They are capable of yielding 5 to 10 gallons per minute to an individual well for needs of a household or for livestock water. Much of the water can be used for irrigation, but care must be taken to determine the compatibility of the water quality with the type of soil to be irrigated.

The vast majority of contaminants in the groundwater of the northern Great Plains are, at present, naturally occurring. But concerns with artificial influence on groundwater quality are increasing. Groundwater supplies that are most susceptible to contamination are shallow, water table aquifers overlaid by sandy soils, with no natural barriers between the top of the water table and the soil surface.

Northern and Central Midwest Regional Aquifer Systems

Within the western area of the Central Lowlands, including parts of Minnesota, Nebraska, Iowa, Missouri, Kansas, and Oklahoma, the principal aquifers consist of surficial deposits and bedrock units. Glacial drift, buried paleovalley alluvial deposits, and valley alluvial deposits form the surficial aquifers; they are mostly unconfined (water table) aquifers. Sedimentary rocks form the bedrock aquifers; they are unconfined near the surface and confined in the deeper formations. Crystalline basement rock provides insignificant quantities of water and effectively forms the base of the groundwater flow system in the region.

Of the glacial deposits, the clay till is generally a poor source of water; yields to wells are small to negligible, and the water can be highly mineralized. Lenses of sand and gravel within the glacial drift yield only small quantities of water because of their limited extent and limited hydraulic connection to other water sources. These may provide for household use or small stock water supplies but are unreliable for public or industrial water supplies.

Valley alluvial aquifers are delineated along and connected hydraulically to surface streams. Aquifers commonly ranging from 10 to 150 feet deep yield 30 to 750 gallons per minute but can exceed 1,500 gallons per minute. Water from both valley and paleovalley alluvial aquifers is suitable for most uses.

The principal bedrock aquifers are, from youngest to oldest, the Cretaceous (Dakota and Niobrara formations) aquifer, the Permian-Pennsylvanian aquifer, the Mississippian aquifer, the Silurian-Devonian aquifer, and the Ordovician (St. Peter-Prairie du Chien-Jordan formations) aquifer. Aquifers in Cambrian rocks underlie the Ordovician formations, but they are used less in the region because of their depth and because they have poorer quality water.

Cretaceous formations extend from central Iowa to the west and southwest into Nebraska and western Kansas where they underlie the High Plains aquifer. The depth of this area's Dakota aquifer and the salinity of its water increase toward the west, thus limiting to eastern Nebraska and western Iowa the area where water for most uses can be obtained. Even here, large concentrations of dissolved solids (often concentrations of iron) limit the use as drinking water.

In western and central Kansas, where localized depletion of water supplies of the High Plains and the alluvial valley aquifers is a concern, the Dakota aquifer is the next available source of groundwater. Presently, however, insufficient information on the quantity, flow system, and quality of Dakota water, as well as on the properties and structure of the rocks of the aquifer, limits widespread planning and development. The state of Kansas is undertaking long-term studies to assess the water resources potential of the Dakota aquifer.

The quality of water in the Dakota aquifer system differs if the aquifer system is recharged locally, whether it has been leached, and how long the water has been in the aquifer. The Dakota's water is harder than water from most other aquifers. Where the Dakota is recharged locally, calcium is the principal constituent; at greater depths, sodium concentrations about equal that of calcium.

Paleozoic rocks underlying large areas of northeast Iowa and the four corners of Iowa, Nebraska, Kansas, and Missouri provide water locally, but they are less important regionally than other aquifers. The rocks consisting largely of limestone, shale, and dolomite are often neglected or bypassed as a source of water because they contain undesirable concentrations of dissolved solids and provide small yields. Because many of these rock systems are at or near the surface and can contain fractures, aquifers are particularly susceptible to surface contamination. Occasionally, channel sandstones within these systems of rock will yield 10 to 100 gallons per minute of water suitable for most uses.

In the northeastern portion of this region (Minnesota and Iowa), the Ordovician and deeper Cambrian aquifers are used extensively and provide large yields of water suitable for most uses. The St. Peter-Prairie du Chien-Jordan aquifer is the uppermost, and thus most used, with the Ironton-Galesville and the Mount Simon aquifers being deeper. There are indications of water flow between the aquifers. They are confined aquifers in the area under consideration, and they receive recharge from structurally higher areas to the east in Wisconsin and Illinois. Potentiometeric surface in the system has declined because of pumping near the recharge areas to the east; simulation indicates this will likely continue. The Ordovician-Cambrian aquifers receive little use in the southwest part of the region (Missouri, Nebraska, and Kansas) because of depth and associated poorer water quality.

Despite the elevated dissolved solids concentration in this system, this water provides the best source of drinking water at some locations. Water in the upper aquifer system is very hard, a calcium-magnesium-bicarbonate type. Concentrations of dissolved solids tend to increase greatly in deeper water, and water becomes a sodium chloride brine in deeper parts of structural basins.

Current Status and Trends of Groundwater

Use of Fertilizers and Pesticides

Although commercial fertilizers and pesticides constitute only a fraction of the source materials that can contaminate groundwater, understanding their use patterns and application trends is part of the key to prevention. After World War II, manufacturers developed the technology to produce nitrogen fertilizer cheaply. As munitions plants were converted to fertilizer plants, fertilizer use increased steadily, nearly quadrupling from 1960 to 1990 (Figure 16.4). Shortly after 1980, nitrogen fertilizer use, measured by the amount of fertilizer purchased, stabilized.

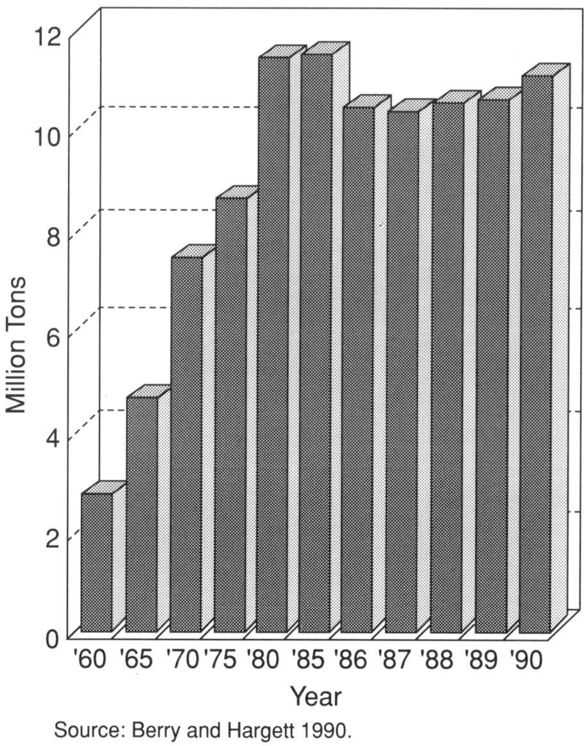

Figure 16.4. U.S. nitrogen fertilizer use from 1960 to 1990

Reduced use of nitrogen and pesticides in 1983 was directly related to a federal supply control program. The payment-in-kind program provided grain from government stocks to farmers who agreed to reduce planted acreage. The corn acreage planted in 1983 was approximately 25 percent smaller than in either 1982 or 1984.

Pesticide use has also increased steadily during the last 20 years (Figure 16.5). In 1982, farmers treated more than 91 percent of their available farmland with pesticides, compared with 74 percent in 1971. The market for pesticides in 1986 had 45,000 products, giving rise to the sale of 1.2 billion pounds of pesticides. Of this amount, about 70 percent was used in production agriculture, 23 percent in forestry, industry, and government programs, and 7 percent by gardeners and homeowners.

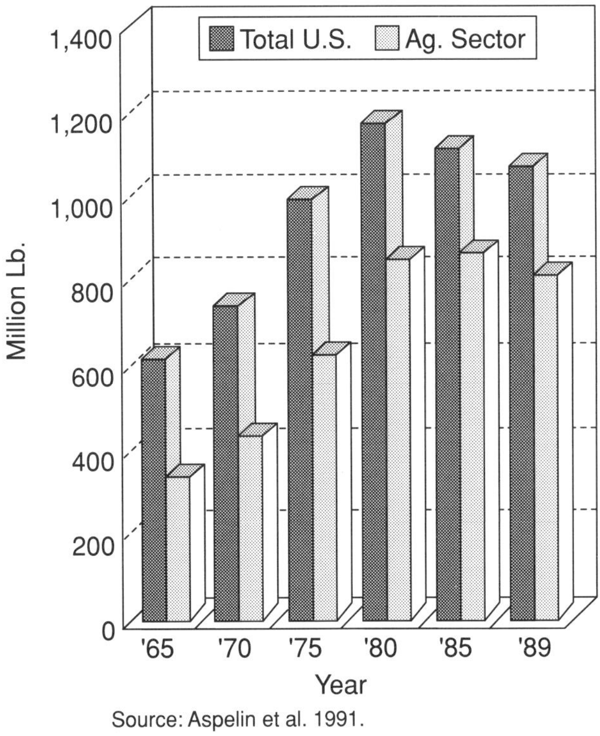

Figure 16.5. U.S. pesticide use from 1965 to 1989

National Well Water Surveys

In the fall of 1990, the EPA completed the five-year National Survey of Pesticides in Drinking Water Wells. The survey was designed to

- Assess the extent and severity of the presence of pesticides and nitrate in drinking water wells nationwide
- Study the relationship of pesticide use and groundwater vulnerability to the presence of pesticides and nitrates in wells

The survey was not designed to provide a representative assessment of the presence of pesticides or nitrates for specific local areas, counties, or states, nor did it assess the presence of agrichemicals in surface or groundwater (USEPA 1990).

According to the survey, about 10.4 percent of the nation's CWS (community water system) wells and about 4.2 percent of the nation's rural domestic wells are estimated to contain at least one pesticide above the minimum reporting limits—most commonly nitrate, atrazine, and the acid metabolites of DCPA—sometimes exceeding EPA's MCL and lifetime health advisory level of 10 milligrams per liter for nitrate. The leading factors associated with contamination of drinking water wells, according to the EPA survey, can be found in Table 16.1.

Another survey, the National Alachlor Well Water Survey (Holden et al. 1992), estimates that less than 1 percent of private domestic wells in the study area have detectable levels of alachlor, metolachlor, and simazine; however, atrazine detections were nearly 12 percent and nitrate was detected in more than 50 percent of the wells. Alachlor use areas, groundwater vulnerability to pesticide contamination, and peanut cropping were factors that determined survey areas. The half-life and likelihood of use of alachlor, cyanazine, atrazine, simazine, and metolachlor support the frequency of their occurrence in this survey.

Nitrates are detected throughout the expanded Great Plains (Figure 16.6). Spalding and Exner (1991) point out that the data in the U.S. Geological Survey's (USGS) Water Storage and Retrieval System are useful to identify broad regions with nitrate problems even though the data are not temporally or regionally representative. Texas, Oklahoma, Kansas, Nebraska, South Dakota, eastern Colorado, and western and northeastern Iowa have areas where nitrate-N concentrations in the groundwater exceeded 3 milligrams per liter (Madison and Brunett 1985). Where nitrate levels near the 10 milligrams-per-liter MCL, both point and nonpoint source contamination are suspected (Holden et al. 1992).

Exner and others (1985) indicate that increased vulnerability to point source contamination occurs in shallow alluvial wells in river and stream valleys and poorly constructed wells near confined animal operations.

Table 16.1. Leading factors associated with contamination of drinking water wells

Community Well Systems		Rural Domestic Wells	
Pesticide Detections	**Nitrate Detections**	**Pesticide Detections**	**Nitrate Detections**
Fertilized pasture and rangeland	Fertilized pasture and rangeland	Market value of crops	Well age
Well depth	Monthly precipitation	Number of beef cows	Monthly precipitation
Other operating wells	Well water pH		Well water pH
	Property farmed		Drainage ditch within ½ mile
			Fertilized pasture and hay land
	Nitrate Concentrations		**Nitrate Concentrations**
	Monthly precipitation		Well depth
	Well water conductivity		Market value of crops
	Total nitrogen fertilizer sales		Surface water with ½ mile
	Well depth		DRASTIC topography score
	Palmer drought index score		Total nitrogen fertilizer sales
	Market value of crops		

Source: Adapted from USEPA 1992.

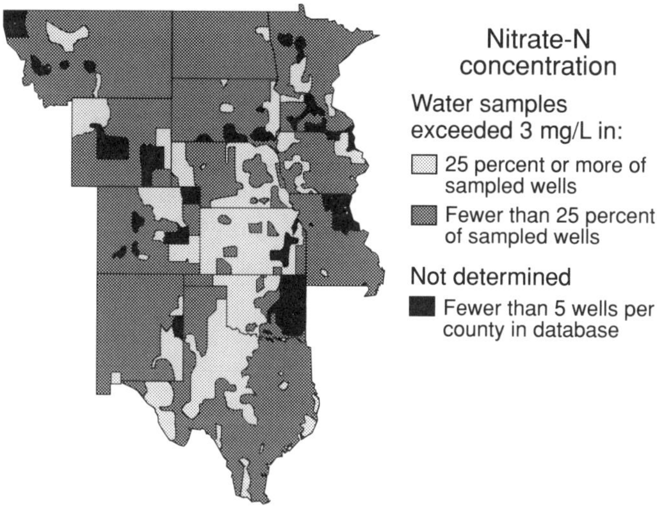

Source: Madison and Brunett 1985.

Figure 16.6. Distribution of nitrate-N concentrations in groundwater

State Surveys and Aquifer Information

The presence and level of materials found in aquifers are related to climate, site characteristics, and land use. The depth from the surface to the aquifer, the type of material overlying the aquifer, the slope of the land surface, and the distance from the point of contamination to the aquifer are site characteristics influencing whether or not chemicals will reach an aquifer (Moore 1988). Cropland, animal operations, and residential areas are some land uses affecting chemical transport to aquifers. Since water drives contaminants to the groundwater, the distribution of precipitation is extremely important in delivery. Precipitation is implicit in many surveys; nonetheless, independent storm intensities and durations are the ultimate determining factors in pollutant delivery to groundwater, especially on sensitive sites.

State surveys indicate some general patterns to contamination in localized areas. Nitrate detections in Iowa were greater in wells more shallow than 50 feet (Kross et al. 1990; Hallberg and Hoyer 1982). Nitrate detections in the Anoka Sand Plains of west central Minnesota were associated with certain land uses: livestock feedlots near wells, cultivated irrigated cropland near wells, and cultivated nonirrigated cropland near wells (Komor and Anderson 1992).

Individual septic systems and irrigation ditches that traverse permeable sand and gravel terraces overlying South Dakota's Rapid Creek aquifer near the Black Hills contribute to the oftentimes poor water quality in the aquifer. Nitrates, total dissolved solids, and fecal coliform are found in parts of the aquifer. The aquifer is less than five feet from the surface in many locations (USDA 1990).

In eastern South Dakota, localized nitrate contamination in the shallow Big Sioux aquifer was found to be primarily from point sources such as leaky fertilizer tanks at distributorships, sewage lagoons, landfills, animal corrals, and possibly septic tanks. A few wells near cultivated fields contained nitrates that were believed to be nonpoint in origin (Spalding and Exner 1991).

North Dakota is emphasizing the sealing of abandoned wells in areas recharging the Icelandic aquifer because of intensive public water use from this shallow aquifer in the northeastern part of the state. Highly permeable materials overlay the aquifer in depths from 2 to 24 feet (USDA 1991).

Elevated nitrate levels in Nebraska's groundwater are, in part, from areas characterized by irrigated corn on well-drained soils less than 50 feet thick (Spalding and Exner 1991). Localized elevated nitrate levels are believed to be from concentrated animal operations, poorly constructed wells, and wells built near abandoned barnyards. Tanner and Steele (1991) have shown that fertilizer leachate contributions alone in some of the shallow alluvial valleys of the glaciated drift region seldom drive aquifer nitrates above the drinking water standard.

In contrast, Gopal (1987) implicated fertilizer leachates, along with industrial wastewater from a fertilizer plant, as sources of nitrate contamination of shallow aquifers underlying coarse-textured soils in the North Canadian River Basin of Oklahoma.

Kreitler and Jones (1975) determined that concentrated feeding operations and terracing of farmland contributed to nitrate contamination exceeding drinking water standards in Runnels County, Texas.

Voluntary and Regulatory Programs, Strategies, and Best Management Practices

Voluntary Action

Several federal agencies and programs have been instituted to encourage voluntary action.

1. The Rural Clean Water Program (RCWP) began in 1992 as a 10-year pilot program designed to assess the effectiveness of agricultural nonpoint source pollution control practices. Those involved in RCWP indicate that voluntary programs and local involvement are necessary to keep the program going. Cost-sharing is necessary to implement expensive best management practices (BMPs), which would otherwise have been cost prohibitive. In order to identify the effects of BMPs used in any project, a moni-

toring strategy must use reliable baseline data, and pre-implementation monitoring is necessary.

So far, one of the most important lessons learned from RCWP is that there may be long lag times from the installation of BMPs to actually realizing water quality improvement. Several projects reported dramatic reductions in phosphorus loading into the lake, yet phosphorus concentration in the lake stayed the same. This is presumably due to the recycling of the phosphorus that is already in the lake.

2. USDA's Water Quality Initiative began in 1990 based on (1) education, technical, and financial assistance; (2) research and development; and (3) database development and evaluation. The objective of the five-year plan is to protect ground and surface water from potential contamination by agricultural chemicals and wastes, especially pesticides and nutrients. Demonstration projects, hydrologic unit area projects, and Agricultural Conservation Program (ACP) special projects were all developed to take positive steps to protect water quality.

3. The Water Quality Incentive Program of the Food, Agriculture, Conservation, and Trade Act of 1990 (FACTA) provides incentive payments to reduce at the source agricultural pollutants, which will enhance and protect surface and groundwater quality. In the Water Quality Incentives Program the producer enters into a three- to five-year agreement to implement specified management practices designed to minimize the generation, emission, or discharge of nonpoint source agricultural pollutants.

4. USGS's Midcontinent Initiative is designed to determine the relationship between natural and anthropogenic factors and the occurrence factors and patterns of atrazine in ground and surface water (USGS 1988). The Midcontinent Initiative supports regional-scale projects that study the transport of herbicides in the atmosphere, surface water drainage system, and the surficial aquifer throughout the Corn Belt states.

5. USGS's National Water Quality Assessment Program (NAWQA) was designed to describe the status and trends in the quality of the nation's ground and surface water resources. This program works in conjunction with the USDA's National Agricultural Statistics Service and the Economic Research Service to develop accurate chemical use information and to compare the use data with herbicide migration through various environmental components.

6. USDA and USGS's Management Systems Evaluation Areas Program (MSEA) resulted from a merger of the 1989 USDA MSEA initiative and part of the USGS Midcontinent Herbicide Initiative. USGS's role in the MSEA projects is to characterize the hydrogeologic framework for the studies and conduct research on the fate and transport of agrichemicals. The 1989 MSEA initiative assessed the basis for public concerns about agricultural pollution and developed a protocol of BMPs to safeguard groundwater in the Midwest from chemical contamination (USDA/USGS/USEPA 1991).

Regulatory Programs

Many western Great Plains states have already imposed regulations on groundwater withdrawal in an effort to reduce drawdown of the aquifers and to raise the water level in the aquifers. In addition, Nebraska has established Special Protection Areas (SPAs), designed by the Nebraska Department of Environmental Quality (NDEQ) (Hutton 1988). Once contamination of a water body has been identified, a natural resource district may request the NDEQ to monitor the groundwater for upward trends of the contaminant or to identify if the concentration exceeds the maximum contaminant level. If either of these two conditions exist, the NDEQ declares the SPA and an action plan is submitted. Penalties can be assessed for noncompliance with the action plan (Ehrman et al. 1990).

Human health and the environment are the focus of the EPA's Pesticides and Groundwater Strategy. If pesticides are identified as having a potential to leach and as having negative effects on human health and the environment, the states are requested to develop a State Management Plan (USEPA 1991).

The EPA actively encourages agencies at all levels to develop partnerships with shared goals and responsibilities. EPA's strategy for protecting the nation's groundwater emphasizes the need for voluntary and regulatory programs to work together to not duplicate efforts.

The National Pollutant Discharge Elimination System Program, established by the EPA, is designed to control point source discharges. In general, states administer this program, which requires a livestock producer to apply for permits when the pollution control agency suspects that the feedlot is causing a pollution problem. Fines are assessed in many states if fish kills are caused by livestock manure polluting the water (Muehling 1992).

Best Management Practices

Water Quality Incentives Program

The ACP Water Quality Incentives Program provides cost-sharing to producers who will implement management practices and systems designed to reduce nonpoint source agricultural pollutants. A resource management plan including management and structural measures to achieve the source reduction of agricultural pollutants will be developed.

Incentive payments under a three- to five-year long-term agreement are given for specified best management practices. Some of the systems related to protecting groundwater include integrated crop management, irrigation water management, waste management, conservation cover, conservation cropping sequence, and well testing.

The integrated crop management system involves managing nutrients from commercial fertilizers and from animal wastes, managing pests through an integrated pest management approach, managing pesticides, using conservation cover, and adjusting cropping sequences to achieve a specific purpose. Incentive payments are not given for structures under this program.

Soil and Pesticide Screening

A first-tier screening procedure for evaluating pesticide losses through the soil profile through leaching and in runoff as adsorbed or in solution is the soil-pesticide interaction screening procedure (Goss and Wauchope 1990). This procedure assumes pesticide loss to a water resource when the pesticide is leached below the root zone or leaves the field boundary in solution or, is adsorbed on sediment suspended in runoff water. Such soil factors as surface horizon thickness, organic matter content of the surface horizon, surface texture, and hydrologic soil group are measured. Pesticide parameters include solubility, soil half-life, and the organic carbon partitioning coefficient. The screening procedure then yields the potential pesticide loss to leaching, the adsorbed loss to runoff, and the solution loss to runoff. An agricultural manager can choose a chemical acording to its rating.

The soil-pesticide interaction procedure has been accepted by the USDA's Soil Conservation Service (SCS) and by the pesticide industry as the best currently available technology to evaluate the relative potential of pesticides to move to groundwater or in surface runoff water (Goss and Wauchope 1990). Although the procedure does not evaluate the actual potential of getting into groundwater, it does indicate the potential to leach below the root zone, where common agricultural practices have little influence. The procedure is commonly found associated with the Pest Management Practice Standard in the Field Office Technical Guide.

Programs, Strategies, and Management Practices to Address Problems

Programs and strategies for maintaining or improving the quality and quantity of our groundwater resources include regulatory approaches and research, investigative, educational, and voluntary programs, either with incentives or without. Perhaps the best strategy to achieve the goals that we have regarding our groundwater resources is through the integration of regulatory, investigative, educational, and voluntary approaches.

Regulatory programs are not the sole solution. The American public is very aware of its environment, and Americans are willing to take action to avoid health hazards from poor quality water or not enough water. Because there will always be flagrant violators, regulations are necessary to enforce sound conservation ethics. Voluntary programs are becoming more popular throughout the states as citizens try to conserve their resources. Citizen monitoring groups are also on the rise. Many farmers are conservation-minded people, even in the absence of cost-sharing programs.

Incentives in farm programs have reshaped the land, the economy, and the way of farming in many instances. Terraces adorn the landscape in the Midwest and reduce the transport of sediment off fields by sheet and rill erosion. Reduced tillage acreage is increasing.

Investigations by state agencies, academic institutions, the USGS, and the EPA have revealed interesting relationships between human activities and the potential for groundwater contamination. These relationships need to be considered in any plan to protect groundwater.

Regulatory programs range from setting standards for drinking water quality to restricting the way a farmer uses a chemical or by denying use of that chemical. Flexibility in regulations or programs is a necessity. Ecosystems do not act like a physical structure; rather, living beings respond to changes in the environment, adapting in a way to preserve themselves or the species.

The microorganisms responsible for particular chemical changes will respond to changes in temperature, pH, moisture, and energy sources. A regulation aimed at one aspect of nutrient or pesticide application without considering the interaction of the elements of the ecosystem is not flexible enough to ensure the livelihood of agriculture. An example is a regulation that prohibits fall application of nitrogen fertilizer based upon nitrate concentrations exceeding the MCL of an important drinking water aquifer. Such restrictive regulation may not keep nitrate from getting to the aquifer at all. A more flexible regulation would be to address the issue from a watershed or aquifer-related viewpoint.

Emphasis on protecting recharge areas from contamination may be the first step in preventing groundwater contamination. Prevention planning should start at a higher level than each individual in the community. The most vulnerable sites should be targeted for more intensive planning and assistance. Nutrient management plans should be written for farmers, ranchers, or urban dwellers and should consider the whole picture rather than one part.

Because nutrient removal is so costly, potential groundwater contamination must be reduced. Flexibility lies in identifying a number of management options, which would allow farmers to alter their current operations according to environmental conditions at different times during the season.

Success in resource planning depends on the partnerships that are built among local people and agencies. Partnerships must be geared toward consensus planning, where each individual has a voice in the planning process. Ultimately, the decision making process may be governed by majority rule. Nevertheless, all individuals in consensus planning bring their ideas before the group. A resident in a small rural town is as important as the farmer or an agency representative. Involving all people in the watershed is a key ingredient for a plan that will ensure that voluntary approaches work. Successful partnerships built upon consensus planning will employ a watershed approach. The degree of success depends on the willingness of the planners to view the entire watershed at once and not as separate tracts of land. Solutions that are lasting have as cornerstones the support of the community in the watershed.

The EPA is adopting the watershed approach in their efforts, as is the SCS. Starting with the big picture at the river basin or watershed level, the planners then work their way down to the farm level, and eventually to individual fields.

The SCS now uses total resource planning, wherein five resources, as well as economic, cultural, and social considerations, are integrated into one plan. The five resources are soil, water, air, plants, and animals. Through an inventory of the land and discussion with the landowner, a conservation management system that addresses all identified resource concerns can be developed.

Then, those involved evaluate the plan's effects on the economic, cultural, or social aspects of the farmer, rancher, or community. Human considerations allow for flexibility in a plan. The likelihood of success is greater if farmers know that the conservation plan will not cost them the farm. The cultures of others are also recognized so that a physical artifact or value of the land to that culture is not destroyed.

Through education, people can develop an understanding that will help them to self-regulate, rather than needing extensive regulations passed by legislation. Programs or strategies must be flexible. Options must be built into them to ensure their just implementation.

Acknowledgments

The authors gratefully acknowledge SCS geologists Marie Marshall, Montana; Carol Reed, North Dakota; Cindy Steele (env. eng.), South Dakota; Mike McCauley and Scott Sumsion, Nebraska; John Littke, Iowa; Don Williams, Missouri; and Gary Kelley, Kansas, for assistance in collecting geologic data. We also appreciate the assistance of Dr. Mary Exner, hydrochemist, University of Nebraska-Lincoln, for identifying sources of information and supplying graphics for groundwater quality; and Tammy Nepple, Janet Bauer, and Brenda Miller for graphics support and text preparation.

References

Aspelin, A. L., A. H. Grube, and V. Kibler. 1991. *Pesticide industry sales and usage: 1989 market estimates.* Washington, D.C.: U.S. Environmental Protection Agency.

Berry, J. T., and N. L. Hargett. 1990. *1990 fertilizer summary data.* Muscle Shoals, Alabama: Tennessee Valley Authority.

Dugan, J., and D. Schild. 1992. *Water-level changes in the High Plains aquifer: Predevelopment to 1990.* Water Resources Investigation Report 91-4165. Lincoln: U.S. Geological Survey.

Ehrman, R. L., M. L. Link, and J. J. Gottula. 1990. Special protection areas: A new nonpoint source management option in Nebraska. *Journal of Soil and Water Conservation* 45(1990):263-264.

Exner, M. E., R. F. Spalding, and C. W. Lindau. 1985. Groundwater contamination and well construction in southeast Nebraska. *Groundwater* 23(1):26-34.

Gopal, B. K. 1987. Investigation of nitrate contamination in shallow ground waters near Woodward, Oklahoma. In *Groundwater quality and agricultural practices,* ed. D. M. Fairchild. Chelsea, Michigan: Lewis Publishers.

Goss, D., and R. D. Wauchope. 1990. The SCS/ARS/CES pesticide properties database: Using it with soils data in a screening procedure. In *Pesticides in the next decade: The challenges ahead,* ed. D. L. Weigmann, proceedings of the Third National Research Conference on Pesticides (Nov. 8-9).

Hallberg, G. R., and B. E. Hoyer. 1982. *Sinkholes, hydrogeology, and groundwater quality in northeast Iowa.* Open-file Report 82-3. Iowa City: Iowa Geological Survey.

Holden, L. R., J. A. Graham, R. W. Whitmore, W. Joseph Alexander, R. W. Pratt, S. K. Liddle, and L. L. Piper. 1992. Results of the national alachlor well water survey. *Environmental Science and Technology* 26:935-943.

Hutton, U. Gale. 1988. Nebraska's special protection area program. In *Planning now for irrigation and drainage in the 21st century: Proceedings of a conference,* ed. Delynn R. Hay. New York: American Society of Civil Engineers.

Komor, S. C., and H. W. Anderson. 1992. Nitrogen isotopes as indicators of nitrate sources in Minnesota sand-plain aquifers. U.S. Geological Survey paper accepted for publication in *Groundwater.*

Kreitler, C. W., and D. C. Jones. 1975. Natural soil nitrate: The cause of the nitrate contamination of groundwater in Runnels County, Texas. *Groundwater* 13(1):53-61.

Kross, B. C., G. R. Hallberg, D. R. Bruner, R. D. Libra, and others. 1990. The Iowa statewide rural well water survey: Water quality data—Initial analysis. Technical Information Series 19. Iowa City: Iowa Department of Natural Resources.

Madison, R. J., and J. O. Brunett. 1985. Overview of the occurrence of nitrate in groundwater of the United States. Water Supply Paper 2275. U.S. Geological Survey.

Moore, John. 1988. SEEPPAGE: A system for early evaluation of the pollution potential of agricultural groundwater environments. Geology Technical Note 5. Chester, Pennsylvania: U.S. Department of Agriculture, Soil Conservation Service, Northeast National Technical Center.

Muehling, A. J. 1992. Comparing livestock waste regulations between Midwestern states. Presentation made at the 1992 International Winter Meeting of the American Society of Agricultural Engineers, Nashville, Tennessee (Dec. 15-18).

Spalding, R. F., and M. E. Exner. 1991. Nitrate contamination in the contiguous United States. In *Nitrate contamination: Exposure, consequences and control,* ed. I. Bogardi and R. D. Kuzelka. NATO ASI Series, Vol. 630. Berlin: Springer-Verlag.

Tanner, D. Q., and G. V. Steele. 1991. *Groundwater quality in the Nemaha natural resources district, southeastern Nebraska.* Water Resources Investigative Report 90-4184. U.S. Geological Survey.

U.S. Department of Agriculture, South Dakota Soil Conservation Service. 1990. *USDA agricultural nonpoint source hydrologic unit area proposal, Lower Rapid Creek Basin, Pennington County, S. Dakota.* Prepared by USDA's Soil Conservation Service and Agricultural Stabilization and Conservation Service; South Dakota's Cooperative Extension Service and Department of Water and Natural Resources; Pennington County Cooperative Extension Service; U.S. Department of the Interior, Geological Survey; Black Hills Resource, Conservation, and Development Area Council; Black Hills Council of Local Governments; and Rapid Valley Conservancy District.

U.S. Department of Agriculture, U.S. Geological Survey, and U.S. Environmental Protection Agency. 1991. Management systems evaluation areas: Program plan (Draft April).

U.S. Department of Agriculture, North Dakota Soil Conservation Service (USDA/NDSCS). 1991. *North Dakota's 1991 proposal for agricultural nonpoint source hydrologic unit area.* Prepared by USDA's Soil Conservation Service, Cooperative Extension Service, and Agricultural Stabilization and Conservation Service; North Dakota Nonpoint Water Quality Task Force; and local sponsors.

U.S. Environmental Protection Agency, Office of Pesticides and Toxic Substances. 1990. *National survey of pesticides in drinking water wells: Phase I report.* Document no. 570/9-90-015. Washington, D.C.: USEPA, Office of Water, Office of Pesticides and Toxic Substances.

———. 1991. *Pesticides and groundwater strategy.* Document no. 21T-1022. Washington, D.C.: USEPA, Office of Pesticides and Toxic Substances.

U.S. Environmental Protection Agency. 1992. *National pesticide survey: Update and summary of Phase II results.* Document no. 570/9-91-021.

U.S. Geological Survey (USGS). 1986. National regional aquifer-system analysis program of the U.S. Geological Survey. USGS Circular 1002. Reston, Va.: U.S. Government Printing Office.

———. 1988. Herbicides in ground and surface water: A midcontinent initiative. Prepared by Michael R. Burkhart, Stephen E. Ragone, E. Michael Thurman, and Charles A. Perry, as part of the Toxic Waste Groundwater Contamination Program. USGS Circular (draft).

Weeks, J., E. Gutentag, F. Heimes, and R. Luckey. 1988. Summary of the High Plains regional aquifer-system analysis in parts of Colorado, Kansas, Nebraska, New Mexico, Oklahoma, South Dakota, Texas, and Wyoming. U.S. Geological Survey, Professional Paper 1400-A. Washington, D.C.: U.S. Government Printing Office.

17. Surface Water: A Critical Resource of the Great Plains

Thomas L. Huntzinger
U.S. Geological Survey

The Great Plains is an area dominated by grassland, cropland, and major river systems that drain the nation's heartland. The hydrologic setting of the area provides a factual foundation for potential water policy and management challenges and how multiple interests may affect management priorities of surface water resources.

This chapter provides technical background necessary for discussion of water resources policy issues within the Great Plains. The discussions, hosted by the U.S. Environmental Protection Agency (USEPA) and U.S. Fish and Wildlife Service, are summarized by state and local water resources representatives as a series of recommendations for water policy priorities and were presented to the governors of the Great Plains states.

The Character of Surface Water

Stream flow characteristics in the Great Plains reflect the climatic transition from western arid and alpine environments of the Rocky Mountains to eastern humid tallgrass prairies and central forests. Intermittent streams in the arid western Great Plains are in contrast to perennial streams in the eastern Great Plains that respond to intense and frequent spring storms. Surface water results from precipitation within a watershed or from shallow groundwater that may or may not originate from precipitation within the watershed. Flooding from alpine and northern High Plains snowmelt frequently occurs in the northern Great Plains.

Sources

Moisture becomes available to the Great Plains by the movement of Pacific, Gulf of Mexico, and Arctic air masses. Movement of moist air concurrently with temperature differences between colliding air masses is the source of precipitation. Pacific air crossing the mountains usually has a drying effect in the fall and results in warming trends

Stanley R. Johnson and Aziz Bouzaher (ed.), *Conservation of Great Plains Ecosystems*, 253-273.

during the winter. Gulf moisture dominates in the summer, and the convergence of this moist Gulf air with cooler Arctic air causes regional precipitation in the spring or winter. The convergence of moist Gulf air and Arctic air is the cause of widespread winter rainfall in the southern Great Plains. The general patterns of these air masses result in a more arid climate (16 versus 36 inches) (Olson 1985) and less runoff (1 versus 10 inches) (Gebert et al. 1986) in the western than in the eastern Great Plains.

Snowpacks typically are 10 to 40 inches deep in the northern Great Plains (Rabenhorst 1981), but these accumulations provide important moisture for the more arid regions. Snowpack in the Rocky Mountains also is important to the Great Plains because snowmelt runoff supplies water for agriculture along the larger river systems during the growing season.

Infiltration of precipitation to the shallow groundwater system can provide the only source of stream flow when no surface runoff occurs. Infiltration is extensive, and small amounts of runoff occur in some sandy regions, such as western Nebraska, where most stream flow is from shallow groundwater. Streams connected only to local or seasonal groundwater usually have quite variable, intermittent flows. Stream flow derived from groundwater may have significantly different temperature and water quality characteristics than stream flow from runoff.

Extremes

Surface water issues commonly are related to management of hydrologic extremes such as floods and droughts. Stream flows that greatly exceed normal conditions are by definition infrequent at a specific site but may be a common occurrence elsewhere, such as in a state or region. Extremes in surface water flow, often destructive socially and economically, also may end or begin critical habitats and biological communities.

Floods result from a combination of soil-moisture conditions, topographic relief, and precipitation intensity. Therefore, floods are affected by both geographic and climatic factors. Geographically, flooding occurs throughout the Great Plains, but occurs more frequently in the areas with larger average precipitation (the eastern part). Other reasons for floods include intense spring rainfall on saturated soils and snowpack that melts over frozen ground, particularly when rain occurs during the snowmelt process in the spring. Snowmelt in alpine areas outside the Great Plains also may cause flooding in large rivers of the Plains.

Flood occurrence is expressed in frequency or probability. Recurrence intervals shown in Table 17.1 are the standard expression of flood occurrence. The probability of a flood with 100-year recurrence interval occurring during a 30-year mortgage is about 25 percent. Every state except North Dakota experienced at least one notable flood during the 1980s, and many Great Plains states experienced two or more (USGS 1991).

Table 17.1. Probability that one or more floods or droughts equal to or more extreme than one of a given recurrence interval will occur during time periods of various lengths

Recurrence interval (years)	Number of years				
	5	10	50	100	500
			percent probability		
2	97	99.9	*	*	*
10	41	65	99.5	*	*
50	10	18	64	87	*
100	5	10	39	63	99.3

Source: USGS 1991.
*Probability greater than 99.9 percent but less than 100 percent.

Droughts result from extended periods of low precipitation, usually accompanied by high temperatures and evapotranspiration. Droughts usually are multiyear in duration and take an extended period to develop. They may have periods of wet conditions, even flooding, within the prolonged dry period, but precipitation occurrences are of smaller magnitude and stream discharge is less than the long-term mean. Nearly every state within the Great Plains experienced notable droughts during the 1980s (USGS 1991).

Water Quality

Since 1972, the U.S. Geological Survey (USGS) has been collecting and analyzing data from a network of sampling stations, called the National Stream Quality Accounting Network (NASQAN) (USGS 1990). Many agencies and organizations collect water quality data, but NASQAN data summaries are readily available on a national and regional scale. Selected chemical constituents that are significant factors for public water supply and aquatic life serve as useful indicators of water quality conditions in the Great Plains. These constituents, commonly related to nonpoint sources of contamination, include dissolved solids, nutrients such as phosphorus and nitrate, pesticides, and sediment.

Major inorganic compounds such as sodium, calcium, magnesium, carbonate, bicarbonate, chloride, sulfate, and other ions make up the dissolved solids found in surface water (Figure 17.1). The presence of these compounds in streams usually results from the natural dissolution of rocks or from human activities, such as water treatment plants and runoff from urban and industrial areas. Irrigation return flows in agricultural areas, such as those along the Arkansas River, also may contribute dissolved solids to streams (Stoner 1985). Public water supplies containing concentrations of dissolved solids more than 500 milligrams per liter are considered impaired by the USEPA. Irrigation water containing concentrations of more than 700 milligrams per liter may require special management, and water containing concentrations of more than 2,000 milligrams per liter is considered unsuitable for irrigation (National Research Council 1973).

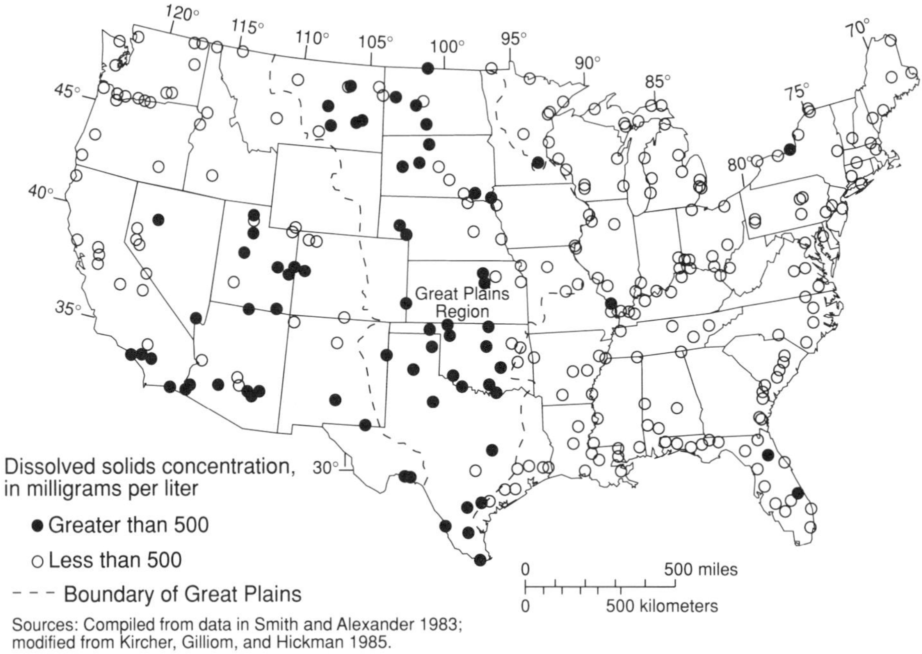

Figure 17.1. Mean annual concentrations of dissolved solids in water from U.S. Geological Survey National Stream Quality Accounting Network stations in the conterminous United States, 1975–81

Phosphorus and nitrate are critical to the life-sustaining potential of surface water. Large concentrations of these nutrients, however, stimulate production of aquatic plants and disturb the balance of the ecosystem. Elevated concentrations of phosphorus result in abnormal aquatic plant growth and a depletion of oxygen in streams. Aquatic habitat is impaired when concentrations of phosphorus are more than 0.05 milligrams per liter in lakes or reservoirs and 0.10 milligrams per liter in streams. Nitrate concentrations larger than 10 milligrams per liter are considered a health risk in public water supplies (USEPA 1992). Concentrations of phosphorus in water from NASQAN stations in the Great Plains for 1975-81 are shown in Figure 17.2. Elevated concentrations of nitrate

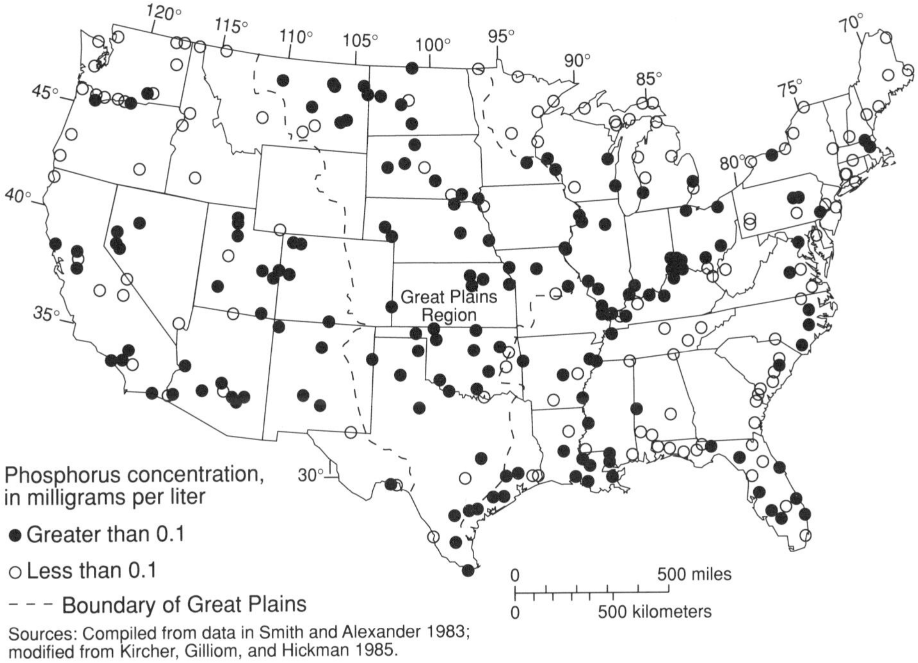

Figure 17.2. Mean annual concentrations of phosphorus at U.S. Geological Survey National Stream Quality Accounting Network stations in the conterminous United States, 1975–81

in some stream reaches may result from fertilizers in runoff, treatment plant effluents, and livestock waste.

The extensive use of pesticides in agriculture has increased production, which benefits the economy of the Great Plains. However, there are concerns about potential effects of pesticides on humans and aquatic organisms. More than 2,500 water samples from about 175 U.S. sites were cooperatively collected by the USGS (1985) and the USEPA, in order to measure 18 organochlorine and organophosphate insecticides. Two of the measured insecticides, lindane and diazinon, were detected in 1.1 to 1.2 percent of the samples. All other insecticides were found in only 0.4 percent or fewer of the samples.

Atrazine was detected in 4.8 percent of the more than 1,300 triazine herbicide samples collected from about 140 U.S. sites. Of the more than 1,700 chlorophenoxy (2,4,5-T, silvex) herbicide samples collected at 180 sites, only 0.2 percent or fewer had detectable concentrations. Some organochlorine insecticides (dieldrin, DDD, DDE) were detected in more than 10 percent of the 1,000-plus bed material samples collected.

A regional study done by Goolsby and others (1991) documented the presence of pesticides (specifically herbicides applied to raw crops) in streams of Iowa, Kansas, Minnesota, Missouri, Nebraska, North Dakota, and South Dakota. In the spring and summer, concentrations of several herbicides exceeded USEPA's maximum contaminant levels in about one-half of the streams sampled. Atrazine, a triazine herbicide used on corn and soybeans, was detected most often and in the largest concentrations.

Sediment, a primary substance present in the rivers of the Great Plains, is a product of natural geologic processes of erosion that transport solid material to the ocean. Human activities such as farming, surface mining, and urban construction often elevate the quantities of sediment in streams. Transport of sediment in streams can result in depositions in stream channels that raise the elevation of the stream, decrease its capacity to convey water, and alter its habitat. Stream bed scour and bank erosion result in the loss of riparian vegetation, degradation of substrate for aquatic organisms, and destruction of cropland and reservoir capacity. Sediment also carries with it other contaminants sorbed to particles or as part of the suspended material.

In the more arid states of the Great Plains, such as South Dakota, Colorado, New Mexico, and parts of Oklahoma and Texas, sediment is transported when surface water flows over a combination of easily eroded sedimentary rocks with little protective vegetation. In these arid states, precipitation is not sufficient to sustain a vegetative cover adequate to control erosion (Meade and Parker 1985). Intense rainfall on steep slopes causes erosion in the more humid areas of the Great Plains. The potential for sediment transport also increases when land is disturbed by farming and urban development. Average concentrations of suspended sediment in rivers in the United States are shown in Figure 17.3.

Sediment load is the mass of sediment transported in streams; this mass, transported from upland, deposits its load at the Gulf of Mexico and helps sustain the coastal region. However, human efforts to control flooding and to maintain upstream water supplies have altered the sediment load downstream. Reservoir construction on the Missouri River in North and South Dakota, for example, has virtually stopped the transport of sediment below major storage reservoirs (Meade and Parker 1985). Other reservoir systems, such as those on the Platte and Kansas Rivers, also decrease downstream sediment transport. Sediment deposition is expected and planned for in reservoir design, but it is a maintenance concern, especially for those reservoirs used as a water supply. Sediment deposition in upstream reservoirs, in combination with altered flows, affects the character of downstream riparian areas. Reservoir controls on the Platte River in central Nebraska have virtually eliminated frequent high flows that transported sediment and bed materials and removed established vegetation from the flood plain. The absence of these frequent high flows has caused woody vegetation to encroach on open sandbar environments, and thus, has affected the ecology of the Platte River (Williams and Wolman 1984).

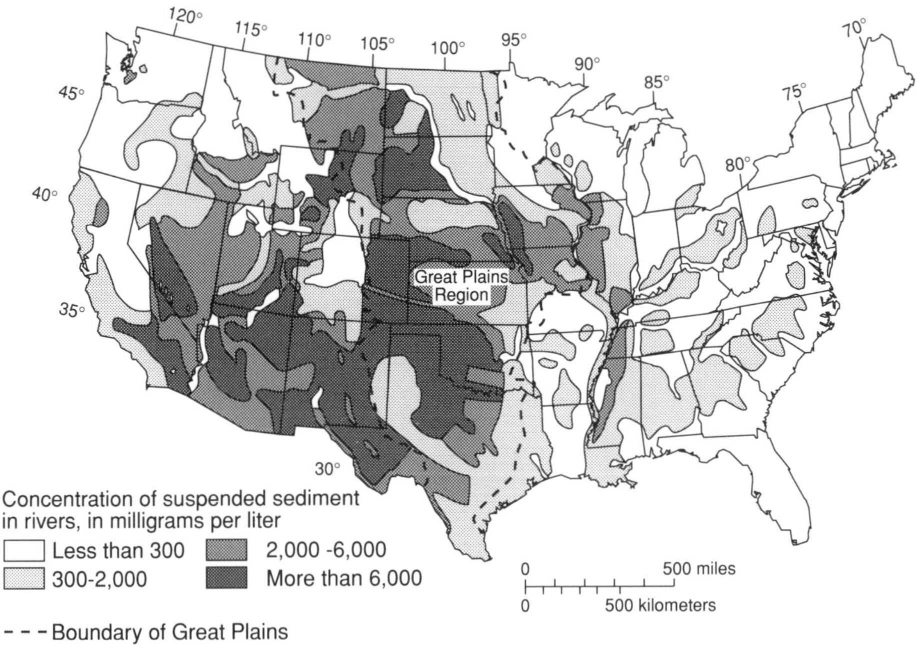

Sources: Modified from Rainwater 1962, plate 3; sediment discharge data compiled by R.S. Parker and R.H. Meade from files of the U.S. Geological Survey, U.S. Army Corps of Engineers, and International Boundary Commission modified from Meade and Parker 1985.

Figure 17.3. Average concentration of suspended sediment in rivers of the conterminous United States

Water Use

Severity of flooding, droughts, and water quality conditions are not always dependent on hydrologic conditions. Deficiencies in supply, as defined by the important or critical demands relative to water availability, often are more reliable measures of water resources priorities. Water demands for various uses in Great Plains states are included in Table 17.2. It should be recognized that only the Dakotas, Nebraska, Kansas, and Iowa are entirely in the Great Plains. The remaining states are only partially in the Great Plains, and water use may not be as uniformly distributed over these states. The percentage of total withdrawal of surface water may be much larger in the mountainous areas than in the Plains areas where groundwater may be used extensively. In-stream uses, such as wildlife habitat, power generation, recreation, and navigation, usually do not diminish stream flows. The schematic in Figure 17.4 shows examples of in-stream uses and their possible seasonal requirements. In-stream uses often have a significant ecological importance in conjunction with other uses, such as power and navigation. Also, water quality conditions generally are important to water use because of potential effects on biological systems both for consumption and habitat.

Table 17.2. Summary by state of freshwater withdrawals for source and use categories in the Great Plains, 1985

		Source					
		Surface Water		Groundwater		Use—Public Supply	
State	Total State Withdrawals	Withdrawals	State Total	Withdrawals	State Total	Withdrawals	State Total
	billion gal./day		percent	billon gal./day	percent	billion gal./day	percent
Colorado	13.500	11,200	83	2.310	17	.737	5
Iowa	2.760	2,090	76	.671	24	.350	13
Kansas	5.670	866	15	4.800	85	.316	6
Minnesota	2.840	2,150	76	.685	24	.473	17
Missouri	6.110	5,470	90	.640	10	.645	11
Montana	8.650	8,450	98	.203	2	.158	2
Nebraska	10.000	4,450	45	5.590	56	.248	2
New Mexico	3.290	1,780	54	1.510	46	.226	7
North Dakota	1.170	1,040	89	.127	11	.069	6
Oklahoma	1.280	707	55	.568	44	.521	41
South Dakota	.674	425	63	.249	37	.080	12
Texas	20.100	12,900	64	7.180	36	2.990	15
Wyoming	6.200	5,700	92	.504	8	.098	2

Table 17.2. Continued

State	Domestic and Commercial		Industrial and Mining		Thermoelectric Power		Agricultural	
	Withdrawals and Deliveries	State Total	Withdrawals and Deliveries	State Total	Withdrawals and Deliveries	State Total	Withdrawals and Deliveries	State Total
	billion gal./day	percent	billion gal./day	percent	billion gal./day	percent	billon gal./day	percent
Colorado	.731	5	.197	1	.123	1	12.500	93
Iowa	.410	15	.301	11	1.810	66	.239	9
Kansas	.316	6	.135	2	.416	7	4.800	85
Minnesota	.583	21	.504	18	1.480	52	.277	10
Missouri	.583	10	.249	4	4.930	81	.347	6
Montana	.173	2	.061	1	.067	1	8.350	97
Nebraska	.224	2	.216	2	2.210	22	7.380	74
New Mexico	.269	8	.085	3	.059	2	2.870	87
North Dakota	.082	7	.015	1	.892	76	.176	15
Oklahoma	.372	29	.317	25	.136	11	.450	35
South Dakota	.109	16	.051	8	.072	1	.507	75
Texas	2.810	14	1.390	7	7.480	37	8.380	42
Wyoming	.123	2	.0165	3	.0238	4	5.670	91

Source: USGS 1990.

Notes: Withdrawal data may be rounded to three significant numbers and might not add to totals because of independent rounding. Percentage of total state withdrawals is calculated from unrounded numbers.

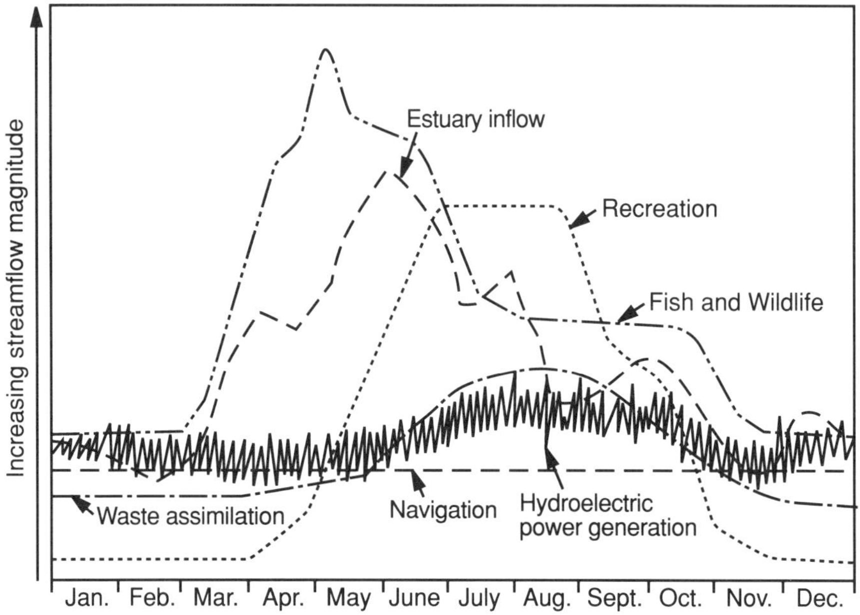

Source: Modified from U.S. Water Resources Council 1978.

Figure 17.4. Schematic of the nature of stream flow requirements for in-stream uses throughout a calendar year

Lakes and Reservoirs

Inland water bodies in the Great Plains contribute significantly to the public water supply and to wildlife habitat. Lakes and reservoirs that make up these inland water sources are of various origins and characteristics. Lakes in the Great Plains originate from geologic processes. The advance and retreat of massive continental glaciers during the Pleistocene Epoch left surficial features and debris that have impounded water to form lakes. Wind-eroded depressions and interdune areas have also formed lakes that are permanent if they receive sufficient runoff or are below the local water table. Lakes also occur in depressions where limestone has dissolved and the surface has slumped and filled with water from precipitation and groundwater. Erosion and accretion in active stream channels form lakes when meanders are cut off, creating oxbow lakes, or when new channels are formed and old ones are abandoned in braided streams. Constructed impoundments or reservoirs occur in all areas of the Great Plains, but they become the dominant water body in the more arid areas. Reservoirs have a combination of riverine and lake characteristics that make them unique systems.

The permanence and health of lakes and reservoirs are dependent on the relative inflows versus the outflows and the presence of oxygen and energy (Wetzel 1983). Precipitation on the water surface, runoff from the surrounding area, and inflows from shallow groundwater are inflows to lakes and reservoirs. Outflows include evaporation and seepage to ground and surface water (either natural drainage or spills from a reservoir). Relative magnitudes of the inflows and outflows may cause lakes to be intermittent as evaporation and seepage exceed seasonal precipitation. In arid areas, water bodies have large fluctuations in depth and shoreline because evaporation rates are large relative to the inflow. Also, lakes that have groundwater inflow have different characteristics than lakes with consistent seepage to groundwater, which have shallow depths and high water temperatures in seasons that are dry and hot, resulting in stress or loss of aquatic organisms.

Partial penetration of sunlight and the retention of heat create temperature gradients in lakes. Warm water is less dense than cooler water and may cause the lake or reservoir to stratify in summer into layers of progressively cooler water from the surface to the bottom. When seasonal temperature changes occur in the fall, cool water at the surface settles, and warmer water rises from the bottom resulting in mixing and circulating of oxygen and nutrients. Thermal stratification may not occur in shallow lakes where sunlight heats the water to the bottom.

Reservoirs may have more complex heat distribution than other lakes because of currents or flow from incoming rivers. Temperature, photosynthesis, and biological activity affect the oxygen content of the water. Oxygen solubility in cold water is significantly greater than it is in warm water. Photosynthesis of aquatic plants generates oxygen that is consumed by bacteria during organic matter consumption, particularly on bottom sediments. The combination of high temperatures and decomposition of extensive amounts of organic matter will stress the aquatic system by disturbing the balance of oxygen in the lake that is critical to the survival of aquatic biota (Wetzel 1983).

Chemical constituents can also affect biological processes in lakes. Salinity is the sum of the concentrations of major ions dissolved in water. Sources of salinity in the environment are found in the rock and soil material near the lake or upstream from the reservoir. Evaporation of lake or reservoir water increases the concentration of major ions.

Nutrients such as nitrogen and phosphorus directly affect productivity in lakes. Nitrogen sources are from both the atmosphere and from land surrounding the lake. Atmospheric sources can be significant in unpolluted lake environments where precipitation adds nitrogen to the lake and where aquatic plants such as blue-green algae add nitrogen to the lake by metabolizing (nitrogen fixation) it from the atmosphere (Wetzel 1983). In agricultural and urban areas, nitrogen applied to the land for plant growth and in municipal and industrial effluent provides a source of nitrogen to receiving lakes.

Phosphorus sources are normally found with the sediment inflow to lakes and reservoirs. This sediment decreases the volume of the lakes, and organic material and nutrients such as phosphorus are transported into the lake from surrounding farmland. Large concentrations of phosphorus usually result in increased productivity of algae and other microorganisms that may deplete oxygen supplies for other organisms in the lake.

In the northern Great Plains, many shallow lakes in glaciated areas and in the Sand Hills of Nebraska (Frey 1963) are supplied by groundwater. Others, such as those in North Dakota, are saline, and many go dry seasonally or during sustained dry periods. Prairie lakes in eastern South Dakota and western Minnesota are less saline than those in North Dakota. The deeper lakes in this area support varied fish populations such as northern pike and bass. The shallower lakes produce bullheads, white suckers, and carp. Large reservoirs constructed on the Missouri River in the northern Great Plains have transformed the river into a continuous series of large impoundments. The drastic changes from a river to lake habitat on the Missouri River have threatened many aquatic species such as the pallid sturgeon (Henry and Ruelle 1992). Many reservoirs also provide water supply and recreation to urban areas, such as those in the Kansas River Basin in eastern Kansas and the Trinity River Basin in Texas.

Water Management

Water management addresses issues related to the needs for water and the protection of life and property from adverse hydrologic conditions. Some management activities involve structural alternatives, and some are nonstructural ones functioning within the natural hydrologic system. Management of water resources occurs at many levels from federal programs to state, local, and individual responsibilities, all of which require cooperation among many levels of government and involve decisions that cross state and other political boundaries.

Flood Control

Flooding, as one of the most destructive natural phenomena in the United States, has been a major challenge for the federal government, resulting in numerous programs and commitments. Flood control structures, such as dams and levees, have been constructed to decrease flooding and flood damage. The U.S. Army Corps of Engineers estimates their flood control structures prevented about $12 billion in damages from 1979 until 1989 (Wingerd and Ming 1991). Rural and urban areas have used federal and local programs to construct and manage small structures for flood control on farms and in urban areas. Urban planning often includes flood detention or retention measures to offset potential increases in flooding caused by urban development.

Nonstructural measures, such as flood plain planning and management, often are used as economical alternatives to structural measures. State and local governments have been encouraged to develop planning and management policies related to urban development in and near flood plains. These policies seek constructive uses for flood plains that enhance traditional urban activities and promote new recreational and educational interests for urban life, such as natural habitat and wildlife preservation.

Drought-Contingency Planning and Implementation

A drought in the context of water management is a water shortage that seriously affects the established economy and lifestyle of the area (Walker et al. 1991). Therefore, a water shortage that has an adverse effect on agricultural productivity, for example, or on the critical threshold of a public water supply may be a more important definition of drought than merely a comparison of current to long-term precipitation or stream flow. Because of the various definitions of drought, management policies may not be specific or focused, and water management related to droughts tends to be more reactive than proactive. Alternatives often are no more than mandates to decrease water use during shortages. Consistent drought-contingency strategies are implemented primarily as disaster aid after a drought becomes a crisis (Walker et al. 1991).

Combining flood control and public water supply often can help moderate the severity of droughts. Water may be stored in flood control reservoirs for water supply. Some refined water-marketing concepts have been implemented in Kansas to utilize flood control storage in federal reservoirs (Kansas Water Office 1993). However, without proper safeguards, water management policies may allow demands on stored flood water to escalate without regard to adequate drought protection. The traditional system of water rights in the western states provides a formal water allocation based on prior appropriation that ensures a consistent priority on available supplies. However, municipalities may develop a more refined list of priority uses within the community's main water right, so that the most valuable demands are met at the expense of the others during droughts. Water management methods used by individual states during droughts are summarized in Figure 17.5.

Power Generation

Power generation sometimes demands large volumes of stream flow, and water storage and head stabilization structures constructed on streams to divert water or to use hydraulic gradients may alter natural flow patterns. However, more than 95 percent of the water diverted for power generation returns to surface water bodies (USGS 1990).

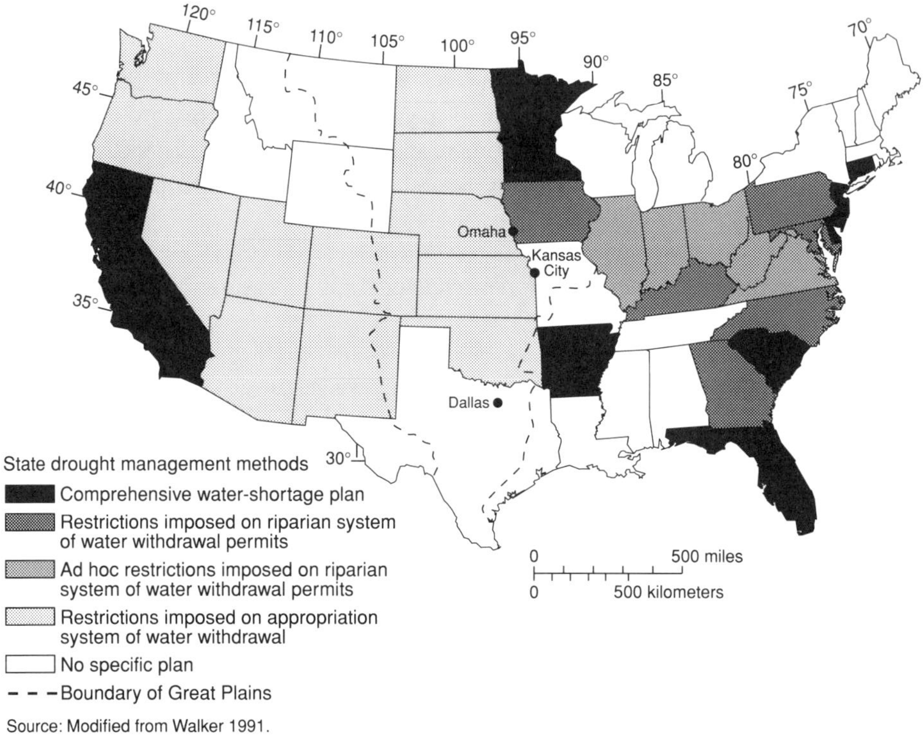

Figure 17.5. State-level methods for managing water resources during droughts

Major changes in aquatic and riparian habitats occur if there is a long distance between the diversion and return-flow points. In addition, water quality conditions may become critical to aquatic life because of less water upstream for dilution.

Water Quality Administration

Water quality has become a major criterion in water resources management. More programs exist within agencies at all levels of public administration than can be described in this chapter. However, some examples of national water quality efforts that relate to issues in the Great Plains are presented here.

Urban Storm Runoff

In November 1990, the USEPA required cities with more than 100,000 population to control pollutants in urban storm runoff. These communities are expected to ensure that their storm runoff meets water quality standards established by the USEPA. The USEPA is also requiring National Pollution Discharge Elimination System (NPDES) permits for large urban storm water discharges to protect receiving streams from contamination. Urban areas that are subject to these storm runoff requirements include municipal streets, catch basins, constructed channels, and storm drains that discharge into receiving streams (Oakley and Forrest 1991). The NPDES permits require 173 cities and 47 counties in the United States to begin applying for permits. Under these guidelines, communities such as Dallas, Texas, and Omaha, Nebraska, have installed water quality monitoring stations operated by the USGS to assist in the design of treatment processes. The permit regulations also pertain to runoff from specific industrial areas, such as treatment plants with effluent discharges larger than one million gallons per day, transportation facilities, landfills accepting industrial waste, recycling centers, and construction sites larger than five acres (Oakley and Forrest 1991). As a result of these regulations, many communities will need technical information and operational resources to obtain permits to discharge runoff water.

U.S. Department of Agriculture Demonstration Projects

As previously mentioned, conventional farming practices for production agriculture are almost totally dependent on chemical fertilizers and pesticides. An intensive monoculture has resulted from the economic motivation to increase yields and maximize operational efficiency in modern agriculture. Fertilizers and pesticides applied appropriately improve crop yield and decrease weed competition, but nutrients and pesticides remaining in soil and plant material may infiltrate to groundwater, be transported to receiving streams with irrigation return flow, or move with precipitation runoff. Agricultural use of herbicides increased more than five times from 1964 to 1982 (Gilliom 1985).

Farm policies of the U.S. Department of Agriculture recognize the need for agricultural producers to conserve soil and water resources and control agricultural chemicals. In selected hydrologic units or watersheds around the country, producers were provided incentives within existing farm programs to implement chemical management and other conservation practices to help control agricultural chemicals. Several agencies —the Soil Conservation Service, the Agricultural Stabilization and Conservation Service, and the Extension Service—put the highest priority on program assistance and solicit participation by producers. Their efforts are intended to cost effectively decrease pollutant loads from agricultural land. The effects are assessed through monitoring water quality changes in receiving streams and in groundwater.

Surface-Mining Reclamation

Coal, sand, and gravel are commodities most commonly obtained from surface mining operations in the Great Plains. Extraction of such materials from surface or near-surface deposits, however, may result in disturbing the drainage, land cover, and ecosystems. Transport of sediment and other contaminants such as trace metals may also result. The percentage of land in the Great Plains that is being surface mined is small, but the disturbance of the surface within a mined area is intense. Agencies, such as the Office of Surface Mining, with responsibility as stewards of public land and private mining activities attempt to control contamination and disturbance of natural habitat and to assist in mine reclamation projects.

Irrigation-Drainage Assessment

Irrigation return flows from agricultural land in the western United States have affected the water quality of receiving streams. Usually, irrigation return flow increases concentrations of common salts that could impair water supplies. Sometimes the return flow contains toxic trace elements such as selenium. Problems related to the presence of toxic trace elements in some wildlife refuges are described in a U.S. Department of the Interior investigation of wetland areas near federal irrigation projects in the West (Feltz et al. 1991).

Return flows from irrigation in the western Great Plains have been managed primarily to ensure an adequate volume of water and to redistribute the flows (Moore et al. 1990). Water quality problems related to irrigation return flows in the Great Plains usually are not as dramatic as in the wetland areas being studied by the U.S. Department of the Interior (Feltz et al. 1991). However, these assessments have increased the awareness of wetland vulnerability to degradation. Consequently, the management of irrigation water from a water quality perspective has been an increasing concern.

National Water Quality Assessment

Objective water quality management decisions by federal, state, and local agencies rely on hydrologic and environmental facts and scientific assessments. A network of stream flow and water quality monitoring stations in the Great Plains provides basic information on many streams. The USGS operates many of these stations and produces interpretive reports assessing the availability and quality of surface water. Through partnerships with state and local agencies and other federal agencies, priorities are decided for program activities in each state.

An important program is the National Water Quality Assessment, which will assess the ground and surface water quality of hydrologic units throughout the country. The program aims to describe water quality conditions and trends, determine the natural and human factors that affect water quality, and relate water quality information to important policy issues, such as quality of sources of public water supplies and nonpoint sources of contaminants. These objectives are approached at two scales—an assessment of water quality in 60 major hydrologic units and a national synthesis of water quality conditions based on results of the assessments in the 60 units. Intensive assessment activities will be undertaken for five to six years in each of the study units. An accompanying, less intense monitoring activity is planned to follow for four or five years before the original assessment is updated (Leahy et al. 1990).

Challenges of Multiple Interests and Priorities

Progress in the stewardship of water resources requires continually adapting to changing needs and increasing expectations. Examples of multiple interests that may be addressed in the Great Plains are sustainable agricultural production, urban and commercial development, and environmental preservation.

Effects of Urban Effluents on Downstream Uses

The challenges of maintaining urban water supplies in arid regions of the Great Plains are well known. More recently, an equally challenging issue has been managing the quality of urban effluents that must meet other urban, agricultural, or in-stream needs. For example, effluents from the Denver area affected the South Platte River, as described in the National Urban Runoff Program (Ellis et al. 1984). These results indicated that concentrations of toxic metals (lead and zinc) exceeded the standards established by the Colorado Water Quality Control Commission in 1980 in all storm runoff samples, while dry weather flows met the standards most of the time. Concentrations of other metals (copper, iron, and manganese) exceeded the standard more than one-half the time during all flow conditions; concentrations of nitrogen compounds met the standards most of the time. Storm runoff and point source discharge had about the same percentages of total flow (14 to 15 percent) in the South Platte River downstream of Denver, except in May when storm runoff was three times the point source discharge. It was found that decreasing contaminant loads of organic carbon, lead, and zinc in the South Platte River would require decreasing contaminant sources, such as urban runoff, irrigation return flows, and point source discharges other than treatment plants (Ellis et al. 1984).

Effects of Agricultural Effluents on Public Water Supplies

Chemicals applied to agricultural land are another contaminant critical to public water supplies. Nonpoint sources of pesticides in the Kansas and Big Blue River basins in northeastern Kansas and southern Nebraska have affected public water supplies, according to Stamer and Zelt (1992). Since pesticides are applied in the spring, many of the pesticides run off the fields during spring storms. Surface runoff from agricultural land ends up in large federal reservoirs on major tributaries of the Kansas River. These reservoirs, designed for flood control and recreation, are now important sources of water supply for urban areas, such as Topeka and suburban Kansas City. Atrazine, the most extensively used herbicide on cropland in the area, was detected during every month of the year at these sampling sites. Median concentrations of atrazine in water samples collected from June through August exceeded 3.0 micrograms per liter. The USEPA's maximum contaminant level for treated/finished publicly supplied drinking water is an average concentration for the year equal to or exceeding 3.0 micrograms per liter (USEPA 1992). Concentrations of atrazine varied seasonally but were consistent in the outflow of the large reservoirs. Atrazine concentrations in the outflow from Perry Lake, a major federal reservoir in northeastern Kansas, exceeded 3.0 micrograms per liter during each of the four seasons. Conventional water treatment does not effectively remove atrazine from finished water (Stamer and Zelt 1992).

Interstate Interests

The Missouri River and its tributaries, as a vast surface water resource that flows through ten Great Plains states, hold the interest of several states. The U.S. Army Corps of Engineers and the U.S. Bureau of Reclamation have for a long time controlled Missouri River flow, and this control has significantly altered native riverine habitat. The original projects authorized flood control, navigation, irrigation, hydroelectric power, and other consumptive uses. But other uses have since been added, including municipal, industrial, wildlife, and recreation uses; pollution control; and reservoir storage for public water supply (MRBIC 1969). The recreational and wildlife areas now support a popular outdoor sporting industry, and business interests in the lower reaches rely on managed releases of water from upstream to transport goods down the rivers. During droughts (e.g., the late 1980s), the multiple interests were increasingly apparent, because the northern states (Montana and the Dakotas) requested the Corps of Engineers to retain water in the upper Missouri to enhance wildlife and increase recreational value. Businesses and municipalities in the lower Missouri, meanwhile, defend the value of navigation and measured reservoir releases (Harkins 1993). Reservoirs have inundated Indian land, making Indian water rights a significant issue.

Maintaining Critical Habitats and Endangered Species

Environmental interests in the Great Plains have an increasing influence on the priorities for use of surface water. The Platte River, like the Missouri River, has major reservoirs and related water management controls. The North Platte River is also controlled by the U.S. Bureau of Reclamation (FERC 1992). Flow of the North Platte River is controlled by large reservoirs from its headwaters in north central Colorado and southeastern Wyoming to the confluence with the Platte River near North Platte, Nebraska. Irrigation and power generation were the major interests in the Platte River Basin from 1900 to the 1930s when the reservoirs were built. The managed flows in the Platte River accelerated the environmental changes occurring on the Platte River downstream. In the late 1980s, the electric power industry that relied on stored water from the Platte River control projects was scheduled for relicensing by the Federal Energy Regulatory Commission. During the relicensing process, environmental interests contested the value that past operational guidelines had assigned to the protection of critical habitat for endangered species. Irrigation and power generation interests currently are defending their long-term investment in diverting stored Platte River water for their use (FERC 1992).

Conclusions

Surface water is a critical resource in the Great Plains. Major river systems provide water for public use, irrigation, wildlife, recreation, and navigation. Stream flow originates from spring storms, snowmelt, and base flows from shallow groundwater. Water management challenges generally are related to extremes in hydrologic conditions, specifically floods and droughts. Water quality conditions also are a major concern, particularly in relation to urban effluents and agricultural drainage. Major river systems crossing political boundaries present special challenges to those responsible for protecting the many competing interests across these boundaries. Increased environmental interests have greatly influenced management decisions in defense of ecological and recreational uses of surface water.

Ongoing programs are addressing many of the challenges facing the Great Plains region. Flood control structures and the flood insurance program reduce flood damage and provide essential information on flood-prone areas so land near waterways can be managed with more confidence. Federal, state, and local governments are increasing their attention to drought-contingency planning by developing more specific and focused programs for adequate water supplies. Some water management plans have combined flood control structures with storage for public water supplies. An understanding of the interaction between in-stream biological needs and the effects of power generation is developing through discussions of relicensing of power-generating facilities. The importance of protecting the health of receiving streams is reflected in the permit guidance and monitoring provided to communities and counties with large storm water effluent discharges. Programs that are designed to improve the quality of runoff from agricultural lands have been implemented. Farm operators and chemical applicators are taking part in educational and management incentives that will decrease the transport of agricultural chemicals to the streams. Increased communication across state boundaries will provide opportunities to make water management decisions that mutually benefit competing interstate interests.

References

Ellis, S. R. 1984. *Analysis of urban storm-runoff data and the effects on the South Platte River, Denver Metropolitan Area, Colorado.* Water Resources Investigations Report 84-4159. Lakewood, Colorado: U.S. Geological Survey.

Federal Energy Regulatory Commission (FERC). 1992. *Draft environmental impact statement, Kingsley Dam Project and North Platte/Keystone Diversion Dam Project.* Federal Energy Regulatory Commission, Office of Hydropower Licensing.

Feltz, H. R., M. A. Sylvester, and R. A. Engberg. 1991. Reconnaissance investigations of the effects of irrigation drainage on water quality, bottom sediment, and biota in the western United States. In *U.S. Geological Survey toxic substances hydrology program: Proceedings of the technical meeting, Monterey, California, March 11-15, 1991,* comps. G. E. Mallard and D. A. Aronson. Water Resources Investigations Report 91-4034. Reston, Virginia: U.S. Geological Survey.

Frey, D. G. 1963. *Limnology in North America.* Madison: University of Wisconsin Press.

Gebert, W. A., D. J. Graczyk, and W. R. Krug. 1986. *Average annual runoff in the United States, 1951-80.* Open-file Report 85-627. U.S. Geological Survey.

Gilliom, R. J. 1985. Pesticides in rivers of the United States. In *National water summary, 1984: Hydrologic events, selected water-quality trends and groundwater resources.* Water Supply Paper 2275. U.S. Geological Survey.

Goolsby, D. A., E. M. Thurman, and D. W. Kolpin. 1991. Herbicides in streams—Midwestern United States. In *Irrigation and drainage: Proceedings of the 1991 national conference, Honolulu, Hawaii, July 22-26, 1991,* ed. W. F. Ritter. New York: American Society of Civil Engineers.

Harkins, J. F. 1993. Missouri River conflict. Written communication from the former director of the Kansas Water Office.

Henry, C. J., and Richard Ruelle. 1992. *A study of pallid sturgeon and shovelnose sturgeon reproduction.* U.S. Fish and Wildlife Service, Fish and Wildlife Enhancement, Pierre, South Dakota.

Kansas Water Office. 1993. Water marketing program. Kansas Water Office Fact Sheet no. 8 (Jan.).

Kircher, J. E., R. J. Gilliom, and R. E. Hickman. 1985. Loads and concentrations of dissolved solids, phosphorus, and inorganic nitrogen at U.S. Geological Survey National Stream Quality Accounting Network stations. In *National water summary, 1984: Hydrologic events, selected water-quality trends and groundwater resources.* Water Supply Paper 2275. U.S. Geological Survey.

Leahy, P. P., J. S. Rosenshein, and D. S. Knopman. 1990. *Implementation plan for the National Water Quality Assessment Program.* Open-file Report 90-174. U.S. Geological Survey.

Meade, R. H., and R. S. Parker. 1985. Sediment in rivers of the United States. In *National water summary, 1984: Hydrologic events, selected water-quality trends and groundwater resources.* Water Supply Paper 2275. U.S. Geological Survey.

Missouri River Basin Interagency Committee (MRBIC). 1969. *The Missouri River basin comprehensive framework study.* Missouri River Basin Interagency Committee Report 3:221-227.

Moore, M. R., W. M. Crosswhite, and J. E. Hostetler. 1990. Trends and associated factors in offstream water use. In *National water summary, 1987: Hydrologic events and water supply and use.* Water Supply Paper 2350. U.S. Geological Survey.

National Research Council. 1973. *Water quality criteria, 1972: A report of the Committee on National Water Quality Criteria, Environmental Studies Board, National Academy of Sciences, and National Academy of Engineering.* Ecological Research Series EPA/R3/73/033. Washington, D.C.: U.S. Environmental Protection Agency.

Oakley, M. M., and C. L. Forrest. 1991. The clock is ticking to comply with new stormwater regulations. *Water Environment and Technology* March: 51-56.

Olson, D. A. (comp.). 1985. *Average annual precipitation in the continental United States, 1951-80.* National Oceanic and Atmospheric Administration.

Rabenhorst, T. D. 1981. *An atlas of the United States.* Department of Geography, University of Maryland, Baltimore County.

Rainwater, F. H. 1962. Stream composition of the conterminous United States. In *U.S. Geological Survey hydrologic investigations atlas,* HA-61 (2 sheets).

Smith, R. A., and R. B. Alexander. 1983. *A statistical summary of data from the U.S. Geological Survey's National Water Quality Networks.* Open-file Report 83-533. U.S. Geological Survey.

Stamer, J. K., and R. B. Zelt. 1992. Distribution of atrazine and similar nitrogen-containing herbicides, lower Kansas River basin. In *U.S. Geological Survey yearbook, 1991.* Reston, Virginia: U.S. Geological Survey.

Stoner, J. D. 1985. Dissolved solids in the Arkansas River basin. In *National water summary, 1984: Hydrologic events, selected water-quality trends and groundwater resources.* Water Supply Paper 2275. U.S. Geological Survey.

U.S. Environmental Protection Agency (USEPA). 1992. *Drinking water regulations and health advisories* (Nov.). Washington, D.C.: USEPA, Office of Water.

U.S. Geological Survey (USGS). 1985. *National water summary, 1984: Hydrologic events, selected water-quality trends and groundwater resources.* Water Supply Paper 2275. U.S. Geological Survey.

———. 1990. *National water summary 1987: Hydrologic events and water supply and use.* Water Supply Paper 2350. U.S. Geological Survey.

———. 1991. *National water summary, 1988-89: Hydrologic events and floods and droughts.* Water Supply Paper 2375. U.S. Geological Survey.

U.S. Water Resources Council. 1978. *The nation's water resources, 1975-2000: Vol. 1, summary.* Washington, D.C.: U.S. Government Printing Office.

Walker, W. R., M. S. Hrezo, and C. J. Haley. 1991. Management of water resources for drought conditions. In *National water summary, 1988-89: Hydrologic events and floods and droughts.* Water Supply Paper 2375. U.S. Geological Survey.

Wetzel, R. G. 1983. *Limnology* (2nd ed.). Philadelphia: Saunders College Publishing.

Williams, G. P., and M. G. Wolman. 1984. *Downstream effects of dams on alluvial rivers.* Water Supply Paper 1286. U.S. Geological Survey.

Wingerd, David, and T. T. Ming. 1991. Flood and drought functions of the U.S. Army Corps of Engineers. In *National water summary, 1988-89: Hydrologic events and floods and droughts.* Water Supply Paper 2375. U.S. Geological Survey.

18. The Missouri River: A Formula for Ecosystem Change

Donald A. Becker and Richard D. Gorton
U.S. Army Corps of Engineers, Omaha District

Historical Perspective

Plains riverine ecosystems have evolved through the intervention of very extreme climatic, geologic, and biotic events. The Pleistocene Epoch played a pivotal role in creating the present day Missouri River. Northeasterly flowing Plains streams such as the Musselshell, Yellowstone, and Cheyenne were dammed by the advancing glacial ice, and flows were diverted southeasterly along the glacial front, forming the present day upper Missouri River (Flint 1957). Plains river systems were further affected by the mid-Holocene warm and dry period (6,000 to 4,000 B.P.), and throughout postglacial time by shorter, but periodic, dry and wet periods. Extremely large floods, exceeding those of this century, occurred on the Missouri River in 1844 and 1881, before most Euro-American settlement. The severe drought of the 1930s, in conjunction with the Great Depression, led to economic and social instability in the Great Plains, creating a setting for the New Deal of the Roosevelt Administration.

Formula for a Regulated River

The vagaries of the Plains climate, linked with flood protection, electrical energy, and transportation needs, were forces that led to a state-federal government partnership for regulation of the Missouri River. Several flood control, navigation, power, and bank stabilization acts stimulated work by the U.S. Army Corps of Engineers on the river, while the Bureau of Reclamation developed plans for irrigation and power projects in the Missouri Basin. A public dispute developed in the basin over how the river should be harnessed for beneficial purposes, so the Pick-Sloan plan was developed by the two agencies. Under this plan, the Corps focused on development of the main stem, while the Bureau was responsible for hydropower development in the tributaries and irrigation projects throughout the basin. Under the aegis of the Pick-Sloan plan, the Bureau and the Corps then developed a public water resources infrastructure in the Missouri

Stanley R. Johnson and Aziz Bouzaher (ed.), *Conservation of Great Plains Ecosystems*, 275–297.

Basin. The large main stem irrigation projects conceived by the Bureau, however, have not been built (Velehradsky 1988).

Purpose and Formulation

As stated earlier, river and basin development emphasized providing flood control, hydropower, water supply, and navigation. Recreation, water pollution abatement, and fish and wildlife resources received little attention from basin interests during plan formulation. These latter resources have received increased attention in more recent years. Benefits from the plan have exceeded earlier expectations and are estimated to approximate \$1.1 billion annually (U.S. Army Corps of Engineers 1992). The ratio of various benefits in the regulated river are shown below. Not completely expressed and disaggregated in this analysis are fish and wildlife and water quality benefits.

Regulated river formula = FC(2.5) + HYD(6) + WS(1) + NAV(0.75) + REC(0.5)

where: FC = flood control, HYD = hydropower, WS = water supply, NAV = navigation, REC = recreation.

Project Development

Main Stem Dams

The locations of dams constructed along the Missouri's main stem are shown in Figure 18.1. Although the dams were originally intended for upstream flood control and downstream navigation, hydropower became an added purpose later. Built between 1937 and 1963 at a cost of \$1.2 billion, these dams were not operable as an integrated system until 1967.

Reconfiguration of the Middle and Lower River

Reconfiguration included extensive work on the channel as well on overbank areas. Initial work began in the late 1800s in order to improve navigation; extensive work on channel improvement started in 1912 after a 6-foot deep channel was authorized by Congress. Later the Bank Stabilization and Navigation Act of 1945 authorized a 9-foot deep channel along 734 miles from Sioux City to the mouth. Work generally progressed in an upstream direction and was completed in 1981. About 19 miles of the river above Sioux City were also altered between 1946 to 1961; it is known as the Kensler's Bend Project. Flood control levees were authorized along the lower river through the 1936 and 1944 Flood Control Acts; however, only 414 miles of federal

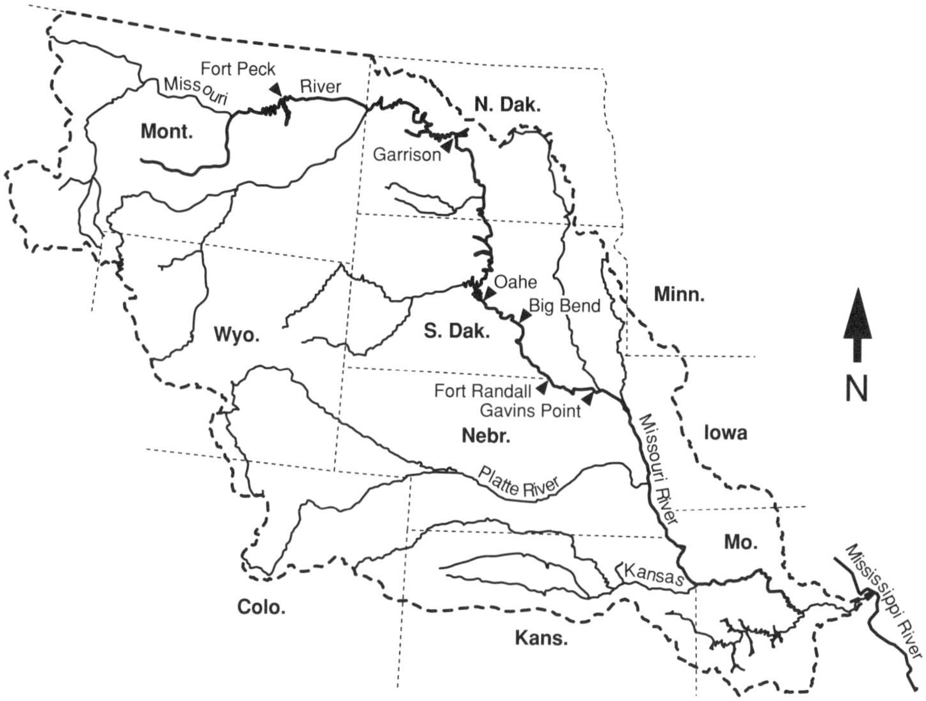

Figure 18.1. Locations of dams along the Missouri River

levees have been actually constructed. The river reach affected by the bank stabilization and navigation project and the main stem levees is shown in Figure 18.2 (Peterson 1988).

The navigation and bank stabilization project converted a wide, partly meandering, partly braided channel into a narrow, deep channel with gentle bendways (Figures 18.3 and 18.4), which shortened the channel by several miles. These alterations were made in order to constrain lateral migration of the river and to use the energy of the flow to scour the channel and maintain a 9-foot navigation depth. Costs of construction were about $460 million (U.S. Army Corps of Engineers 1976).

River Effects and Responses

First-order Changes

First-order changes include direct responses to dam closure and river training work. Impoundment of the upper river transformed much of a shallow, sand-bottomed channel and its broad flood plain into several large prairie lakes. In the upper river (Sioux

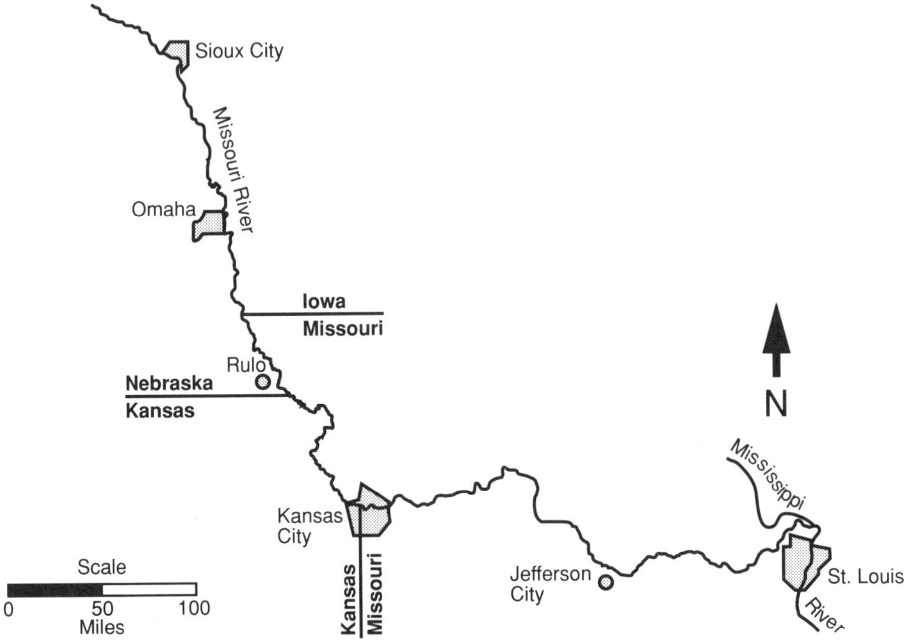

Figure 18.2. Map of the study area—Sioux City, Iowa, to St. Louis, Missouri

City to Fort Peck Dam) about 990,000 acres of bottomland, including 760 miles of channel, were converted to water storage areas. Five river segments, totaling 395 river miles, remained in a more or less free-flowing and unconstrained state. As indicated, hydraulic conditions in the lower river changed drastically; in addition, levees prevented flooding of adjacent flood plain land along several river reaches. In the lower river, about 166,000 acres of river surface area were lost, areas that included sandbars and some vegetated islands (U.S. Army Corps of Engineers 1981).

Second-order Changes

Second-order changes are indirect changes from dam and channel operation.

Flow Pulse and Hydroperiod

Historic runoff patterns in the upper river typically involve two peaks each year: a snowmelt runoff in March-April and a rise in June from combined mountain snowpack and Plains rainstorms. River flows are now regulated by successive reservoir releases. Flow control is high in the interlake reaches and in the reach extending from Gavins

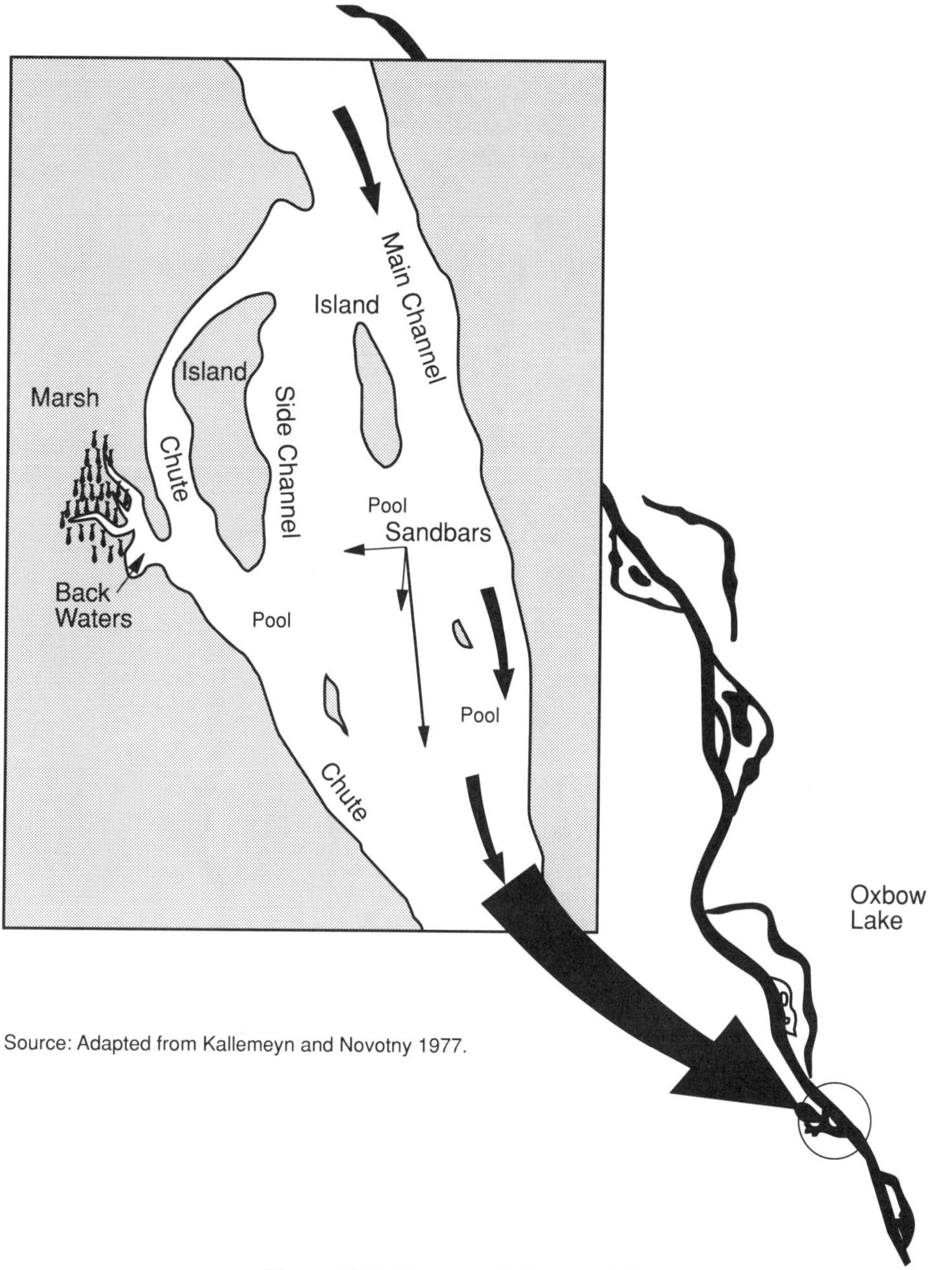

Source: Adapted from Kallemeyn and Novotny 1977.

Figure 18.3. Diagram of characteristics in an unchannelized Missouri River

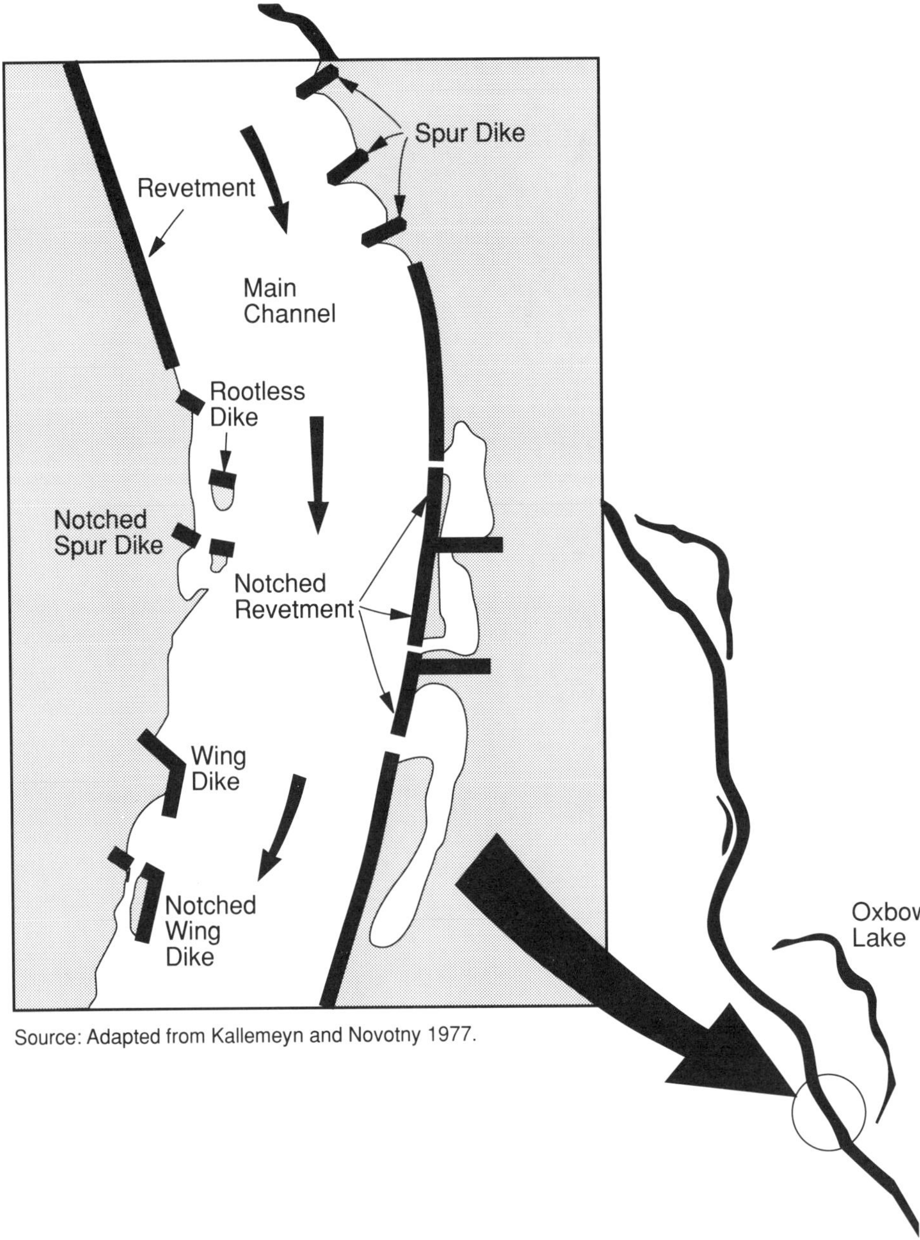

Source: Adapted from Kallemeyn and Novotny 1977.

Figure 18.4. Diagram of structures in the channelized Missouri River

Point Dam to the Platte River. Tributary inflows dampen the effects of regulation downstream, but the effects of regulation often extend to the mouth, especially during low flood periods. Figures 18.5 and 18.6 compare the historical flow pattern with the regulated pattern at seven main stem locations.

Releases from the upstream dams during the snowmelt period tend to be quite low, while the highest flows occur either in winter (Fort Peck and Garrison dams), or in mid- to late summer (Oahe, Fort Randall, and Gavins Point dams). This pattern reflects how the system is operated for flood control, hydropower operations, and navigation. The magnitude of peak flows during spring and early summer are greatly dampened below each dam. Another effect of regulation is to raise mean and low flow levels. Historically, flows under 10,000 cubic feet per second were common in late summer, fall, and winter in the lower river. Under regulation, flows less than 10,000 cubic feet per second rarely occur even during droughts (U.S. Army Corps of Engineers 1993b).

Channel Bed Change

Changes in channel beds have been a problem below Gavins Point Dam to near Omaha over the past 25 years. Because the river channel lacks firm bed controls, clear water releases from the dam tend to pick up sediment from the river bed. As a result, there has been a gradual bed lowering of several feet since the dam was closed. In reaches unconstrained by bank protection, as in the Gavins Point to Ponca reach, channel widening also has occurred. As a result of degradation and widening, the channel can now carry a larger volume of water, and the frequency of overbank flooding is very low. Lowering of the main stem channel has also affected the channel beds of several downstream tributaries (U.S. Army Corps of Engineers 1991a).

Third-order Changes

Third-order changes result from biological and chemical events that are acting on or exerting feedback effects upon the first- and second-order events.

Upper, Free-flowing Reaches

Various studies (Reily and Johnson 1982; USFWS 1987), and general observations (U.S. Army Corps of Engineers 1989, 1993a) indicated alterations in wetland and riparian habitats along river segments below the main stem dams. The reduced levels of flooding and overbank scour and erosion are implicated. While there are no quantitative data available, it is roughly estimated that at least 105,000 acres of bottomland habitat have been affected. The rate of cottonwood regeneration, which requires fresh soil and adequate water for germination and survival, appears to be less than mortality. Over time, stands become open, decadent, and susceptible to invasion by exotic and other species that are less biologically productive.

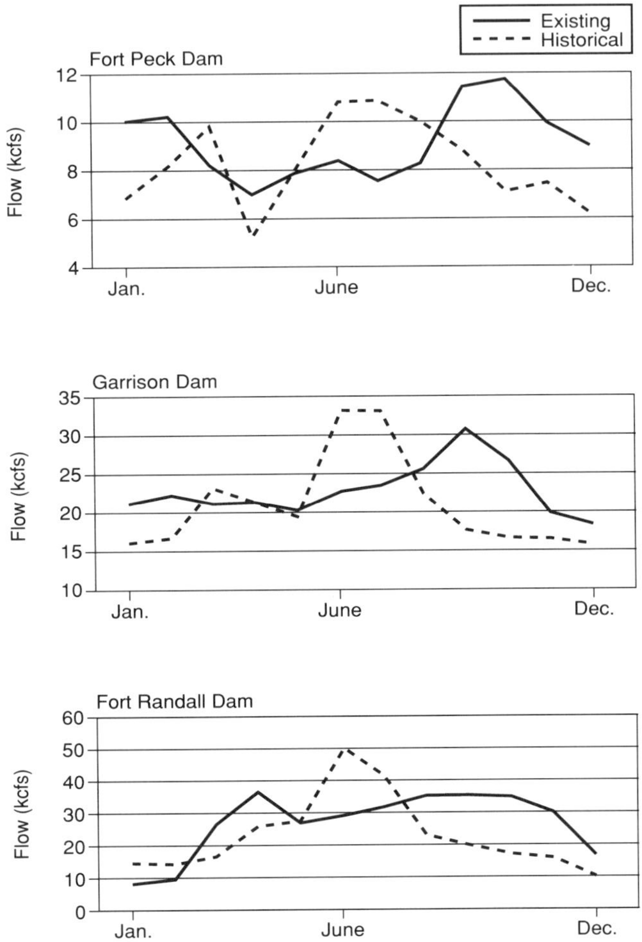

Source: U.S. Army Corps of Engineers 1993b.

Figure 18.5. Existing and historical flows below Fort Peck Dam, Garrison Dam, and Fort Randall Dam as estimated by a model of hydrological conditions

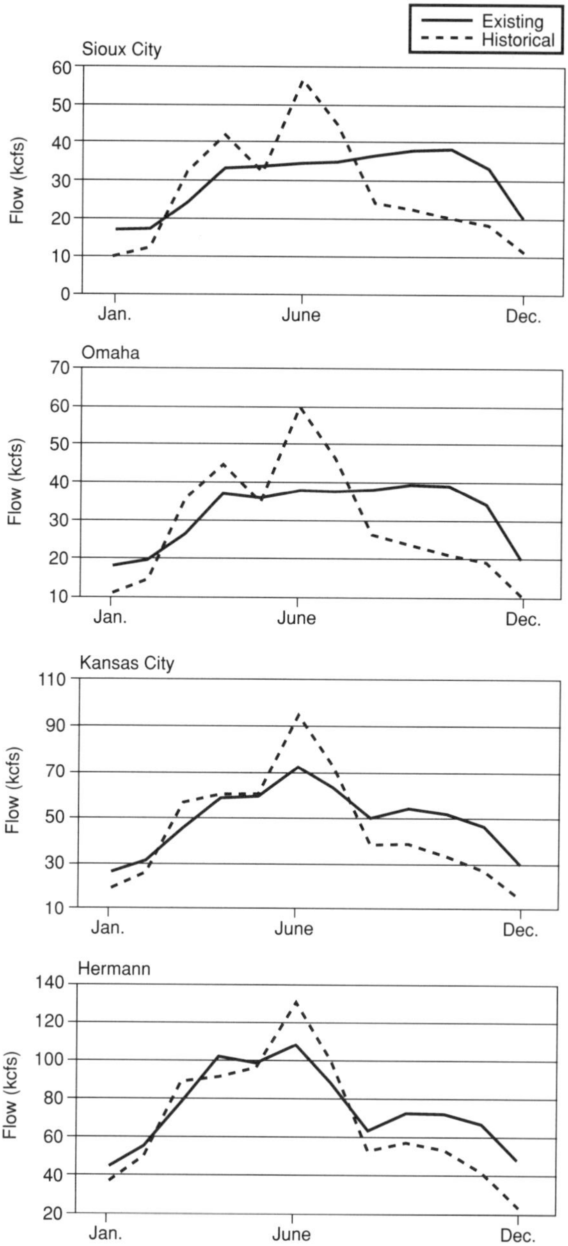

Source: U.S. Army Corps of Engineers 1993b.

Figure 18.6. Existing and historical flows below Sioux City, Omaha, Kansas City, and Hermann as estimated by a model of hydrological conditions

Lower River Meander Belt

Meandering and associated erosion and deposition processes no longer occur along the river, and flooding has been greatly reduced because of flow regulation, bed degradation (above Omaha), and levee construction. As a result, the number and diversity of wetland and riparian communities within this zone of the river have been degraded and continue to decline as a result of surface water loss, reduced groundwater levels, siltation, or actual conversion to cropland and other uses. Also, woodland encroachment into open or marshy wetlands has been a major problem in most areas. The estimated amount of habitat loss is about 309,000 acres. This does not include land lost directly because of channel constriction and realignment (U.S. Army Corps of Engineers 1981).

Aquatic Community Diversity

A continuous riverine system has been broken up into a more diverse hydrosystem, consisting of five large prairie lakes, four interlake river segments, and at least two distinct river segments below the most downstream dam (Gavins Point). The last interlake segment, the Fort Randall reach, has recently been designated as a National Recreation River. The segment from Gavins Point Dam to Ponca was also given this designation in 1978.

Flow releases from each of the lakes vary as do the biological responses in downstream reaches (U.S. Army Corps of Engineers 1989). These responses are also mediated or influenced by the geographic location (ecoregion), geological setting, shoreline configuration, and mean depth. A major biological response to creation of the Great Prairie Lakes along the upper Missouri has been a westward shift of prairie waterfowl migrations. In the lower river there has been an adverse response because aquatic habitat diversity (especially shallow, turbid, warm water areas) has decreased, leading to lower levels of aquatic production. Overall, plankton communities in the system have increased in diversity and abundance, however, because of development in the upper river lakes and subsequent movement downstream with releases from the dams. On the negative side, the dams have blocked spawning migrations of native river fish to suitable aquatic areas.

Water Quality

With stream flow regulation, the potential for water quality degradation has been reduced, since higher flows are now stored for release during low flow periods. Flows during fall and winter now rarely drop below 10,000 cubic feet per second, so mean concentrations of potential pollutants are lower during these periods than under preregulation conditions.

Since sediment trapping efficiency of the upstream reservoirs is high, most of the sediment load in the lower river now comes from downstream tributaries. Sediment concentrations now typically vary from 200 to 1,500 parts per million during the navigation period, although temporary increases during spring and summer flooding typically may exceed 20,000 parts per million (U.S. Army Corps of Engineers 1976).

Light transmissivity in the water column is relatively high in upper river segments: Fort Peck Dam to Milk River, Garrison Dam to Bismarck, Oahe Dam to Lake Sharpe, Fort Randall to Niobrara, and Gavins Point Dam to Sioux City. The euphotic zone extends to the channel bottom in shallower areas, which enables the establishment of aquatic beds and periphyton in the river where suitable substrate, nutrient, and hydraulic conditions allow. Below Sioux City, light transmissivity becomes progressively lower. Low light levels and stressful hydraulic conditions in the channel do not allow aquatic bed vegetation to develop. In the reservoirs, depth of light penetration is generally high with the exception of Lewis and Clark Lake where inflowing river currents, wind turbulence, and active shore erosion tend to keep sediment particles in suspension. Depth of the euphotic zone in the other main stem lakes generally exceeds 10 feet (Benson 1968).

Water quality in the main stem lakes is generally good. Only occasionally do heavy metals, pH, pesticides, phosphate, and fecal coliform bacteria exceed state and federal standards. While the reservoirs tend to be oligotrophic to mesotrophic in production, seasonal algal blooms can occur at Lake Oahe, Fort Peck, and Lake Sakakawea during the summer months. During the mid-1980s, an algal bloom in Lake Sakakawea stretched over 100 miles but had no apparent negative impact on fisheries or recreation (Anderson 1989). At Lewis and Clark Lake, phosphate levels are high enough to create nuisance algal blooms, but higher turbidity levels and reservoir flushing rates apparently arrest aquatic production. Existing water quality data are too limited to establish any eutrophication trends, but there appears to be a correlation between phosphorous content of the reservoirs and phytoplankton production, suggesting that phosphorous may be limiting production in the lower lakes (U.S. Army Corps of Engineers 1993c).

Carbon Fixation

Patterns and levels of carbon flow in regulated rivers are not well defined but are highly influenced by dam operations. In the main stem lakes, due to reduced turbidity levels (Neel et al. 1963), fixation of carbon now occurs within a moderately deep water column occupying a volume of about 6.9 million acre-feet and covering 990,000 acres of river valley. Phytoplankton data suggest that the hydro system is basically autotrophic above Gavins Point Dam, since comparison of pre- and post-impoundment phytoplankton data show that numbers are 10 times higher after impoundment (Cowell 1970; Benson 1968). This tentative conclusion is supported by carbon studies on a large reservoir in Texas where 77 percent of the organic matter was found to originate from primary sources (Lind 1971). Recently, concern has been expressed that the main stem dam system, by interrupting inputs from sediment and detrital sources on the flood plain (Hesse et al. 1989), has actually lowered secondary production in the lower river. However, while localized effects may exist below Fort Randall and Gavins Point Dam, other carbon mitigating factors appear to exist because:

- The release of large quantities of plankton (9,000 metric tons annually) into the river below Gavins Point Dam affects plankton assemblages downstream as far as the Missouri border (Hudson and Cowell 1966; Repsys and Rogers 1982)
- Recovery of sediment and detrital material levels in the lower river takes place (Kallemeyn and Novotny 1977; Slizeski et al. 1982)
- Total organic carbon in the water column is considerably higher than averages for freshwater systems described by Hem (1970)

Species Sensitivities

As a result of habitat alteration, changes in the system have been widespread. In the main stem lakes and tailwaters, the native warm water fish communities have shifted to mixed communities containing cool and cold water species. Species considered to be predators and sight-feeders are also more common than detrital and bottom feeders. As a result, 66 species of fish are found in the reservoirs, and species richness has increased in the entire river system since regulation (Pflieger and Grace 1987). A large and significant sport fishery industry exists in the lakes as well as in tailwater river areas below each of the six main stem dams. While fishing is important in downstream river areas, pressure is less because of numerous factors, including access, species availability, and catchability (U.S. Army Corps of Engineers 1993b).

Several native riverine fish have been declining in abundance as a result of habitat alteration, blockage or access to habitat, overfishing, and hybridization (U.S. Army Corps of Engineers 1993b). Some fish have been listed or proposed for federal listing, including the pallid sturgeon, lake sturgeon, sturgeon chub, sicklefin chub, paddlefish, and blue sucker. The latter two fish may not be "listed" because they are present, at least at low levels, in all of the lakes and in most river reaches. The blue sucker is reportedly increasing downstream of Sioux City (Atchison et al. 1986). Lack of spawning habitat and nursery areas are believed to be primary factors in the decline of the sturgeon, while loss of turbid river habitat may restrict the small chubs (Berry 1988). Turbid habitat found in several Plains streams, such as the Little Missouri, Cheyenne River, White River, and Yellowstone drainage, may provide suitable refuge for these small fish.

The least tern and piping plover are sensitive to regulated conditions since they depend upon open scoured bars for nesting habitat and fledging of young. Loss of sandbar habitat has been widespread along both the upper and lower river within the historic breeding range of these birds (U.S. Army Corps of Engineers 1991b).

Hydraulic conditions in the navigation channel and low amounts of slack-water habitat also limit production of other native riverine fish, especially the blue catfish and buffalo fish. Macroinvertebrate production is also low, being largely confined to slack-water areas or rock-rubble areas along the channel (Kallemeyn and Novotny 1977). Also fish in the navigation channel have access to less zooplankton than fish in the main stem lakes or in the tailwater areas.

Mother Nature's Formula for the Missouri River

Natural energy releases along streams and rivers are typically mild or moderate for long periods of time, but occasionally the energy release rate is explosive and catastrophic. In an unregulated river, the release includes floods and attendant phenomena such as ice jams and channel evulsions. In addition, the river valley is affected by windstorms, blizzards, and ice storms. Late- or early-season drought and fire, as additional natural disturbances along Plains rivers, also direct or influence the nature of the aquatic and riparian community.

The natural formula (NF) along the river includes the following dynamics:

$$NF = PF + ER + AG + SE + BC$$

where: PF = peak flow, ER = erosion, AG = aggradation, SE = seasonality, BC = biologic or chemical activity.

Natural Mitigation on a Regulated System

Great Prairie Lakes

Wave action along the margins of the main stem lakes induces the formation of erosional and depositional features that mimic those found on natural lakes and seashores. Erosion of headlands and littoral drift have resulted in barrier beaches across some of the smaller embayments, partially isolating these from the main lake. Aquatic populations that develop in these protected ponds are different from those found in the main lake. Well defined beaches, often with extensive cobble and rubble, are also developing along parts of Fort Peck, Garrison, Oahe, Big Bend, Sharpe, and Fort Randall reservoirs. These beaches, when inundated, provide spawning areas for certain fish species such as walleye and sauger. Shore birds, including the least tern and piping plover, prefer barren, gravelly shorelines. At Lake Sharpe, beaches are sufficiently high at some locations to prevent shoreline erosion during periods of moderate wave activity, and patches of aquatic bed and emergent vegetation have developed within many of the smaller embayments.

Main Stem Deltas

At the upper ends of reservoirs, water surface slopes decrease, resulting in loss of velocity and settling of sediments in the water column. Channel beds aggrade, resulting in higher water surface profiles and adjacent water-logging of overbank land. These processes, acting together, have created more than 100,000 acres of wetlands in the six main stem lakes. While much of this acreage is characterized by shallow, turbid water or monotypic stands of emergent or scrub-shrub vegetation, about 50 percent has a diverse mix of open water, aquatic bed, emergent, wet meadow, and woody habitat. Large numbers of waterfowl and other aquatic birds use these areas, as well as upland or big game during the severe winters. Fur-bearers and native riverine fish are able to utilize shallow areas, side channels, and "hemi" marshes to meet important habitat requirements (U.S. Army Corps of Engineers 1989). Sensitive fish species, such as the paddlefish, use the delta area at Lake Sakakawea as they move from the lake to spawning grounds in the Yellowstone River. Turbid waters of the delta area also are used by the pallid sturgeon (Harberg 1992) and possibly by the sturgeon chub and sicklefin chub as well. In addition, the delta areas have considerable potential for fishery and waterfowl enhancement projects because of their reliable water supplies and available land.

Biotic Influences

New habitats in the Missouri River system have been exploited by several aquatic species that formerly were either uncommon or absent; several species of plankton have invaded the system (Morris et al. 1968; Repsys and Rogers 1982). New plants observed in the system include hybrid swarms of cattail (narrow-leaf, broadleaf), purple loosestrife, tamarisk, and various pondweeds (U.S. Army Corps of Engineers 1989). While probably always present in riverine backwaters, large mayflies developed very productive populations in the bottoms of several lakes. Fish assemblages changed in both the upper and lower river. The large, open reservoirs favored lake species or adaptable river species such as gizzard shad, emerald shiner, perch, drum, walleye, northern pike, and crappie. Further, state fish and game agencies have introduced several cold water species into the system: rainbow smelt, lake whitefish, cisco, lake trout, brown trout, chinook salmon, and rainbow trout. Other cool and warm water species include white bass, tiger musky, smallmouth bass, and grass carp (U.S. Army Corps of Engineers 1993c). Although the lower river has changed less than the upper river, factors such as overfishing, escapement from main stem and tributary lakes, and competition have also affected the fish assemblage. Certain species such as red shiner, emerald shiner, freshwater drum, river shiner, gizzard shad, skipjack herring, and white bass have increased in the river while carp, river carpsucker, blue catfish, bigmouth buffalo, sauger, flathead chub, and Plains minnows have decreased (Pflieger and Grace 1987; Funk and Robinson 1974).

Adding Fish and Wildlife Benefits to the Formula

Federal-State Management Initiatives

Various projects for fish and wildlife have been implemented under the regulated river formula. Along the main stem lakes, many subimpoundments and game management areas exist, and lake and tailwater fisheries have flourished.

Large federal wildlife refuges for waterfowl production and other purposes are located at Fort Peck Lake, Lake Sakakawea, and Lake Oahe. In addition, North Dakota maintains large, multipurpose game management areas along Lake Sakakawea and at the upper end of Lake Oahe. Large wildlife areas have been developed along Lake Francis Case and Lewis and Clark Lake by South Dakota, while Nebraska has developed a large game management area upstream from Lewis and Clark Lake.

In the lower river during the 1960 to 1970 period, Iowa acquired and maintained several backwater areas, but until recently other states have not pursued habitat development projects. Lack of federal funding for the mitigation plan has significantly hindered all states, but in the 1980s there was more interest in and capability to develop river projects.

Early on, popular cool water fisheries were developed in the main stem lakes. Later, cold water fisheries were added in the three large upper lakes. Naturally spawning populations of rainbow trout have developed in the tailwaters of Fort Peck Dam; other populations of cold water fish in the upper lakes or tailwaters are maintained by stocking. Diverse tailwater fisheries for warm and cool water species have also developed below Fort Randall and Gavins Point dams because of an abundance of food and variable hydraulic and substrate conditions. More recently, a smallmouth bass population has been established below Fort Randall.

In recent years, the Missouri River Natural Resources Committee (MRNRC) has developed tailwater flow recommendations, and these guidelines are being followed, to the extent practicable, at several dams. At Fort Peck, flow releases are timed to provide for sauger and rainbow trout spawning and rearing and to maintain riffle habitat. At Oahe Dam, releases are made to flush and maintain riverine habitat and to improve fishing and other forms of river recreation. Spring season releases from Fort Randall Dam are timed for sauger and paddlefish spawning and rearing when these are compatible with flood control operations. Also, the MRNRC has recommended a three-year intrasystem regulation plan to improve lake fisheries in the upper three reservoirs. The system would be unbalanced to achieve this goal; for example, each reservoir would be subject to a “drawdown” and a “filling” during a three-year period (U.S. Army Corps of Engineers 1991c).

The Role of Federal Environmental Legislation

Fish and Wildlife Coordination Act of 1958

Between 1950 and 1970, federal environmental legislation had some influence on fish and wildlife development along the river. The Fish and Wildlife Coordination Act of 1958 was enacted when Oahe Dam, Big Bend Dam, and the Bank Stabilization and Navigation project were being constructed, so it was necessary to consider mitigation of fish and wildlife resources affected by the project. Accordingly, a Lake Oahe/Sharpe Mitigation Plan was developed by 1963, and a Fish and Wildlife Mitigation Plan was developed for the lower river by 1981 (U.S. Army Corps of Engineers 1981). Unfortunately, the Lake Oahe/Sharpe Plan, despite several reformulations, still has not received federal approval. The lower river mitigation plan has received public support for many years, but only recently have funds been provided. This plan provides for restoration work on nearly 30,000 acres of bottom land at a cost of $67.6 million. Land acquisition and habitat development work is in progress at several locations.

Other Federal Legislation

Since 1970, several federal laws have had subtle and indirect influences on operation of the system: the National Environmental Policy Act of 1969, the Federal Water Pollution Control Act (FWPCA) of 1972 and its subsequent amendments, the Endangered Species Act of 1973 as amended, the Streambank Erosion Control and Demonstration Act of 1974, and the National Parks and Recreation Act of 1978. Under the FWPCA, point source loading of municipal effluents has been drastically curtailed. This, along with improved stream flow, has virtually eliminated dissolved oxygen problems in the river. Dredge and fill authority has expanded federal jurisdiction along the river during the past 20 years, especially for activities affecting wetlands. However, wetlands are still being lost along the river due to inadequate early season flows, bed degradation, and drainage.

The Endangered Species Act has heightened awareness of habitat loss in the system for federally listed species. Measures are being taken to avoid sensitive shore birds when altering flow releases from dams during spring and summer. However, curtailing the early season flows changes lower river stages, which reduces sandbar scouring and leads to subsequent loss of habitat. Artificial measures (such as vegetation clearing and dredging) are being evaluated to improve reproductive success. While the Endangered Species Act can affect operation of the Missouri River system, it cannot effectively address broader river fishery, wetland, and riparian habitat management issues.

Designation of two different segments of the Missouri as National Recreation Rivers (Gavins Point Dam to Ponca in 1978 and Fort Randall to Lewis and Clark Lake in 1991) could be viewed as protection for the environmental resources along the river. However, sandbar, wetland, and riparian habitat has continually degraded in the 1978-

designated reach since land use has not been regulated and flow levels and stages are too low to scour sandbars, wetlands, and adjacent riparian areas. Master planning, initiated in 1993 in both of these reaches, attempts to update the Gavins Point Dam reach and to develop the initial management plan for the Fort Randall reach.

The Streambank Erosion and Control Demonstration program of the late 1970s, while limited to only the Garrison and Gavins Point river reaches, effectively illustrated that highly erodible streambanks could be protected in an aesthetic manner while improving bank aquatic habitat.

One Good Idea

An innovative program developed by the Corps was the dike-notching effort in the lower river. Completed in the early 1980s, various dikes and revetments were notched in order to create quiet slack water for fishery habitat. Fishery managers have noted improved fish populations and sport fishing along the river. Concern has recently developed, however, that the small slack water areas are filling with sediment and some of the values gained will be lost.

Water Resources Development Act (WRDA) of 1986

One section of this act provides for restoration of fish and wildlife resources lost during federal projects. Utilizing this authority and funding, the Corps and the Papio-Middle Missouri Natural Resource District have restored Boyer Chute along the lower river, and projects at Bullard Bend, Nebraska, and Hidden Lake, Nebraska, are also being evaluated. The Corps' attempts to develop a waterfowl management project in the delta at the upper end of Lake Sakakawea, however, were unsuccessful due to design, high costs, and policy restrictions on cost-sharing. The project was built faster and cheaper under the aegis of the Mitchell Bill (U.S. Fish and Wildlife Service funding), together with design and construction expertise provided by Ducks Unlimited, a private conservation organization (Buse 1993). Recent changes in cost-sharing policies under WRDA, however, may reduce some constraints.

Master Water Control Manual Review

The states of the upper and lower basin requested a review of the manual for main stem system operation. A major goal of the review is to reduce resource management conflicts and identify trade-offs. Through this effort, several environmentally based (EQ) flow regimes have been formulated and evaluated, along with various regimes focusing on national economic development (NED) benefits and the regulated river formula. One EQ alternative was a split-season flow option, which provides for both an early and a late navigation period. This option releases an early-season, high-flow pulse in the system, which mimics a natural hydroperiod. A wetland simulation model indicated that various EQ and NED flow options maintained more wetlands over the simulation period than under existing conditions (U.S. Army Corps of Engineers 1993a). This suggests that wetland acreage can be improved without significantly affecting other main stem uses of water.

Regulated River Formula Status and Long-term Issues

Main stem flow regulation has varied little since the dams were filled about 30 years ago. So far, there have been only small losses of hydropower benefits from release constraints required under the Endangered Species Act. Modified operations, including flow release to scour sandbar habitat for sensitive shore birds, need not necessarily result in significant NED losses unless wise and prudent management and recovery efforts are thwarted by interagency conflicts, interference from political interests (especially legal actions), or fiscal constraints. Agencies with resource management responsibilities and concerns along the river, especially the Corps, the U.S. Fish and Wildlife Service, and numerous state resource agencies, need to continue to resolve issues as they arise and to seek practical solutions. In these situations, the Missouri River Basin Association can provide assistance with focusing state concerns and moderating conflicts between state and federal interests.

Channel bed changes within some river reaches, but especially at Williston, Bismarck, Niobrara, and below Gavins Point Dam, have been a chronic problem since they have affected flooding, water supply intakes, recreation, and fish and wildlife habitat. These are being addressed to the extent they can be within existing Corps authority. For example, a sediment study, focusing on degradation and aggradation problems, is in progress along the river between Niobrara and Ponca, Nebraska. Hydraulic alternatives for improved sediment management will be examined in order to find a way to alleviate these concerns.

Water Use Conflicts

Conflicts occur among functions, within functions, and spatially or geographically. The discussion here only presents examples of actual or potential conflicts and is not intended to be comprehensive.

Among Functions

A pulsed release of spring flow for environmental purposes could conflict in varying degrees with the flood control requirement of constrained spring flow or the need for navigation flow in summer and fall. Also, during droughts, reducing the navigation flow would adversely affect recreation and fish and wildlife development along the lower river that depend on a late-season flow. The Master Water Control Manual Review is evaluating various flow options to determine the magnitude and location of these conflicts. As indicated previously, a very serious conflict exists below Gavins Point Dam, where there has been habitat degradation from the effect of inadequate spring flow and channel instability.

Within Functions

Reduction of flows in spring and summer to avoid flooding of sensitive shore birds is not compatible with the need to supply adjacent wetland and riparian habitats with an early season water supply necessary for fish reproduction and waterfowl use. Such reduction also encourages woody vegetation to grow in the acute channel area, creating a long-term elimination of existing chutes and backwaters, and attachment of sandbars and islands to the flood plain. A coordinated, long-term approach to manage sensitive species is needed so that other habitats can also be maintained along the river. Habitat model simulations conducted during the master study indicated that EQ or higher spring flows benefit all aquatic communities along the river, including sandbar habitat.

Spatial

Conflicts between upper and lower basin interests relate to location along the river and their policies on development and management of fish and wildlife and recreational resources. The upper basin states have adapted several programs to the existing water manual and are managing large wildlife areas and subimpoundments according to existing reservoir water elevations. Similarly, Iowa and, more recently, Missouri and Nebraska have been developing wildlife and recreation areas along the lower river, assuming existing levels and patterns of flow. Maintaining higher reservoir levels during drought periods in the upper river, however, would curtail navigation flow in the lower river and adversely affect lower basin wildlife and recreation.

Environmental Additives and the Formula

Through public input received during annual reviews under the Master Water Control Manual Review, numerous NED and EQ alterations in water management were identified (U.S. Army Corps of Engineers 1992). An approach is sought that will satisfy both upper basin and lower basin interests, including actions to alleviate environmental, endangered species, bed instability, and recreational problems along the river.

With this in mind, what future direction should environmental programs take along the river? These could be: (1) along historical lines, exploiting new opportunities that arise within the regulated river environment; (2) along more natural lines, restoring lost environmental conditions; or (3) both approaches. These questions are only broadly addressed here.

Formula Alteration

Through the Master Water Control Manual Review, which in 1993 released a preliminary draft of its environmental impact statement for public review, interested groups and individuals have the opportunity to comment on various NED and EQ alternatives. While EQ options generally increase the value of environmental resources, several NED options also considerably improve environmental values.

The U.S. Fish and Wildlife Service, during the Master Water Control Manual Review, formally stated that fish and wildlife resources should be given co-equal status with other resources during operation of the river system (USFWS 1993). While this can be accomplished in resource planning studies, implementation of actual changes may affect basin economic interests and may represent a negotiated political decision. Additional congressional approval or authorization may be required.

Other major modifications to the river system might include structural work at dams to provide fish passage, construction of sills or dikes in degraded river reaches to raise water levels, or subdivision of the upstream ends of reservoirs in order to reduce shoreline erosion and provide large, protected areas for fish habitat improvement and harbors. Aggradation and degradation problems associated with the headwaters of some reservoirs will need to be evaluated thoroughly, also possibly requiring some major structural modifications.

Formula Adaptation

In the upper river, there are numerous opportunities to protect or enhance oxbow, backwater, and side channel wetlands, to restore sandbars for sensitive species of shore birds, to protect scenic shorelines from unnecessary development, and to develop habitat for fish and waterfowl along some of the larger Plains tributaries. Using river ice to scour sandbars and other habitat under regulated flow conditions has not been attempted but may be feasible.

In the main stem lakes of the Dakotas, fish spawning and rearing habitat in such tributaries as the Heart, Cannonball, Cheyenne, and White Rivers could be improved for river-running fish such as paddlefish, sturgeon, catfish, and sauger. In the main stem lakes, small barrier reefs could also be constructed in shallow areas, providing not only quiet water and aquatic bed habitat for game fish and waterfowl but also protecting recreation areas and curtailing shoreline erosion. Further actions could be taken to protect ecologically sensitive shorelines and fish spawning areas, to retain sediment in upper delta areas, and to protect woody draw and grassland habitat in the uplands around the reservoirs. Opportunities to enhance delta areas are numerous, including such options as island construction, silt-dredging from old oxbows, development of pumps and diversion canals, installation of small water control structures, and manipulation of vegetation.

In the lower river, more attention could be directed to habitat conditions in some of the larger tributary rivers, including the Big Sioux, Little Sioux, Nishnabotna, and Big Nemaha. This is because the potential for restoration of the main stem is limited by engineering constraints, small-scale projects, and high costs. Low flows are a problem on several tributary streams, however, and can only be alleviated through multipurpose projects that provide flow regulation and protection of existing flows. Some projects may materialize near tributary confluences with the main stem if tributary problems can be related to main stem operations (e.g., the Corps' bed control structure on the Little Sioux River, Iowa, which restricts fish movement).

Institutional Changes and the Formula

Experience shows that broadly based federal environmental legislation does little for the river: Only resource studies, innovative project planning, a cooperative interagency spirit, and timely and adequate funding are really effective. The Missouri River Basin Association can assist with bringing groups together to discuss mutual needs, while the MRNRC and private conservation groups can assist in focusing on specific needs and developing specific environmental restoration measures, strategies, and projects. In addition, partnerships need to be strengthened with nontraditional groups, including Native American tribes, private corporations, and private conservation groups in order to protect and improve the river's resources. Further, we need to loosen institutional bonds that tend to escalate project costs and to reduce capability to provide prompt and effective technical assistance and designs. We have observed that intergovernmental partnerships have led to several good environmental projects in the Missouri Basin, particularly within the Prairie Pothole Region.

In conclusion, the Corps needs direction from the public so that finite resources, namely water and the river valley lands, can be developed in a balanced manner. Whether the formula is changed or adapted, decisions will ultimately be driven by the economic and physical constraints of the present system.

References

Anderson, J. 1989. *Missouri River reservoirs wetland study.* Internal memorandum from engineering division to planning division (July 18), U.S. Army Corps of Engineers, Omaha District, Omaha, Nebraska

Atchison, G. J., R. W. Bachmann, J. G. Nickum, J. B. Barnum, and M. B. Sandheinrich. 1986. *Aquatic biota associated with channel stabilization structures and abandoned channels in the middle Missouri River.* Technical Report E-86-6. U.S. Army Corps of Engineers.

Benson, N. G. 1968. *Review of fishery studies on Missouri River main stem reservoirs.* Research Report no. 71. North Central Reservoir Investigations, Bureau of Sport Fisheries and Wildlife. Washington, D.C.: U.S. Government Printing Office.

Berry, C. R. 1988. Rare fish in the upper Missouri River Basin. In *A review of the Missouri River main stem system operation: Symposium proceedings.* Topeka, Kansas: Missouri Basin States Association.

Buse, R. L. 1993. Personal communication of the chief of the Project Development Branch, Omaha District, U.S. Army Corps of Engineers.

Cowell, B. C. 1970. The influence of plankton discharges from an upstream reservoir on standing crops in a Missouri River reservoir. *Limnology and Oceanography* 15:427-441.

Flint, R. F. 1957. *Glacial and Pleistocene geology.* New York: John Wiley and Sons.

Funk, J. L., and J. W. Robinson. 1974. *Changes in the channel of the lower Missouri River and effects on fish and wildlife.* Aquatic Series no. 11. Missouri Department of Conservation, Jefferson City.

Harberg, M. 1992. Personal communication of the environmental resource specialist, Missouri River Division, U.S. Army Corps of Engineers, Omaha, Nebraska.

Hem, J. D. 1970. *Study and interpretation of the chemical characteristics of natural water.* Water Supply Paper (2nd ed.). U.S. Geological Survey.

Hesse, L. W., G. R. Chaffin, and J. Brabander. 1989. Missouri River mitigation: A system approach. *Fisheries* 14:11-15.

Hudson, P. L., and B. C. Cowell. 1966. Distribution and abundance of phytoplankton and rotifers in a main stem Missouri River reservoir. *Proceedings of the South Dakota Academy of Science* 45:84-106.

Kallemeyn, L. W., and J. F. Novotny. 1977. *Fish and fish food organisms in various habitats of the Missouri River in South Dakota.* Document no. OBS-77/25. U.S. Fish and Wildlife Service, Columbia, Missouri.

Lind, O. T. 1971. The organic matter budget of a central Texas reservoir. In *Reservoir fisheries and limnology,* ed. G. E. Hall. Special publication no. 8. Washington, D.C.: American Fisheries Society.

Morris, L. A., R. N. Langemeier, T. R. Russell, and A. Witt, Jr. 1968. Effect of main stem impoundments and channelization upon the limnology of the Missouri River. *Transactions of the American Fisheries Society* 97:380-388.

Neel, J. K., H. P. Nicholson, and A. Hirsch. 1963. *Main stem reservoir effects on water quality in the central Missouri River, 1952-1957.* U.S. Public Health Service, Region VI, Water Supply and Pollution Control, Kansas City, Missouri.

Peterson, D. K. 1988. Missouri River main stem development. In *A review of the Missouri River main stem system operation: Symposium proceedings.* Topeka, Kansas: Missouri Basin States Association.

Pflieger, W. L., and T. B. Grace. 1987. Changes in the fish fauna of the lower Missouri River, 1940-1983. In *Community and evolutionary ecology of North American stream fishes,* ed. W. J. Matthews and D. C. Heins. Norman: University of Oklahoma Press.

Reily, P. W., and W. C. Johnson. 1982. The effects of altered hydrologic regime on tree growth along the Missouri River in North Dakota. *Canadian Journal of Botany* 60:2410-2423.

Repsys, A. J., and G. D. Rogers. 1982. Zooplankton studies in the channelized Missouri River. In *The Middle Missouri River: A collection of papers on the biology with special reference to power station effects,* ed. L. W. Hesse, G. L. Hergenrader, H. S. Lewis, S. D. Reetz, and A. B. Schlesinger. Norfolk, Nebraska: Missouri River Study Group.

Slizeski, J. H., J. L. Anderson, and W. G. Dorough. 1982. Hydrologic setting, system operation, present and future stress. In *The Middle Missouri River: A collection of papers on the biology with special reference to power station effects,* ed. L. W. Hesse, G. L. Hergenrader, H. S. Lewis, S. D. Reetz, and A. B. Schlesinger. Norfolk, Nebraska: Missouri River Study Group.

U.S. Army Corps of Engineers. 1976. *Draft environmental statement, continuing construction and O & M.* Missouri River Bank Stabilization and Navigation Project, Missouri River Division, Omaha, Nebraska.

———. 1981. *Fish and wildlife mitigation plan.* Report and final EIS. Missouri River Bank Stabilization and Navigation Project, Missouri River Division, Omaha, Nebraska.

———. 1989. *Missouri River wetlands: New resources in main stem reservoirs.* Unpublished report. Omaha District, Omaha, Nebraska.

———. 1991a. *Investigation of channel degradation, 1991 update, Missouri River Gavins Point Dam to Platte River confluence.* Omaha District, Omaha, Nebraska.

———. 1991b. *Biological assessment, Missouri National Recreational River.* Omaha District, Omaha, Nebraska.

———. 1991c. *Annual operating plan and summaries, 1990-91 operations.* Missouri River Division, Omaha, Nebraska.

———. 1992. *Initial evaluation report, Missouri River Master Water Control Manual Review and Update Study.* Missouri River Division, Omaha, Nebraska.

———. 1993a. *Wetlands and riparian resources, Vol 7 F: Environmental studies, Missouri River Master Water Control Manual Review and Update Study.* Missouri River Division, Omaha, Nebraska.

———. 1993b. *Riverine fisheries, Vol. 7 C: Environmental studies, Missouri River Master Water Control Manual Review and Update Study.* Missouri River Division, Omaha, Nebraska.

———. 1993c. *Reservoir fisheries, Vol. 7 A: Environmental studies, Missouri River Master Water Control Manual Review and Update Study.* Missouri River Division, Omaha, Nebraska.

U.S. Fish and Wildlife Service. 1987. *Draft Fish and Wildlife Coordination Act Report.* Pierre-Fort Pierre Area Studies, Pierre, South Dakota.

———. 1993. *Interim fish and wildlife report on the Missouri River Master Water Control Manual Review and Update Study.* North and South Dakota Field Office, Bismarck, North Dakota.

Velehradsky, J. E. 1988. Federal laws, policies and regulations for system operation. In *A review of the Missouri River main stem system operation: Symposium proceedings.* Topeka, Kansas: Missouri Basin States Association.

Energy and Mineral Resources

19. Energy and Mineral Resources of the Great Plains

Kenneth S. Johnson and Charles J. Mankin
Oklahoma Geological Survey

The Great Plains, for a long time considered to be America's bread basket, also is a major contributor to the nation's mineral and energy supply. The U.S. portion of this vast, open region that extends from south Texas to the northwestern part of Canada has provided a significant portion of the mineral and energy needs of this country. The sparse population, the relatively easy access, and, most important, the geology of the region gave rise to the discovery and development of huge reserves of crude oil, natural gas, and coal.

Equally important is the development of the nonfuel minerals of the region. The industrial minerals were important in developing the urban centers of the Great Plains. The use of clays and shales in the manufacture of bricks and tile, shale and limestone in the production of Portland cement, sand and gravel for concrete and roads, and gypsum for wallboard and plaster are a few of the many applications of industrial minerals of this region.

The discovery of valuable ores of both precious and industrial metals gave rise to local population centers and created centers of wealth. Lead and zinc deposits in the tri-state area of Kansas, Oklahoma, and Missouri produced a thriving industry. The discovery of gold in Colorado resulted in a large influx of people to the western edge of the Great Plains and generated substantial wealth in that part of the region.

Today, the Great Plains continues to be a major source of mineral and energy supply for the United States. The Great Plains contains about 25 percent of U.S. land area (including all 50 states) and about 12 percent of the population, yet it produced 39 percent of the nation's mineral and energy wealth during 1991. The value of mineral and energy production in the Great Plains during 1991 was almost $50 billion (Table 19.1), whereas that of the nation was slightly more than $127 billion.

Of the $50 billion generated by mineral and energy production in the Great Plains in 1991, 43 percent came from the production of petroleum, 34 percent from natural gas, 8 percent from coal, and 15 percent from nonfuel minerals.

A discussion of the major energy and nonfuel minerals that are produced in the Great Plains follows. It should be noted that the information on quantities and values for each of the commodities is based upon state totals. It was a difficult task to separate

Stanley R. Johnson and Aziz Bouzaher (ed.), *Conservation of Great Plains Ecosystems*, 301-322.

the part of each commodity that was produced only in the Great Plains part of a state. The portion of a commodity that should be attributed to the Great Plains in several of the states was estimated, with the result that, in most instances, the state totals did not differ markedly from the part allocated just to the Great Plains part of the state. The largest uncertainty is in the production of natural gas in Texas and New Mexico: If the Permian basin of Texas and New Mexico is included in the Great Plains, then the difference is small; if not, the discrepancy for natural gas is more than 50 percent for those two states.

Table 19.1. Value of mineral and energy production, by state for various commodities, in the Great Plains states during 1991

State	Petroleum	Natural Gas	Coal	Minerals	Total
		millions of dollars			
Colorado	626	403	396	338	1,763
Iowa	0	0	8	344	352
Kansas	1,129	774	11	366	2,280
Minnesota	0	0	0	1,019	1,019
Missouri	3	0	63	880	946
Montana	356	86	411	690	1,543
Nebraska	110	2	0	89	201
New Mexico	1,363	1,422	500	986	4,271
North Dakota	674	89	232	17	1,012
Oklahoma	2,118	3,166	53	276	5,613
South Dakota	32	1	0	290	323
Texas	13,461	10,071	657	1,265	25,454
Wyoming	1,732	824	1,568	929	5,053
Totals	21,604	16,838	3,899	7,489	49,830

Sources: EIA 1992c; API 1993.

Petroleum

Petroleum (including crude oil and natural gas liquids) production began in the Great Plains with the discovery of crude oil at Florence, Colorado, in 1863. The Florence oil field, located on the western edge of the Great Plains, is the oldest continuously producing oil field west of the Mississippi River. In 1945, Florence was combined with the Cañon City field, creating the Florence-Cañon City oil field. Cumulative production from the combined field was more than 15 million barrels through 1990 (International Oil Scouts 1991).

Discovery of the Florence field was followed in 1889 with discoveries of petroleum in Kansas, Missouri, and Texas. Petroleum was discovered in 1891 in Indian Territory (now eastern Oklahoma) and in Wyoming in 1894. Petroleum was discovered in the remaining Great Plains states in the early to middle part of the twentieth century.

The Great Plains region has contributed substantially to the nation's petroleum production. Of the 614,000 producing oil wells in the United States, more than 60 percent are located in the Great Plains states (API 1993).

Of these, Texas ranks first with one-half, Oklahoma second with one-fourth, and Kansas third with one-eighth of the producing oil wells in the Great Plains (Table 19.2).

Table 19.2. Number of producing oil wells in the Great Plains in 1991, by state

State	Number	Percent
Colorado	5,818	1.50
Kansas	48,628	12.55
Missouri	846	0.22
Montana	3,750	0.97
Nebraska	1,751	0.45
New Mexico	17,638	4.55
North Dakota	3,486	0.90
Oklahoma	101,438	26.17
South Dakota	154	0.04
Texas	192,459	49.66
Wyoming	11,620	3.00
Total	387,588	100.01

More than 70 percent of the 388,000 producing oil wells in the Great Plains are classified as stripper wells (wells that produce less than ten barrels of oil per day) (API 1993) (Table 19.3).

Within the Great Plains, the states that have produced oil for the longest period of time have more stripper wells. For example, more than 90 percent of the producing oil wells in Kansas are classified as stripper wells, whereas only 12 percent are so classified in South Dakota.

In the Great Plains, the average stripper well production is about three barrels per day. Consequently, production costs for these wells are high, ranging from $9 to $12 per barrel, exclusive of taxes and royalty payments. With low world oil prices ($18 to $20 per barrel), there has been a marked increase in the rate of plugging and

Table 19.3. Number of stripper wells in the Great Plains in 1991, by state, and percentage of stripper wells to total wells in that state

State	Stripper Wells	Percent of Total Wells in State
Colorado	4,432	75.18
Kansas	44,959	92.45
Missouri	701	82.86
Montana	2,962	78.99
Nebraska	1,190	67.96
New Mexico	15,085	85.53
North Dakota	1,264	36.26
Oklahoma	72,821	71.79
South Dakota	19	12.34
Texas	129,332	67.20
Wyoming	3,030	26.08
Total	275,795	

abandoning of these high-cost wells in the Great Plains. In 1991, 12,132 wells were plugged and abandoned in the region; that was almost 2,000 more than in the previous year and amounted to 70 percent of all wells plugged and abandoned in the United States that year.

The 1991 petroleum production for the Great Plains region was 1.14 billion barrels, or 63 percent of the U.S. total of 2.71 billion barrels (API 1993). This production distribution among the Great Plains states is shown in Figure 19.1.

As previously mentioned, some of this production is from portions of the included states that lie outside the Great Plains. For example, production from the Texas Gulf Coast and the Texas offshore districts is included, as is production from the western parts of New Mexico, Colorado, and Wyoming. Despite the difficulty in separating the production from different districts within a state, we estimate that about three-fourths of Texas's production is in the Great Plains, as is more than 90 percent of the New Mexico, Colorado, and Wyoming production. Thus, the quoted numbers are reasonable representations of the oil production and related activities for the Great Plains.

Cumulative production of petroleum through 1991 from the Great Plains, as shown in Figure 19.2, is 115 billion barrels, or almost two-thirds of the U.S. total of 182 billion barrels (API 1993; IPAA 1990).

Texas has provided 64 percent of the cumulative oil production from the Great Plains, followed by Oklahoma with 14 percent, and Kansas, New Mexico, and Wyoming, with about 6 percent each.

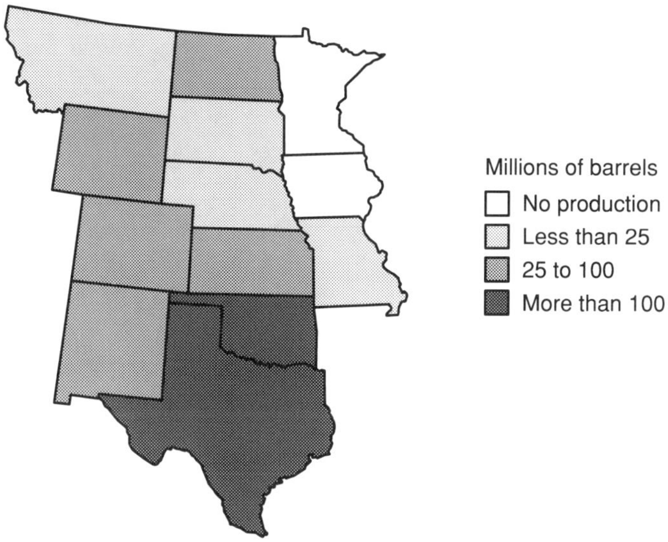

Source: API 1993.

Figure 19.1. Petroleum production in the Great Plains states for 1991

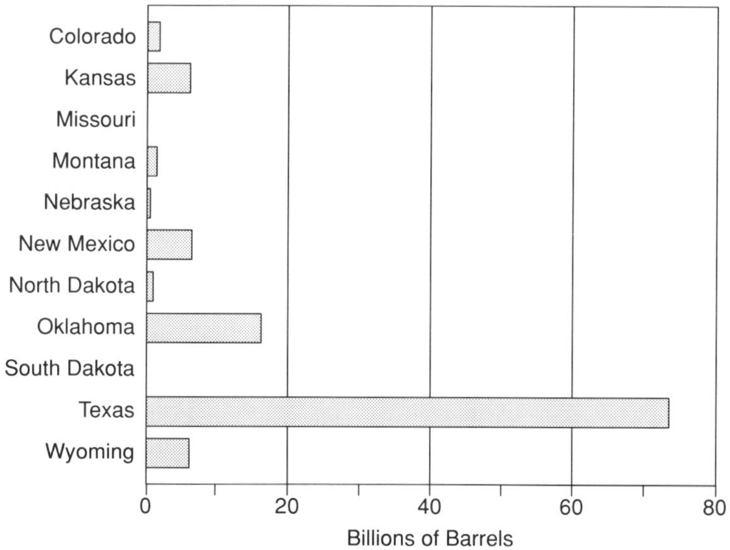

Sources: API 1993; IPAA 1990.

Figure 19.2. Cumulative petroleum production for the Great Plains states through 1991

About 10 billion barrels of reserves remain in existing fields in the Great Plains (EIA 1992a). That is about 40 percent of the remaining reserves in the United States, estimated to be almost 25 billion barrels. At the end of 1991, petroleum reserves had declined in the United States by 1.6 billion barrels from the previous year, a decline of 6 percent. During the same period, reserves in the Great Plains declined by 416 million barrels, a decline of 4 percent.

While the Great Plains states have produced a large amount of petroleum since discovery of the Florence field in 1863, and the general consensus of the petroleum industry is that most of the large fields have been found, the region does have the opportunity to produce significant quantities of additional petroleum. Only about one-third of the oil present in a field at the time of discovery will be extracted using current recovery technologies. This means that when the stated reserves in existing fields have been extracted, these oil fields in the Great Plains will still contain about twice as much oil (about 230 billion barrels) as was produced previously.

Thus, the 230 billion barrels of unproduced petroleum that remain in the known fields of the Great Plains is a major economic opportunity for the region. If only 20 percent of that oil could be recovered, it would represent nearly twice the present U.S. reserves. Petroleum recovery specialists (petroleum geologists and reservoir engineers) who have reviewed field conditions in this region believe that between 15 and 30 percent of the remaining oil-in-place could be recovered with currently available technologies.

To realize this opportunity, several things have to happen. First, because of the high operating costs in the United States, a stabilized price for crude oil is needed. At present, the world price of crude oil is established by the Organization of Petroleum Exporting Countries (OPEC). Because of OPEC's ability, through increases in production quotas, to drop the price to levels below U.S. domestic production costs, few operators are willing to risk the financial resources needed to employ improved recovery methods. A stabilized price would eliminate that risk, leaving only the technical risk of employing the proper recovery technology.

A substantial investment in human resources will have to be made to obtain and analyze field information for reservoir characterization and simulation studies. Such investigations are necessary to determine where the remaining oil resides in individual fields. Then pilot studies need to be developed in selected fields to verify the appropriate recovery methods. Finally, through the application of improved recovery methods, together with better field management practices, the ultimate recovery of an additional 30 to 60 billion barrels of oil from known fields in the Great Plains would be possible.

While the opportunity for finding new, large fields in this region is considered to be small, most oil industry experts agree that a large number of small fields may yet be found. With knowledge gained through the studies of existing fields, such information and methods of recovery could be applied to new, smaller fields; this would enable recovering a larger percentage of the oil-in-place during primary recovery, thereby improving the economics for small fields.

Therefore, while petroleum production in the Great Plains is declining, the opportunity does exist now to significantly expand its contribution to the economy of the region.

Natural Gas

Natural gas is rapidly becoming the primary energy commodity produced in the Great Plains. Accompanying the decline in both price and production of petroleum since 1985, the value of natural gas to the region has declined by almost one-half ($35 billion versus $19 billion). Despite the decline of well-head price of natural gas during this period, production has increased. Therefore, the decline in value of natural gas to the region has been much less dramatic ($25 billion versus $18 billion). Currently, the projected increases in demand for natural gas indicate that it probably will become the primary energy commodity produced in the Great Plains within the next two years.

The early history of natural gas development is not well documented in the United States, much less in the Great Plains. In the early history of petroleum development, the discovery of natural gas was considered to be a liability. The market value was limited to local consumption. Because most of the discoveries in the Great Plains were in areas of sparse population, any natural gas produced generally would exceed local demand. It was not until the mid-1940s when interstate pipelines were developed to convey the commodity to distant markets that natural gas became an important national energy commodity.

Some information is available on the discoveries of natural gas in the Great Plains. Such discoveries commonly were associated with the discovery of crude oil, and the gas was classified as associated natural gas; that is, it was either dissolved in the crude oil, or was a gas cap in the same reservoir with petroleum.

The earliest report of marketed production of natural gas in the Great Plains is in Kansas in 1882. Five years later, marketed production of natural gas began in Missouri, followed by Texas and Wyoming in 1889. Colorado began producing natural gas in 1893, and the remaining gas-producing states of the Great Plains began producing natural gas in the early to middle twentieth century (IPAA 1990).

Natural gas production is becoming a more important component of the economy in the Great Plains (Table 19.4). Of the 277,000 natural gas and condensate wells in the United States, 45 percent are located in the Great Plains (API 1993). More than 90 percent of the natural gas wells of the region are located in four states: Texas (38 percent), Oklahoma (23 percent), New Mexico (16 percent), and Kansas (14 percent).

Table 19.4. Number of producing natural gas wells in the Great Plains in 1991, by state

State	Number	Percent
Colorado	5,562	4.44
Kansas	17,948	14.34
Missouri	6	0.00
Montana	2,802	2.24
Nebraska	12	0.01
New Mexico	20,021	16.00
North Dakota	100	0.08
Oklahoma	28,216	22.55
South Dakota	54	0.04
Texas	47,615	38.05
Wyoming	2,821	2.25
Total	125,157	100.00

However, the number of producing gas wells is a poor measure of the quantity of production, for the average annual production per well by state ranges from 530 million cubic feet (mcf) in North Dakota to 3 mcf in Missouri (Table 19.5). The range in production per well is even more dramatic within a state. For example, in Oklahoma the range in production per well is from more than 10 billion cubic feet (bcf) to less than 10 mcf per year. Thus, it would take more than 1,000 small gas wells to equal the production of one very large well.

There are four major natural gas producing areas in the Great Plains: the Permian basin of west Texas and southeastern New Mexico; the Anadarko basin of western Oklahoma and the Panhandle of Texas; the Arkoma basin of eastern Oklahoma; and the Guymon-Hugoton gas field of western Kansas, the Panhandle of Oklahoma, and the Panhandle of Texas. The Guymon-Hugoton gas field is the largest in the United States; it has an estimated ultimate recovery in excess of 75 trillion cubic feet (tcf).

The amount of natural gas produced from the Great Plains for 1991, 11.26 tcf (API 1993), was 61 percent of the U.S. total of 18.52 tcf. That production was distributed among the Great Plains states as shown in Figure 19.3.

Texas (6.33 tcf), Oklahoma (2.15 tcf), and New Mexico (1.04 tcf) produced 85 percent of the natural gas from the Great Plains in 1991. Wyoming ranked fourth with 777 bcf, followed by Kansas with 565 bcf, and Colorado with 286 bcf. The remaining five states collectively produced 107 bcf of natural gas in 1991.

The cumulative production of natural gas from the Great Plains through 1991 was 472 tcf (Figure 19.4); this is 61 percent of the cumulative production for the United States (775 tcf).

Table 19.5. Average annual natural gas production in 1991, by state

State	Number of Wells	Production	Production/Well
		bcf	mcf
Colorado	5,562	286.00	51
Kansas	17,948	565.00	31
Missouri	6	0.02	3
Montana	2,802	52.00	19
Nebraska	12	0.80	67
New Mexico	20,021	1,038,00	52
North Dakota	100	53.00	530
Oklahoma	28,216	2,154.00	76
South Dakota	54	0.90	17
Texas	47,615	6,334.00	133
Wyoming	2,821	777.00	275
Total	125,157	11,260.72	90

Notes: bcf = billion cubic feet; mcf = million cubic feet.

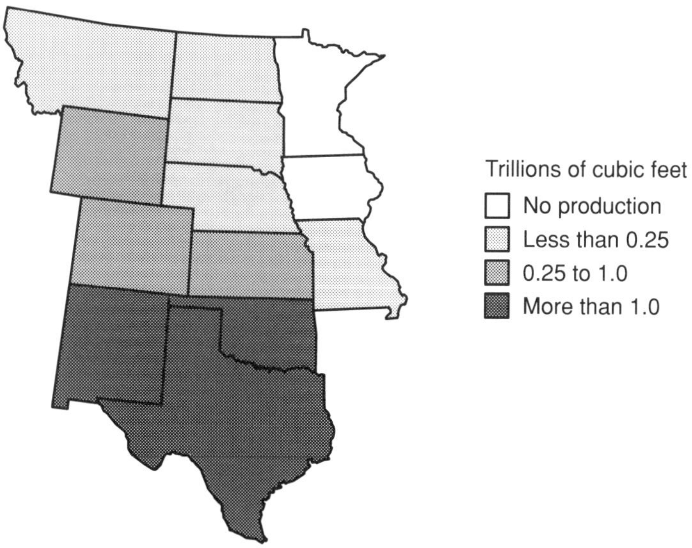

Source: API 1993.

Figure 19.3. Production of natural gas from the Great Plains states for 1991

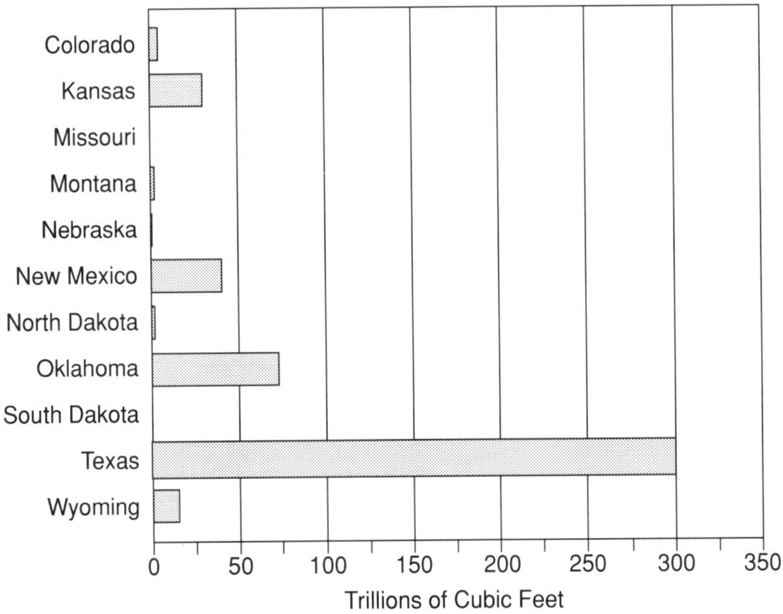

Sources: IPAA 1990; API 1993.

Figure 19.4. Cumulative production of natural gas from the Great Plains states through 1991

The Great Plains is not only an important producer of natural gas but the region also is a major consumer (Table 19.6). More than one-half of the natural gas produced in the Great Plains is consumed in the Great Plains. Texas is the largest producer and also is the largest consumer. In 1991, Texas consumed 3.6 tcf of natural gas, 2.8 tcf of which was used in industrial processes and for generating electricity (EIA 1992b).

The reserves of natural gas in the Great Plains at the end of 1991 were about 115 tcf, almost two-thirds of the total reserves for the United States (175 tcf) (EIA 1992a). Therefore, the Great Plains has a well-developed production and transportation infrastructure for natural gas and has a strong reserve base to support future activities.

There are significant opportunities for expanding natural gas use in the Great Plains. Growth in the use of compressed natural gas for transportation would be substantially enhanced if President Clinton's proposal for a "Btu tax" were enacted in its present form. Likewise, proposed changes in electric power generation will provide increased opportunities for the use of natural gas in gas turbines, and perhaps even in base-load generating systems. Finally, increased air quality standards for major cities in the Great Plains will push industrial and commercial operations in the direction of increased natural gas use.

Coal

Both bituminous and subbituminous coal and lignite are produced in the Great Plains. In this chapter, the data reported are for all forms of coal combined. While differences in energy content are recognized, those differences are not considered to be significant for this document.

Information on the early production of coal from this region is sparse. In the middle to latter part of the nineteenth century, coal produced in the Great Plains states was primarily for local consumption. Information is limited on the earliest date of coal production for several of the states in the Great Plains (Table 19.7). Coal production was reported from Missouri prior to 1840 (Keystone Manual 1992; Finkelman et al. 1990).

Table 19.6. Consumption of natural gas in the Great Plains in 1991, by state

State	Bcf
Colorado	260.51
Iowa	233.18
Kansas	369.24
Minnesota	314.48
Missouri	256.33
Montana	45.40
Nebraska	115.81
New Mexico	218.76
North Dakota	39.79
Oklahoma	570.15
South Dakota	26.25
Texas	3,583.68
Wyoming	97.29
Total	6,130.87

Table 19.7. Year of first coal production in the Great Plains, by state

State	Year
Colorado	1864
Iowa	1840
Kansas	1869
Missouri	pre-1840
Montana	1807
New Mexico	1882
North Dakota	unknown
Oklahoma	1872
Texas	unknown
Wyoming	unknown

Most of the coal and all of the lignite are surface mined in the Great Plains. The number of mines and volume of production for each of the producing states is shown in Table 19.8 (EIA 1992c).

The number of operating coal and lignite mines in the Great Plains states is only 4 percent of the total number of operating mines in the United States (3,022 mines). However, the volume of production is 36 percent of the U.S. total (995.984 million

Table 19.8. Number of mines and coal and lignite production in the Great Plains in 1991, by state

State	Number of Mines	Production (million short tons)
Colorado	21	17.834
Iowa	3	0.344
Kansas	3	0.416
Missouri	5	2.304
Montana	9	38.237
New Mexico	7	21.518
North Dakota	8	29.530
Oklahoma	22	1.841
Texas	15	53.825
Wyoming	31	193.854
Total	124	359.703

short tons) (EIA 1992c). Thus, the number of operating mines is not a good measure of the volume of coal production. The largest operating coal mines in the United States are located in the Great Plains states. In Wyoming, the largest coal-producing state, the average production per mine is more than 6 million short tons per year, whereas the average for West Virginia, ranked second in production, is 250,000 short tons per year. Kentucky, ranked third in production, has an even smaller production per mine at 190,000 short tons per year.

Cumulative coal production data are not readily available for the Great Plains states, but data on the demonstrated reserves indicate that coal and lignite are in abundant supply (Table 19.9). The R/P ratio is the reserve base divided by the 1991 production for each state. Thus, the ratio provides a measure of the number of years that coal or lignite could be produced in each state assuming the production remains constant at the 1991 level. Thus, the Great Plains theoretically could continue at its present rate of production for 676 years.

Not only are the Great Plains states important producers of coal and lignite, they are also major consumers. In 1991, the consumption of coal and lignite in the Great Plains was 290 million short tons (Table 19.10), which was one-third of the total consumption of coal in the United States. The Great Plains states consume more than 80 percent of the coal and lignite produced in the Great Plains, and virtually all of this is used for electric power generation.

Table 19.9. The demonstrated reserve base of coal and lignite and the reserve-to-production (R/P) ratio for each of the Great Plains states in 1991

State	Reserve Base (million short tons)	R/P Ratio
Colorado	16,986	952
Iowa	2,191	6,368
Kansas	977	2,350
Missouri	6,004	2,606
Montana	119,967	3,137
New Mexico	4,457	207
North Dakota	9,627	326
Oklahoma	1,589	863
Texas	13,333	248
Wyoming	68,006	351
Total	243,137	676

Table 19.10. Consumption of coal and lignite by Great Plains states in 1991

State	Million Short Tons
Colorado	16.22
Iowa	18.74
Kansas	14.88
Minnesota	16.99
Missouri	25.77
Montana	10.55
Nebraska	8.86
New Mexico	12.86
North Dakota	28.60
Oklahoma	16.35
South Dakota	2.86
Texas	92.06
Wyoming	25.15
Total	289.89

Nonfuel Minerals

Nonfuel mineral resources are widely distributed in the Great Plains (Figures 19.5-19.8), and they are critical to the well-being of the people and the infrastructure of the society. These minerals are the raw materials for the manufacturing, construction, and chemical industries, and they are a part of almost all consumer goods.

It is impossible to describe and assess the known and suspected resource base of all the nonfuel minerals in such a large region as the Great Plains, but it is possible to discuss the resources that are currently being mined or produced. Production figures commonly available do not separately list production from the various parts of each state, so it is necessary to discuss the total mineral production of a state if any part of it is within the Great Plains (Table 19.11). Nonfuel minerals commonly are divided into two classes, metallic minerals and nonmetallic minerals.

Metallic Minerals

Metallic mineral resources are those rocks and deposits from which one or several metals can be extracted by mining and/or metallurgical processes. Metals produced in the Great Plains states include: precious metals, such as gold, silver, platinum, and palladium; base metals, such as copper, lead, and zinc; and other metals, such as beryllium, iron, magnesium, molybdenum, and uranium.

Precious Metals

The precious metals are so named because of their historical importance as media of exchange (currency), or their exceptionally high value. Gold, silver, platinum, and palladium are the precious metals produced in the Great Plains states (Figure 19.5). Most of this production is in the far western part of the states, such as in western Montana, Colorado, and New Mexico, and technically this is largely in the Rocky Mountains. The precious metals most commonly are present in igneous rocks, and igneous rocks typically are in the core of mountain ranges. Some of the precious metal production is accompanied by coproduction of base metals.

Gold is the leading nonfuel mineral produced in Montana and South Dakota (Table 19.11), but it also is important in Colorado and New Mexico. Gold is used mostly in jewelry and art but also is used by the electronic industry and in dentistry. Silver production is highest in Montana, but it also is significant in Colorado, New Mexico, and South Dakota. About half of the U.S. silver is used in the manufacture of photographic products, and much of the remainder goes to the electronic industry, electroplating, and jewelry. Platinum and palladium are produced only in Montana, but

Table 19.11. Value and types of nonfuel minerals produced in the Great Plains states in 1991

State	Value (thousands of dollars)	Rank in U.S.	Percent of U.S. Total	Principal Minerals, in Order of Value
Colorado	338,405	31	1.12	Sand and gravel, molybdenum, cement, stone, and gold
Iowa	344,032	29	1.14	Stone, cement, sand and gravel, and gypsum
Kansas	365,879	25	1.21	Helium, salt, stone, cement, and sand and gravel
Minnesota	1,018,950	9	3.37	Iron ore, sand and gravel, stone, and industrial sands
Missouri	880,351	12	2.91	Lead, cement, stone, and lime
Montana	690,297	17	1.95	Gold, copper, cement, and silver
Nebraska	89,392	44	0.30	Cement, sand and gravel, stone, and lime
New Mexico	985,510	10	3.26	Copper, potash, sand and gravel, and molybdenum
North Dakota	17,366	48	0.06	Sand and gravel, lime, and gem stones
Oklahoma	275,525	35	0.91	Stone, cement, iodine, sand and gravel, industrial sand, and gypsum
South Dakota	289,922	34	0.96	Gold, cement, sand and gravel, and stone
Texas	1,264,661	7	4.19	Cement, stone, magnesium metal, and sulfur
Wyoming	929,176	11	3.08	Soda ash, clays, helium, and cement
Totals	7,489,466		24.46	

Source: U.S. Bureau of Mines 1992.

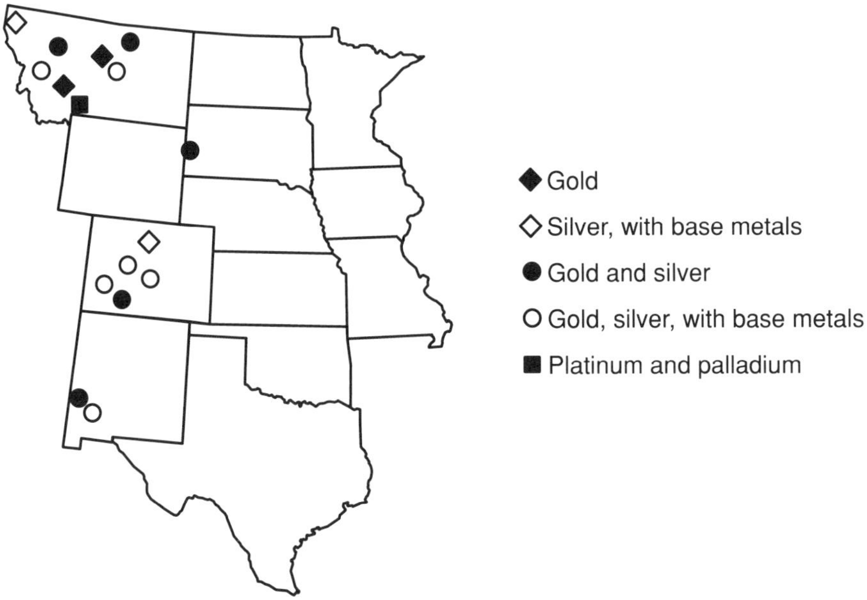

Source: U.S. Bureau of Mines 1992.

Figure 19.5. Major precious metal-producing areas in the Great Plains states

they are a significant commodity in that state. Platinum-group metals are used mainly in the automotive, electronic, electrical, dental, and medical industries.

Future production of precious metals, and additional resources that may be discovered and brought into production, is largely limited to the Rocky Mountains and Black Hills portions of the Great Plains states.

Base Metals

Base metals are any of the common metals that are chemically active, including copper, lead, and zinc. These metals commonly are found in igneous rocks in the mountainous regions of the west, but there also are significant deposits in some of the sedimentary rocks that underlie most of the Great Plains (Figure 19.6).

Copper is the leading nonfuel mineral produced in New Mexico, is the second leading mineral in Montana, and is also produced in Colorado (Table 19.11). Copper is consumed mainly in building construction, electrical and electronic products, industrial machinery, transportation, and consumer goods. Lead and zinc are produced only in

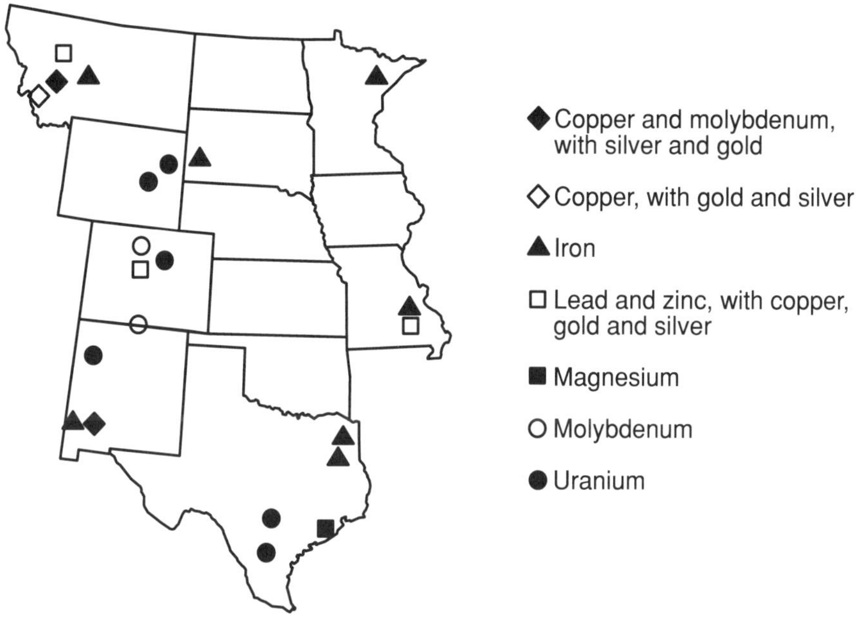

Source: U.S. Bureau of Mines 1992.

Figure 19.6. Major base metal-producing areas in the Great Plains states

Missouri, among the Great Plains states; lead is the leading nonfuel mineral in Missouri, whereas zinc is of lesser importance. Lead is used mainly by the transportation industry (including batteries, fuel tanks, solder, seals, and bearings) and in the electrical, electronic, and communication industries. Zinc is used primarily in construction materials and also in the transportation, machinery, and electrical industries.

Future production of base metals will undoubtedly continue from both the Rocky Mountains and the sedimentary rocks elsewhere in the Great Plains. Rock types favorable for hosting base metals are widespread in the Great Plains, but more exploration must be carried out to find those areas where mineralizing fluids have encountered these potential host rocks and have formed economic deposits.

Other Metals

Other metals in the Great Plains are produced from a variety of igneous, sedimentary, and metamorphic rocks, as well as from sea water. Significant levels of production occur both in the mountain regions of the west and in the Plains region.

Iron is the leading "other metal" being produced in the Great Plains: It is the major nonfuel mineral in Minnesota (Table 19.11) and also is produced in Missouri, Montana, New Mexico, and Texas (Figure 19.6). Molybdenum, used in making various machinery, electrical, transportation, and chemical items, is an important resource being produced in Colorado and New Mexico. Magnesium is recovered from sea water in Texas by an electrolytic process, and it is one of that state's most important nonfuel minerals (Table 19.11). Magnesium is used mainly in aluminum-based alloys, as well as for castings and wrought-magnesium products, desulfurization of iron and steel, and as a reducing agent in nonferrous metals production. Uranium is a metal, though it is used mainly as a fuel by the nuclear industry. Uranium is produced in Colorado, New Mexico, Texas, and Wyoming.

Additional sources of all these metals are quite diverse, ranging from rocks to sea water, and continued exploration is needed to discover new reserves.

Nonmetallic Minerals

Nonmetallic mineral resources include any rock, mineral, or other naturally occurring substance that is of economic value, excepting metallic ores and mineral fuels. Such minerals, also commonly called industrial minerals, are vital to our economy: Generally they must be available in large quantities and be fairly close to market areas in order to supply the manufacturing, construction, and chemical industries at a reasonable cost. For purposes of this discussion, nonmetallic minerals will be divided into two categories: construction minerals, such as cement, gypsum, sand and gravel, and stone; and other minerals, such as barite, phosphate, salt, and sulfur.

Construction Minerals

Construction minerals are widespread in the Great Plains, and the major ones are produced in most of the states. These minerals and products include cement, clays and shale, gypsum, perlite, pumice, sand and gravel, and stone (Figure 19.7).

Cement, which is made from limestone and clay or shale, is the leading construction material produced in Nebraska and Texas and is among the top commodities in eight other Great Plains states (Table 19.11); cement is not produced in Minnesota or North Dakota because both lack limestone resources that can be mined.

Common clays and shale, used mainly to manufacture brick, tile, and cement, are a widespread resource that is being mined in each of the Great Plains states. Specialty clays are used for light-colored brick, refractory materials, pottery, ceramics, and as an additive in oil-well drilling fluids. These clays, which include bentonite, kaolin, fuller's earth, fire clay, and ball clay, are mined in only about half of the Great Plains states.

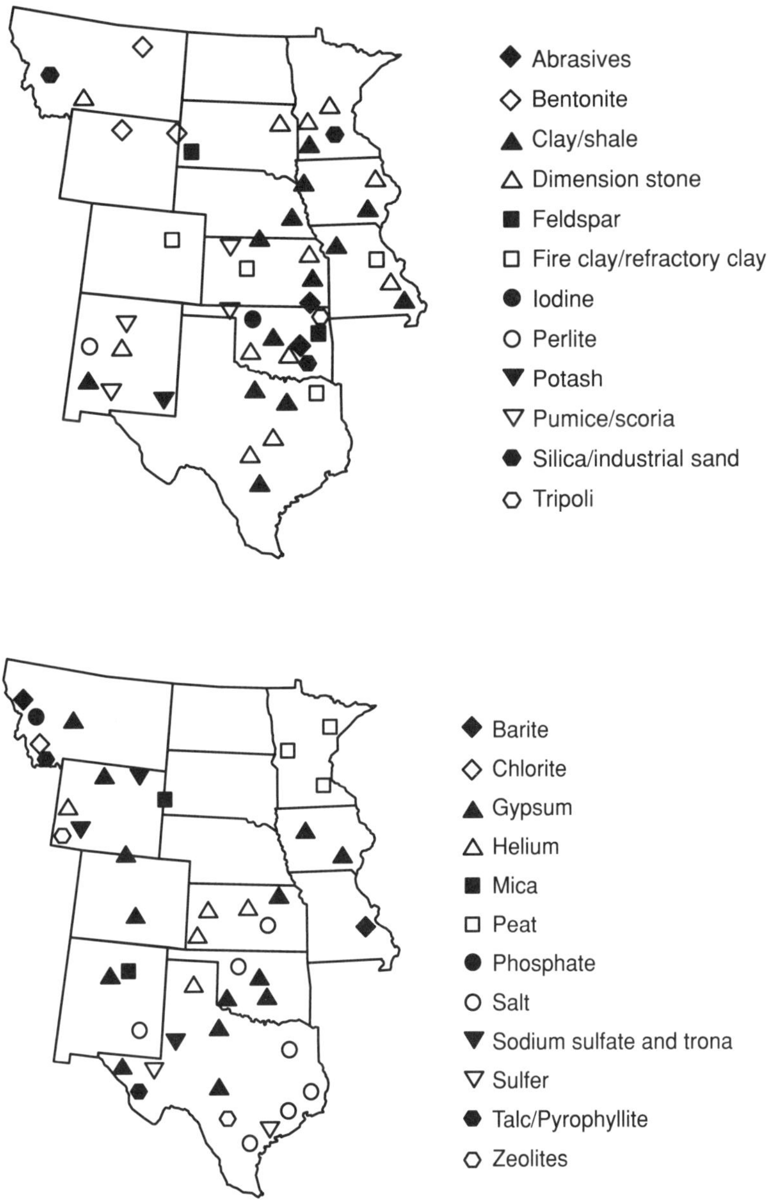

Source: U.S. Bureau of Mines 1992.

Figure 19.7. Major nonmetallic mineral-producing areas in the Great Plains states

Only in Wyoming are clays one of the leading mineral commodities, and this is because of the abundance of high-value bentonite, which is used by the petroleum industry in drilling fluids.

Gypsum is used mainly in the manufacture of wallboard, but other important uses are in producing plasters and cement. Gypsum is being mined in eight of the states (Figure 19.7) and is one of the major commodities in Iowa and Oklahoma (Table 19.11).

Perlite and pumice are glassy volcanic materials used in the construction industry. Pumice is a natural lightweight material, and perlite can be made to expand or "pop" into a lightweight material upon controlled heating. Both materials then can be used as lightweight aggregate in concrete blocks, thus reducing the weight of structures substantially. Perlite and/or pumice are produced in Colorado, Kansas, Oklahoma, and New Mexico (Figure 19.7).

Sand and gravel, suitable for construction, are nearly ubiquitous in the Great Plains states; they are so widespread and widely developed that their production areas would appear to blanket the region if noted on a map. Deposits of sand and gravel typically are associated with modern and ancient rivers and streams that crossed the region; therefore, these deposits occur mainly in flood plains of modern rivers and in the terrace deposits left in ancient river courses. Sand and gravel are produced in all 13 of the Great Plains states, are among the leading commodities in nine of the states, and are the leading construction minerals in Colorado and North Dakota (Table 19.11).

Stone resources consist of limestone, dolomite, granite, rhyolite, and other well-indurated rock materials. Stone either is cut or slabbed, and (maybe) polished, to be sold as dimension stone, or it is simply crushed into various sizes and marketed as crushed stone, mainly for aggregate. Stone is the leading nonfuel mineral being produced in Iowa and Kansas and is one of the leading minerals in seven other states; it also is produced to a lesser degree in the four remaining states of the region (Table 19.11).

Other Nonmetallic Minerals

The remaining nonmetallic minerals are a varied group of substances that occur in solid, liquid, and gaseous forms. Some of these minerals are extracted primarily for their chemical properties, while use of the others is based mainly on their physical properties.

Minerals used mainly for their chemical properties include feldspar, industrial sand and gravel, iodine, lime, peat, phosphate rock, potash, salt, soda ash, sodium sulfate, and sulfur. These are used in glass-making, fertilizers, metals processing, soaps, detergents, and in the manufacture of chemicals. Feldspars, industrial sand and gravel (high-purity silica sands), and soda ash (trona) are used mainly in glass manufacture.

Feldspar also is used in making pottery and other ceramic goods; silica sands' second major use is as a foundry sand; and soda ash is also used to manufacture chemicals, soaps, and detergents. Feldspar is produced in Oklahoma and South Dakota, whereas silica sands are produced in about half of the Great Plains states and are one of the leading minerals in Minnesota and Oklahoma (Table 19.11). Most of the U.S. soda-ash production is from Wyoming, and it is Wyoming's leading mineral.

Fertilizers consume 50 to 90 percent of the production of phosphate rock, potash, and sulfur, and the manufacturing of chemicals consumes most of the remainder of the production of these minerals. Among the Great Plains states, phosphate rock is produced only in Montana; potash is produced only in New Mexico, where it is one of the leading minerals; and sulfur is produced only in Texas, where it also is one of the leading minerals.

The remaining minerals that are used mainly for their chemical properties are diverse. Iodine, with Oklahoma as the sole U.S. producer, is used in animal feed supplements, catalysts, inks and colorants, pharmaceuticals, and disinfectants. Lime is produced in ten states in the region and is a leading mineral in Missouri, Nebraska, and North Dakota; lime is used mainly in steel making, flue-gas desulfurization, and paper and pulp manufacturing. Peat, produced in the region only in Iowa, Minnesota, and Montana, is used mostly as an organic source for soil improvement, potting soil, and nurseries. Salt, which is widespread in the Great Plains and is produced in Kansas, New Mexico, Oklahoma, and Texas, is used primarily in the chemical industry, as well as in de-icing roads and in the food and agriculture business. Salt is one of the leading minerals produced in Kansas. Sodium sulfate is used mainly in manufacturing soap and detergents, pulp and paper, textiles, and glass; it is produced only in Texas in Great Plains states.

Minerals used mainly for their physical properties include barite, gemstones, helium, mica, talc and pyrophyllite, tripoli, and vermiculite. Barite, used almost entirely as a weighting agent in oil and gas well drilling fluids, is produced in Missouri and Montana. Gemstones are produced in all 13 states of the region and are a leading mineral in North Dakota; gemstones include precious (sapphire) and semiprecious stones in some states, and freshwater mussel shells for producing cultured pearls in other states. Helium gas, produced along with natural gas in Colorado, Kansas, New Mexico, Texas, and Wyoming, has cryogenic applications, as well as use in welding, pressurizing, and purging. Helium is the leading mineral produced in Kansas and is also one of the leading minerals in Wyoming. Scrap mica is produced in New Mexico and South Dakota; it consists of small particles of mica used in joint cement, paint, roofing, oil well drilling fluids, and rubber products.

Talc and pyrophyllite, which are mined in Montana and Texas, are ground up for various uses. Talc is used in ceramics, paint, paper, and roofing; pyrophyllite is used in ceramics, refractories, and insecticides. Tripoli is a lightweight, high-purity silica mineral that is ground to a powder and used as an abrasive or buffing compound; tripoli is produced only in Oklahoma in the region. Vermiculite, produced only in Montana, is used in agriculture and as insulation and a lightweight aggregate.

Conclusions

The Great Plains states have contributed substantially to the mineral and energy wealth of the United States. Given the reserves and estimated undiscovered resources in the region, the Great Plains can continue to contribute significantly to meeting the nation's mineral and energy needs in the coming century.

However, for that to happen, there will have to be an improved public understanding of the need for such commodities and the value to this nation of obtaining much of those needs from domestic supplies. Unless that occurs, the United States will continue to rely more on foreign supplies for its mineral and energy needs, thus exporting jobs and placing an important part of the economy in others' hands. In addition, technology must continue to advance to allow further extraction of minerals from the Great Plains.

References

American Petroleum Institute (API). 1993. *Basic petroleum data book,* 13(1). Washington, D.C.: American Petroleum Institute.

Energy Information Administration (EIA). 1992a. *U.S. crude oil, natural gas, and natural gas liquids reserves, 1991 annual report.* Washington, D.C.: U.S. Department of Energy.

———. 1992b. *Natural gas annual, 1991.* Washington, D.C.: U.S. Department of Energy.

———. 1992c. *Coal production, 1991.* Washington, D.C.: U.S. Department of Energy.

Finkelman, Robert B., Friedman, S. A., and Hatch, Joseph R. (ed.). 1990. *Coal geology of the interior coal province, western region.* Reston, Virginia: Environmental and Coal Associates.

International Oil Scouts Association. 1991. *International oil and gas development,* Vol. 61. Austin, Texas: Mason Map Service, Inc.

Independent Petroleum Association of America (IPAA). 1990. *The oil and natural gas producing industry in your state* 60(8). Washington, D.C.: Independent Petroleum Association of America.

Keystone Manual. 1992. *1992 Keystone coal industry manual.* Chicago: Maclean Hunter Publishing Co.

U.S. Bureau of Mines. 1992. *Mineral commodity summaries, 1993.* Washington, D.C.: U.S. Bureau of Mines.

Agriculture in the Great Plains

20. Productivity of Great Plains Soils: Past, Present, and Future

G. A. Peterson
Colorado State University
C. V. Cole
U.S. Department of Agriculture

Introduction to Soils and Climate

Soils are products of the interaction of climate, organisms, time, topography, parent material, and human intervention in management. Soil properties of the Great Plains have been predominantly shaped by the climate variable. However, cultivation and grazing pressures in the region have been significant factors in the most recent 150 years. Soil management, as it interacted with climate and soils, has affected the region's productivity; the effects of those management practices will continue to affect soil productivity in the future.

Aandahl (1982) stated that "the uniqueness of the Great Plains results from its climate, which provides enough moisture for growing crops most years, although the threat of drought is ever present." Productivity of soils in the region depends greatly on the amount and timing of the precipitation. Productivity (total plant biomass produced per year) of the soil has always been a function of the climate, long before planned management of the Plains began.

The soil property soil organic carbon is greatest in the northeast (coolest and wettest area) and declines toward the warmer and dryer regions of the southwest, as illustrated in Figure 20.1 (Cole et al. 1989). Weathering rates are slow because of low precipitation (30 to 100 centimeters per year) (Figure 20.2), leaving soils that are very fertile in bases such as calcium, magnesium, and potassium. Over many years of grassland vegetation, large amounts of nitrogen, phosphorus, and sulfur accumulated in organic forms in conjunction with carbon. Cultivation depleted these organic nutrients simultaneously with soil carbon loss in cropland soils (Figure 20.1).

Precipitation effectiveness, however, is greatly modified by the temperature variable (Figure 20.3). Mean annual temperature varies from 4°C to 22°C from the Canadian border to southern Texas. Obviously, the variation in mean annual temperature is directly

Stanley R. Johnson and Aziz Bouzaher (ed.), *Conservation of Great Plains Ecosystems*, 325–342.

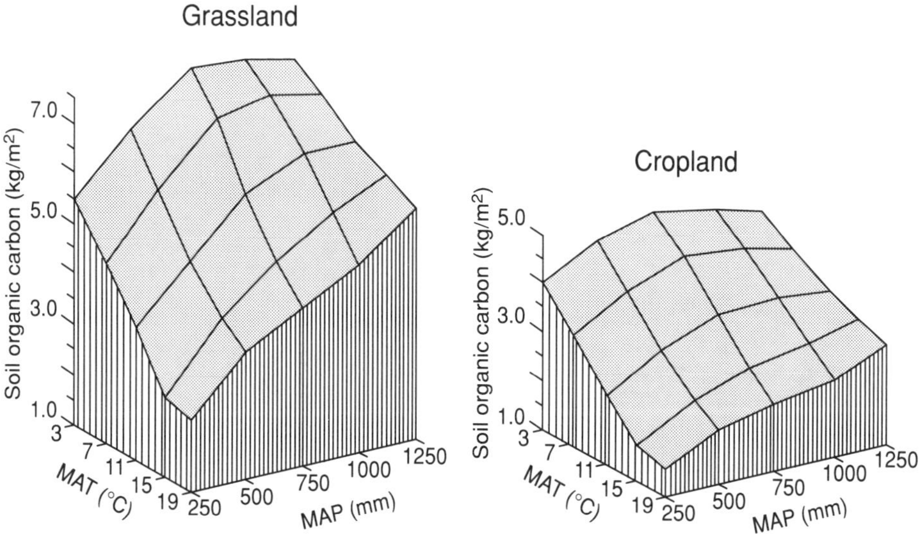

Figure 20.1. Simulated levels of soil organic carbon as a function of mean annual temperature (MAT) and precipitation (MAP) in the Great Plains: native grasslands on fine-textured soils, and cropland on fine-textured soils after 40 years of cultivation in wheat fallow

related to latitude and growing season length, which varies from 100 to 133 days north to south. Mean annual temperature, in turn, greatly affects the amount of water used in evapotranspiration.

Seasonal evaporation varies from 60 to 200 centimeters per year from north to south across the region (Figure 20.4). Where potential evapotranspiration is low, water from rainfall or snowmelt is more likely to enter and remain in the soil and percolate through the soil. The latter process speeds mineral weathering and allows a more distinct soil profile to develop. In like fashion, reduced evapotranspiration increases the potential for plant production because more water is available in the root zone and a given amount of plant production requires less total water. The latter reduces the probability of crop failure due to moisture stress. Reduced air temperatures result in cooler soils and plant residues returned to the soil decompose more slowly. The latter reduction allows soil organic matter to accumulate to a greater degree in northern latitudes than in the south.

Because of the many possible combinations of precipitation, temperature, and evaporation, a wide range of soil properties is found across the region. Greatest weathering of minerals and greatest organic matter accumulations are found in northern latitudes with highest precipitation and lowest evaporation. In contrast, the least weathered soils

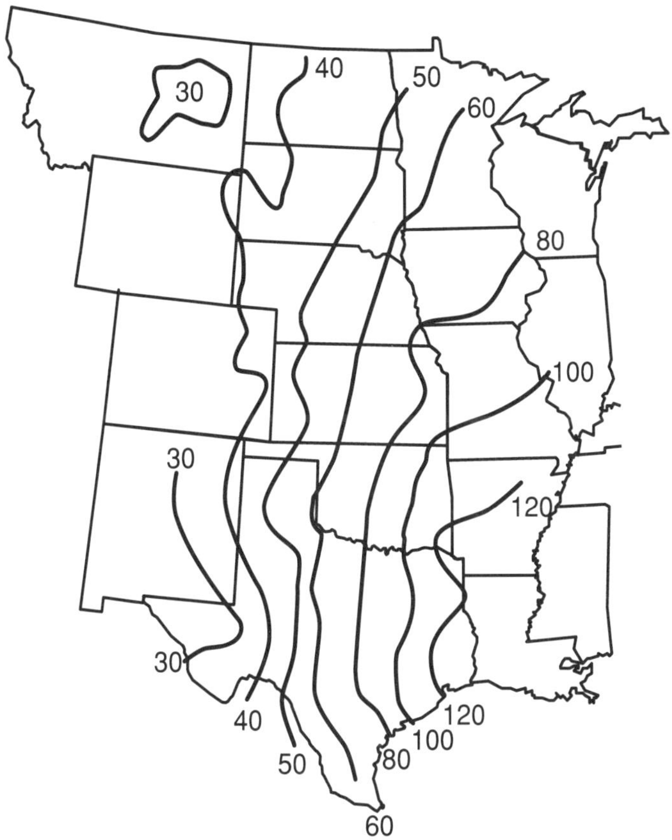

Figure 20.2. Mean annual precipitation (cm) in the Great Plains

are found where precipitation is at a minimum, and air temperatures and evaporation are the highest (i.e., in the southern areas of the region). Jenny (1941) showed that across a rainfall interval of 30 to 100 centimeters, the thickness of the surface soil free of lime (calcite) increased 5 centimeters for each additional 2.5 centimeters of annual precipitation. This feature characterizes the general weathering pattern of Great Plains soils and illustrates the great dependence of soil properties on climate.

Classification of soils into groups, based on measurable soil properties, as discussed above, is important to management decision making. Principal soils of the Great Plains are classified as Mollisols, Entisols, Aridisols, Vertisols, and Ustalfs:

> The Mollisols, formed under grass and forbs, have dark-colored surface horizons high in organic matter and bases, and they are the most extensive soils. In some landscapes, all of the soils are Mollisols. Those in the north, on the youn glacial tills, are thinner

and less developed—usually lacking argillic horizons. In general, to the south, they [soils] become older, more developed, and thicker. Entisols, the very young soils, are on the hilly and steep slopes, in very sandy tracts such as the Sandhills of Nebraska, and in the floodplains of streams with recent alluvium. Aridisols, the very dry soils, are in the western portions of the Great Plains. The driest area is in the southeastern part of New Mexico, extending into Texas. Most of the Vertisols, clayey soils with high shrink-swell potentials, are in the Blackland Prairies and Coastal Plains of Texas. Most of the Ustalfs, which formed under trees and have light-colored surface horizons, are in the Cross Timbers region of Texas and Oklahoma. Other Ustalfs, which formed mainly under grasses and have darker-colored surface horizons, are in the southern Great Plains, especially the southwestern part. These have sandy and fine sandy loam surface horizons and are too low in organic matter to be Mollisols (Aandahl 1982).

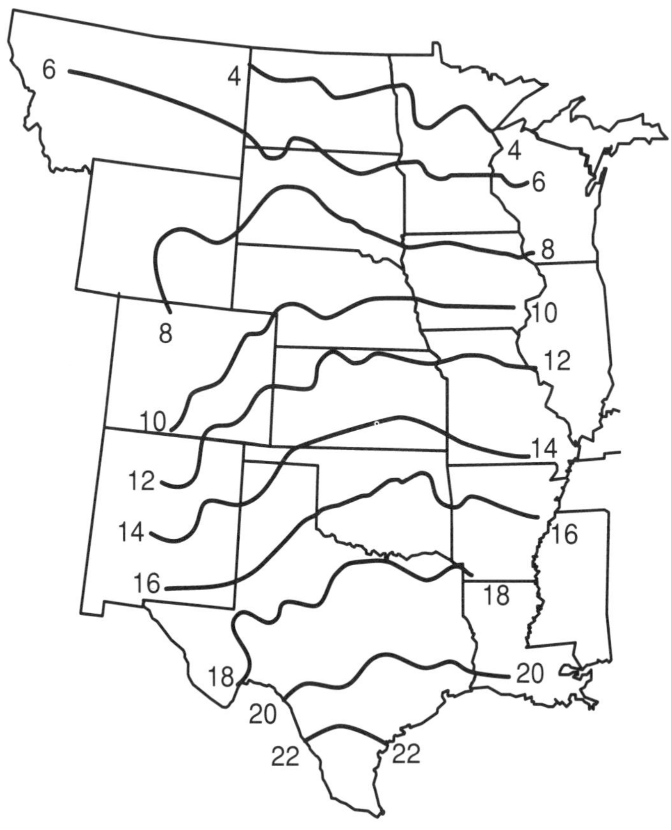

Figure 20.3. Mean annual temperature (°C) in the Great Plains

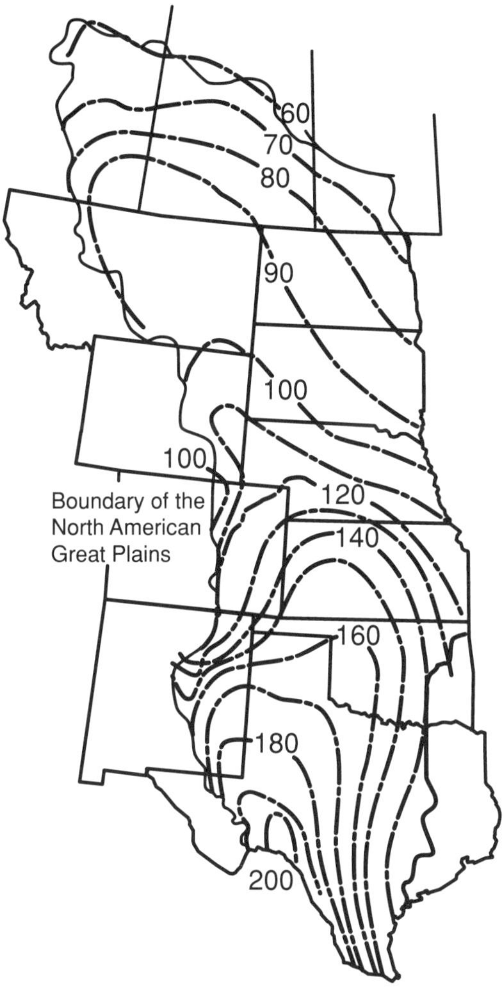

Figure 20.4. Mean annual evaporation in the Great Plains

Soil Productivity, 1850 to 1940

Shifts from Native Vegetation to Crops

Managed agriculture started in the Great Plains about 150 years ago. The first use of the region was for cattle grazing, which originated in Texas before the Civil War. By 1876, cattle were raised in all Great Plains states (Stewart 1990). The Federal Homestead Act, passed in 1862, allowed large land tracts to pass into private ownership, which ended open grazing in many areas. By the 1870s cultivated agriculture was well established in the eastern portions of the Great Plains. Land was plowed from the native vegetation and mixed farming began, oftentimes with techniques brought from the more humid European or eastern United States—techniques that were not suitable for the arid conditions of the Plains. Serious management mistakes caused great financial and social hardship to such an extent that federal and state experimental stations were established through the regions in the early 1900s to study the effects of different rotations and tillage practices on crop production under dryland conditions and to develop better management practices.

High wheat prices following World War I stimulated more sod breaking, and the most rapid conversion of grassland to cultivated agriculture occurred between 1915 and 1925. Serious wind erosion problems arose as a result of continued mismanagement and prolonged drought conditions in the 1930s in the Plains states. Entire communities collapsed because of the boom-to-bust problem that accompanied the Dust Bowl menace.

Drought is not new to the world, and particularly to the Great Plains. Dendrologists have shown that periods of drought as long as 38 years occurred before cultivated agriculture began in the region (Weakly 1962). Furthermore, these drought episodes had profound effects on the soil itself. Movement of soil by wind in these extended periods reshaped entire landscapes. Obviously, modern management strategies must account for the possibility of a water deficit at some time in every crop year. Efficient storage of precipitation, retention of that water against evaporation and weed use, followed by efficient use by crop plants, are major concerns of sustainable agriculture in the Plains.

Techniques used by the early settlers failed because they did not allow efficient storage and retention of water. In fact, their primary tillage tool, the moldboard plow, promoted inefficient capture and retention of precipitation. Soil was completely disturbed, and surface cover was destroyed, destabilizing the soil by reducing aggregate size and hastening organic matter depletion. As the mixing and grinding of the implements reduced aggregate size, it also stimulated oxidation of organic matter. To further complicate the situation, the open, bare surfaces left by these tillage techniques were susceptible to wind and water erosion.

Productivity Assessment

Soil organic matter greatly influences soil productivity from both physical and chemical aspects. Physically, it acts as a binding agent in soil structure formation, which directly affects soil aggregate stability. Aggregate stability is critical to maintaining macropore space in surface soil and subsequent water infiltration. Medium- to fine-textured soils with low organic matter content do not effectively resist raindrop impact. Intense rainfall or prolonged rain even at low intensity destroys surface aggregates and "sealing over" of surface soil pores occurs. Infiltration is reduced to almost nil and runoff potential increases. Obviously, erosion can follow if topographic conditions permit. At this point not only is soil loss imminent, but potential water storage in the soil profile is drastically reduced, decreasing potential plant productivity in subsequent weeks and months. Prolonged problems of this nature compound because lower plant productivity means less carbon being added to the soil and thus permitting further declines in soil organic matter.

Soil organic matter also is the storehouse for most of a soil's nitrogen-supplying capacity. Losing surface horizons to erosion by wind or water immediately reduces the nitrogen-supplying capacity of a given soil because the surface horizon is the zone of maximum organic accumulation. Obviously, if nitrogen supply becomes deficient for plant growth, productivity will be reduced unless outside nitrogen sources are added to the system.

When the system changed from native vegetation with little crop removal—except a part of the biomass by livestock—to cultivated agriculture with annual removal of grain or forage, this greatly altered the nitrogen budgets in the Great Plains. Organic matter retention in soils is only possible if the correct proportions of carbon, nitrogen, and phosphorus are present. When tillage stimulates carbon oxidation, nitrogen is mineralized and is used by the crops. When nitrogen is removed in cropping there is less opportunity to hold carbon in the soil organic matter.

Because soil organic matter directly influences critical chemical and physical properties of soil, it can be used as a sensitive index of soil productivity if used in comparison to the amount of organic matter present in the native condition at that particular site. Total quantities of nitrogen and phosphorus, the two most limiting plant nutrients in the Great Plains, were usually in a state of equilibrium with the climate and biology of the environment when cultivation began. Nitrogen was being consumed or lost by the soil-plant system at the same rate it was being added, while soils were in their native condition before settlers arrived. Nitrogen mineralized from the soil was absorbed by plants rapidly enough to minimize leaching potential in all but the wettest years.

The phosphorus available to plants was a function of phosphorus mineralized from organic matter plus the solubility of phosphate minerals. Cycling phosphorus through the soil pool reduced the opportunity for reactions with the soil mineral pool and thus kept phosphorus availability to plants at a reasonable level for the plant community on a given soil. As discussed earlier, native soil organic content was primarily a function of climate in the Great Plains, and so relative organic matter gains or losses within a climatic area can provide a useful productivity index within that area.

Removing nitrogen and phosphorus by animal grazing and nutrient losses through erosion were minimal under native conditions. Soil cultivation radically changed the equilibrium conditions in the native soils. Perturbations of the soil surface by tillage simultaneously exposed the soil surface to increased erosion, and the stirring decreased soil aggregate size (Peterson and Westfall 1990). New organic surfaces were exposed to mineralization by stirring, and both nitrogen and phosphorus were released from organic pools at faster rates than in noncultivated systems. Removing nitrogen and phosphorus from the system in grain and forage crops also became more important than under native systems grazed by undomesticated animals. Furthermore, cultivated crops, particularly under dryland conditions, did not return as much carbon to the soil as did the native plants. This stimulated nitrogen and phosphorus conversion to mineral forms because the steady-state carbon-to-nitrogen and carbon-to-phosphorus ratios were no longer possible.

Jenny (1941) was one of the first to quantify the cultivation impact in nitrogen and soil organic matter (Figure 20.5). Note that it required 35 to 50 years for a new steady-state to be established. A nitrogen budget calculated by Peterson and Vetter (1971) showed that erosion and crop removal accounted for most of the nitrogen losses in a typical cultivated system, but that significant quantities of nitrogen were still unaccounted for. Lamb and others (1985) conclusively demonstrated that some of the rapidly mineralized nitrogen in a newly cultivated soil was leached below the root zone of winter wheat even in areas with less than 400 millimeters of annual precipitation. Olson and colleagues (1974) also have documented that dryland farming with the wheat-fallow system has resulted in significant quantities of nitrogen being leached below the 180-centimeter depth. Irrigation of these same soils accelerated leaching losses to soil zones below the rooting depths of most crop plants, except alfalfa, which can recover nitrogen from as deep as 5 meters in permeable soils.

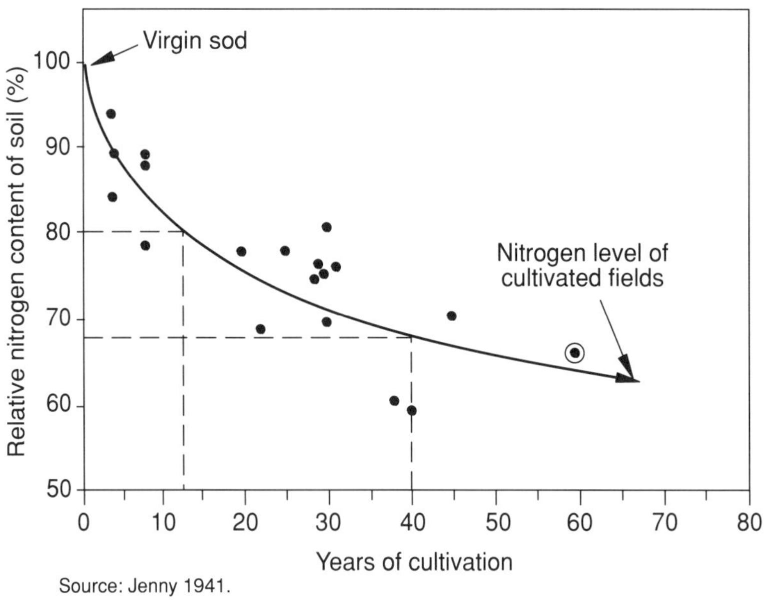

Figure 20.5. Decline in soil nitrogen content with time of cultivation

Phosphorus is used by cultivated crops in smaller quantities than is nitrogen and is not subject to leaching. However, phosphorus is lost from soil-plant systems by erosion processes when soil cover is removed and aggregate size is reduced by tillage. Tiessen and others (1982) showed that there were substantial losses of total phosphorus attributable to erosion, and that relatively large losses came from the organic phosphorus fraction. This is particularly serious in calcium-dominated systems such as those found in arid and semiarid areas.

Soluble phosphorus, whether released in organic matter mineralization or by mineral dissolution, does not remain in solution but precipitates as inorganic forms of low solubility, which are less available to plants. The net result is that a soil that was capable of supplying adequate phosphorus for the native plants, and even to cultivated crops in the early years immediately following sod breaking, has a decreasing capacity to supply phosphorus needs as time under cultivation increases. This change results from phosphorus removal in harvested plant parts and conversion of mineralized phosphorus to unavailable mineral forms.

The upshot is that fertilizer supplements of both nitrogen and phosphorus begin to be needed as time under cultivation increases. For example, the transformation from a nitrogen- and phosphorus-rich condition to a deficient condition occurred in the Great Plains over a period of about 35 to 50 years (Sander and Peterson 1968). Soils broken from the native sod around 1915 to 1920 had nutrient deficiencies by 1965. The time interval was shortened in irrigated areas because crop removal of nutrients, especially nitrogen, was greater. Dryland areas of the Plains did not require inorganic fertilizers until well into the 1960s. Grain protein content declined more quickly than yield and was the first indicator of a nitrogen deficiency problem (Sander and Peterson 1968).

Organic matter depleted under the intense cultivation practices of the pioneer farmers and productivity went with it. Haas and others (1957) and Peterson and Vetter (1971) have documented loss of carbon and nitrogen from surface soil ranging from 35 to 50 percent. Actual amounts of carbon and nitrogen loss range from 10,000 to 20,000 kilograms per hectare and 500 to 1,000 kilograms per hectare for carbon and nitrogen (Peterson and Vetter 1971). Anderson and Peterson (1973) report an exponential decline in corn yields for years after sod breaking that closely paralleled the decline in total soil nitrogen content. Soil nitrogen content declined 49 percent and soil productivity (indexed by corn grain yield) declined 71 percent during the 28 years after sod breaking.

Productivity of Great Plains soils declined precariously in the early years after sod breaking because of inappropriate management techniques. Fortunately, producers and research scientists have been able to stem the degradation and even improve soil productivity since World War II.

Soil Productivity, 1940 to 1990

Present Management Techniques

Duley and Russel (1939) were among the first to recognize the need for using crop residues as soil covers during noncrop periods and seedling growth stages when crop canopies are incomplete. Their primary concerns were erosion control and moisture conservation, which they addressed by developing a sweep machine that would kill weeds but leave most of the crop residue on the soil surface. Their technique also disturbed the soil less and indirectly began to affect nitrogen and carbon cycling, although they were not particularly aware of it at the time. By 1954, agronomists had found that minimum tillage promoted soil tilth and conserved organic matter (Melsted 1954). Furthermore, the positive interaction of improved tilth, soil organic matter, and productivity was being recognized.

Effects of Management Changes on Soil Properties

In the early 1900s the combination of low-yielding varieties, complete straw removal at harvest, and intensive cultivation caused serious losses of organic carbon and available nutrients. These changes were associated with detrimental physical conditions, poor water conservation, and accelerated erosion. After 1940, management practices were modified to include higher yielding varieties, application of nitrogen and phosphorus fertilizers, no straw removal at harvest, and stubble mulch tillage. Analysis based on statistical and mechanistic models by Cole and others (1989) indicated that these management changes arrested soil carbon and nitrogen losses and that soil carbon levels stabilized.

Hobbs and Thompson (1971) showed that adequate fertilization of soil that had lost a significant amount of surface horizon organic matter would often restore the soil to full productivity. Lueking and Schepers (1985) reported that a combination of irrigation and nitrogen fertilizer on corn grown on coarse-textured soil in the Nebraska Sandhills actually increased soil organic matter. They observed increases of 7 percent for soil carbon weight and 5 percent for nitrogen weight in the upper 30 centimeters of the soil as compared with native soil conditions before irrigation and fertilization. It is important to note that the soil they were studying was very low in organic matter, even in its native condition, because of coarse soil texture and limited precipitation.

Power and Myers (1989) reported that adopting minimum tillage methods, which leave more cover on the soil during vulnerable times, does reduce soil erosion and associated nutrients. Interactions of tillage and fertilization are the primary management keys to understanding what will happen to organic matter levels and subsequent productivity. Blevins and others (1984) and Lamb and colleagues (1985) have documented the close link between the amount of tillage and nitrogen conservation (organic matter). Work by the Lamb group (1985) showed that after 12 years of cultivation of newly broken prairie soil, losses of soil nitrogen from the 0- to 30-centimeter depth were 3 percent for no-till, 8 percent for stubble mulch, and 19 percent for plowing. It is critical that tillage be reduced if organic matter and productivity are to be maintained and improved.

Changes in Productivity

The development and adoption of systems that minimize erosion potential, decrease soil disturbance, enhance water storage, and encourage more efficient plant water use have stabilized and begun to improve soil productivity in the present. Greb (1979) summarized the effects of tillage management changes on the productivity of the wheat-fallow system (Table 20.1).

Table 20.1. Progress in fallow systems and wheat yields, Central Great Plains Research Station, Akron, Colorado

Years	Systems	Annual Precip.	Fallow Water Storage	Fallow Efficiency	Yield	Water Use Efficiency
		mm	mm	percent	kg/ha	kg/ha/mm
1916-30	Maximum till, plow, harrow	439	102	19	1,070	1.21
1931-45	Shallow disk, rod weeder	401	112	24	1,160	1.42
1946-60	Stubble mulch, sweeps, rod weeder	416	137	27	1,730	2.05
1961-75	Stubble mulch, sweeps, herbicides	389	157	33	2,160	2.76
1976-90	Minimum tillage (projection)	411	183	40	2,690	3.23

Source: Adapted from Greb 1979.

Note that water use efficiency (WUE) more than doubled from 1916 to 1975 in conjunction with changes from maximum tillage systems to stubble mulch practiced with minimum tillage. Duley and Russel (1939) expressed the opinion,

> When all these things are considered, it would seem that the proper utilization of crop residues on the surface of the soil may offer a simple and practical method for reducing runoff and erosion. By increasing infiltration and reducing evaporation losses, a more efficient use may be made of rainfall, which appears to be absolutely necessary for the maintenance of a stable and permanent agriculture in the Great Plains.

They were truly prophets in their own time and Greb's summary data bear it out: WUE increased from 1.21 (1916-30) to 2.76 (1961-75) kilograms/hectare/millimeter of water by decreasing tillage and keeping a crop residue cover on the soil surface.

Soil Productivity in the Next Century

Potential Management for the Future

Hanway (1976) viewed the future of agriculture in the Great Plains in a very perceptive manner. His vision was:

> The Great Plains has greatly expanded and stabilized production of crop and livestock products in the last 40 years (1935–1975). Agriculture in this region has advanced with increased irrigation, improved range and livestock management practices, better crop varieties, expanded use of fertilizer, and more timely and effective cultural practices. Irrigation has been a primary driver in the expansion of agricultural production in the plains. *However,* available water supplies and the rising cost of energy will set limits on further expansion. Inadequate recharge to prevent decline of water tables will cause some areas now irrigated to revert to rainfed agriculture.

The dust storms of the "dirty '30s" were directly caused by tillage systems that bared the soil. Duley and Russel (1939) pioneered the concept of subsurface tillage to control weeds while leaving the soil surface covered with crop residues. They called their method *stubble mulch farming.* Their system greatly decreased soil erosion potential but had inconsistent effects on soil water storage.

Hanway (1976) further commented regarding tillage:

> Is tillage necessary? No, it is not, at least not to the extent now used because *tillage wastes water*. Herbicides can replace most, if not all, tillage involved in dryland production systems in the Great Plains. Effective use of herbicides will reduce weed control problems and avoid those water losses directly due to tillage operations. In the Great Plains, where the primary limiting factor of production is water, substituting herbicides for tillage is critical to improved production and efficiency.

In addition, Hanway recognized that tillage choices are often based more on tradition than on scientific principles. He suggested that economic advantage for one system over another was the only thing that would ensure change. He concluded:

> Tillage, as generally employed in even the best systems, increases erosion potential by both wind and water and increases evaporation of water that should be saved for plant use. Any tillage above a scientifically determined essential minimum is excessive and undesirable. Tillage is following the horse as an outmoded element in Great Plains agriculture.

Expected Changes in Productivity

In 1976, Hanway identified five deterrents to adoption of water conserving practices. They involved issues ranging from an inadequate research base to inadequate planting equipment to deep-seated tradition. At the beginning of the twenty-first century most of these obstacles have been overcome, and now, 17 years after Hanway's analysis, the benefits of reduced tillage through increased productivity per unit of precipitation should be in the reaping. Unfortunately, not all of the potential benefits he believed could come about have developed. Residue management is slightly improved and erosion potential reduced, but increased soil productivity and economic stability are not as apparent. Traditional wheat fallow, a mainstay risk-reducing production system, has not and cannot efficiently use the additional water stored with reduced and no tillage systems.

Greb (1979) reported increasing WUE values with stepwise improvement in tillage systems (Table 20.1). His database ended in 1979, a point where the maximum possible fallow efficiency had been nearly reached, because fallow water storage was now at the upper limits of the soil's water-holding capacity. Since the potential yield of a wheat crop is determined as much by the amount of spring precipitation as it is by the amount of stored soil water at planting, any additional small gains in stored water will not result in substantial wheat yield increases. Data from Wicks and Smika (1973) demonstrate this situation: A tilled wheat fallow system with 127 millimeters of stored water at planting produced 2,670 kilograms per hectare of grain, while a no-till treatment that conserved 274 millimeters of water under the same climatic conditions produced 3,150 kilograms per hectare (five-year means). More than doubling stored soil water only increased grain yield 18 percent. Production per millimeter of stored water was 21 kilograms/hectare/millimeter of water with tillage and only 11.5 kilograms/hectare/millimeter with no tillage. Obviously, the extra stored water was not used efficiently. The long 14-month fallow period between wheat crops may not be necessary if good water conservation techniques, like reduced and no tillage, are used.

Rotations including summer crops provide an excellent opportunity to take advantage of our new capabilities to store water efficiently and increase WUE substantially in the Great Plains. Rotations with summer crops were the missing pieces in Hanway's perspective for the future of the Plains. The benefits of crop diversity that accompany rotation farming are well known around the world but were not feasible in the Plains before improvements in water conservation techniques. Peterson and Westfall (1990) report that three- and four-year rotations, like wheat-corn fallow (WCF) and wheat-corn-millet fallow (WCMF), have doubled annual grain production in northeastern Colorado, compared to wheat fallow (WF).

WF = 1,170 kg/ha/yr; WCF = 2,240 kg/ha/yr; WCMF = 2,430 kg/ha/yr

Grain production also has doubled using the same total amount of precipitation with no supplemental irrigation. In addition, the amount of biomass carbon returned to the soil also has doubled.

McGee (1992) studied these same extended rotations over a soil and climatic gradient in eastern Colorado and reported annual grain WUE of 2.8, 4.7, and 5.9 kilograms/hectare/millimeter for WF, WCF, and WCMF rotations (Table 20.2). Comparing these WUE values with those reported by Greb (1979) (see Table 20.1) shows that McGee's WF is similar to Greb's 1961 to 1975 value but does not quite meet the 1990 projection for WF. Extended rotations, possible because of the excellent water conserving features of no-till practices, create large productivity gains far beyond Greb's projected values for WF. Rotations of WC(S)F and WC(S)MF resulted in WUE values of 4.7 and 5.9 compared with 2.8 kilograms/hectare/millimeter for WF.

Soil organic matter, a characteristic used as a soil productivity index, also has increased under the intensified crop rotations. Wood and others (1991) demonstrated that, just four years after initiating no-till techniques, a no-till WCMF rotation had significantly higher carbon and nitrogen contents in the 0- to 10-centimeter soil layer than did WF practiced with no tillage. Wood's group (1990) further showed that after 3.5 years of no tillage, potential carbon and nitrogen mineralization, carbon turnover, and relative nitrogen mineralization were 61, 39, 36, and 43 percent greater under WCMF than WF. Differences in potential carbon and nitrogen activity between rotations were due to greater surface organic carbon concentrations under WCMF than WF, which were related to cumulative plant residue additions over the 3.5-year time period. The overall conclusion was that both the soil organic carbon and nitrogen were very sensitive

Table 20.2. Annual water use efficiency as affected by cropping system, climate, and soil gradients in eastern Colorado

	Yield WUE		
Soil Position	WF	WC(S)F	WC(S)MF
		kg/ha/mm	
Sterling (Lowest ET)			
Summit	2.8	4.1	5.0
Sideslope	2.8	5.0	6.2
Toeslope	3.0	5.9	7.2
Mean	2.9	5.0	6.1
Stratton (Medium ET)			
Summit	3.4	4.2	4.8
Sideslope	2.6	4.0	4.4
Toeslope	4.0	6.7	7.7
Mean	3.3	5.0	5.6
Walsh (Highest ET)			
Summit	2.0	2.9	4.3
Sideslope	2.3	3.4	5.2
Toeslope	2.6	6.2	8.3
Mean	2.3	4.2	5.9
Crop Rotation Means	2.8	4.7	5.9

[L.S.D. Site × Soil = 7; Site × Rotation = 5; Soil × Rotation = 4]

Notes:
ET = evapotranspiration
WF = wheat fallow
WC(S)F = wheat-corn(sorghum) fallow
WC(S)MF = wheat-corn(sorghum)-millet fallow

to change in cultural practices and were increased by increasing cropping intensity under no-till management.

Metherell (1992), using the CENTURY model (Parton et al. 1992), simulated the effects of long-term use of no-till crop rotations in eastern Colorado on soil productivity and soil organic matter with various climate change scenarios. With ambient CO_2, using a full range of historic weather data, the simulations predicted that the intensive crop rotations would increase crop productivity and soil organic matter over time, especially in the coolest parts of eastern Colorado. Metherell also predicted that with elevated CO_2 levels and higher temperatures, soil organic matter would continue to increase if wheat remained in the rotation. At higher temperatures, the more intensive rotations with corn, sorghum, or millet still benefited productivity and soil organic matter accumulated, but to a lesser degree because these C-4 species responded adversely to the higher summer temperatures, decreasing carbon return to the soil.

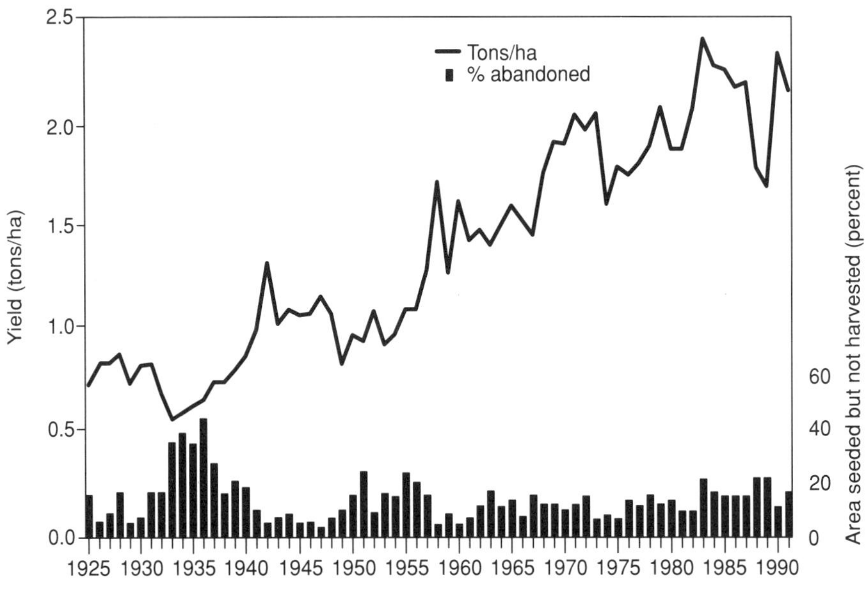

Source: Jenny 1941.

Figure 20.6. Wheat yields and seeded land abandoned in the Great Plains from 1925 to 1991

Conclusions

Current predictions indicate that the Great Plains will sustain higher levels of productivity over the long term when management practices incorporate high yielding varieties, minimum soil disturbance, replacement of nutrients removed in harvest, and residue conservation (Cole et al. 1988). Adaptability of agriculture to stressful climatic variations is evident in the Great Plains, as illustrated in Figure 20.6. Notwithstanding four droughts and their impacts, wheat growers and associated scientists have more than doubled yields since 1925. Increased water conservation using no-till technology makes more intense crop rotations possible, which should further enhance productivity compared to the wheat production data reported in Figure 20.6. Management systems that incorporate more intense crop rotations can increase net income to the producer by 30 percent annually compared to standard tilled wheat fallow farming (Peterson et al. 1992). It is encouraging to note that economic incentives do exist for adopting these soil improving systems, but current policies may limit their adoption.

Overall, the future for soil productivity in the Great Plains appears to be bright, and these trends should continue.

References

Aandahl, A. R. 1982. *Soils of the Great Plains.* Lincoln: University of Nebraska Press.

Anderson, F. N., and G. A. Peterson. 1973. Effects of continuous corn (Zea mays L.), manuring, and nitrogen fertilization on yield and protein content of the grain and on the soil nitrogen content. *Agronomy Journal* 65:697-700.

Blevins, R. L., M. S. Smith, and G. W. Thomas. 1984. Changes in soil properties under no-tillage. In *No-tillage agriculture,* ed. R. E. Phillips and S. H. Phillips. New York: Van Nostrand-Reinhold.

Cole, C. V., I. C. Burke, W. J. Parton, D. S. Schimel, D. S. Ojima, and J. W. B. Stewart. 1988. Analysis of historical changes in soil fertility and organic matter levels of the North American Great Plains. In *Challenges in dryland agriculture,* proceedings of the International Conference on Dryland Farming, Aug. 15-19, Amarillo, Texas.

Cole, C. V., J. W. B. Stewart, D. S. Ojima, W. J. Parton, and D. S. Schimel. 1989. Modelling land use effects of soil organic matter dynamics in the North American Plains. In *Ecology of arable land,* ed. M. Clarholm and L. Bergstrom. Netherlands: Kluwer Academic Publishers.

Duley, F. L., and J. C. Russel. 1939. The use of crop residues for soil and moisture conservation. *Agronomy Journal* 31:703-709.

Greb, B. W. 1979. Reducing drought effects on croplands in the west-central Great Plains. USDA-ARS Information Bulletin 420. U.S. Department of Agriculture, Agricultural Research Service.

Haas, H. J., C. E. Evans, and E. F. Miles. 1957. Nitrogen and carbon changes in Great Plains soils as influenced by cropping and soil treatments. USDA Technical Bulletin 1164. Washington, D.C.: U.S. Government Printing Office.

Hanway, D. G. 1976. Cultural advances in integrated dryland production systems. In *Conservation tillage: Proceedings of Conservation Tillage Workshop, Ft. Collins, Colorado, Aug. 10-12*. Great Plains Agricultural Council Bulletin 77. Fort Collins, Colorado.

Hobbs, J. A., and C. A. Thompson. 1971. Effect of cultivation on the nitrogen and carbon contents of a Kansas Argiustoll. *Agronomy Journal* 63:66-68.

Jenny, H. 1941. *Factors of soil formation*. New York: McGraw-Hill Book Co.

Lamb, J. A., G. A. Peterson, and C. R. Fenster. 1985. Wheat fallow tillage systems' effect on a newly cultivated grassland soils' nitrogen budget. *Soil Science Society of America Journal* 49:352-356.

Lueking, M. A., and J. S. Schepers. 1985. Changes in soil carbon and nitrogen due to irrigation development in Nebraska's Sandhill soils. *Soil Science Society of America Journal* 49:626-630.

McGee, E. A. 1992. Water use and storage efficiency in no-till dryland agroecosystems. M.S. thesis, Colorado State University, Ft. Collins, Colorado.

Melsted, S. W. 1954. New concepts of management of Corn Belt soils. *Advances in Agronomy* 6:121-142.

Metherell, A. K. 1992. Simulation of soil organic matter dynamics and nutrient cycling in agroecosystems. Ph.D. dissertation, Colorado State University, Ft. Collins, Colorado.

Olson, R. A., J. H. Muir, R. W. Wesely, G. A. Peterson, and J. S. Boyce. 1974. Accumulation of inorganic nitrogen in soils and waters in relation to soil and crop management. In *Effects of agricultural production on nitrates in food and water with particular reference to isotope studies*. AEA-PL-539/3. Atomic Energy Agency, Vienna, Austria.

Parton, W. J., B. McKeown, V. Kirchner, and D. Ojima. 1992. *Users guide for the CENTURY model*. Colorado State University, Fort Collins, Colorado.

Peterson, G. A., and D. J. Vetter. 1971. Soil nitrogen budget of the wheat-fallow area. *Farm, Ranch, and Home Quarterly* 17(Winter):23-25.

Peterson, G. A., and D. G. Westfall. 1990. Sustainable dryland agroecosystems. In *Conservation tillage: Proceedings, Great Plains Conservation Tillage Symposium*. Great Plains Agricultural Council Publication 131. Bismark, North Dakota: Great Plains Agricultural Council.

Peterson, G. A., D. G. Westfall, and A. D. Halvorson. 1992. Economics of dryland crop rotations for efficient water and nitrogen use. In *Proceedings of the Great Plains Soil Fertilizer Conference*, March 3-4, Denver, Colorado.

Power, J. F., and R. J. K. Myers. 1989. The maintenance or improvement of farming systems in North America and Australia. In *Soil quality in semiarid agriculture: Volume I*, ed. J. W. B. Stewart, proceedings of an International Symposium on Soil Quality, June 11-16, Saskatoon, Canada.

Sander, D. H., and G. A. Peterson. 1968. Western Nebraska wheat protein content dropping. *Farm, Ranch, and Home Quarterly* 15(2):10-11.

Stewart, B. A. 1990. Limits, bounds, and future of conservation tillage in the Great Plains. In *Conservation tillage: Proceedings, Great Plains Conservation Tillage Symposium*. Great Plains Agricultural Council Publication 131. Bismark, North Dakota: Great Plains Agricultural Council.

Tiessen, H., J. W. B. Stewart, and J. R. Bettany. 1982. Cultivation effects on the amounts and concentrations of carbon, nitrogen and phosphorus in grassland soil. *Agronomy Journal* 74:831-835.

Weakly, H. E. 1962. History of drought in Nebraska. *Journal of Soil and Water Conservation* 17:271-274.

Wicks, G. A., and D. E. Smika. 1973. Chemical fallow in a winter wheat-fallow rotation. *Journal of the Weed Science Society of America* 21:97-102.

Wood, C. W., D. G. Westfall, G. A. Peterson, and I. C. Burke. 1990. Impacts of cropping intensity on carbon and nitrogen mineralization under no-till dryland agroecosystems. *Agronomy Journal* 82:1115-1120.

Wood, C. W., D. G. Westfall, and G. A. Peterson. 1991. Soil carbon and nitrogen changes on initiation of no-till cropping systems. *Soil Science Society of America Journal* 55:470-476.

21. Agricultural Systems and Technologies of the Great Plains

Melvin D. Skold
Colorado State University and the Great Plains Agricultural Council

The Great Plains is a vast and diverse area. Its 10 states, covering almost one-third of the U.S. land mass, are mostly semiarid. Because of its size and geographic definition, the region has dramatic extremes in elevation, temperature, precipitation, and in wind direction and velocity. The growing season varies from 110 to more than 300 days. Annual precipitation ranges from 12 to more than 30 inches. The semiarid character results in important parts of the region being on the extensive margin between cropping and rangeland grazing. Further, year-to-year variability in precipitation gives rise to strategies for which risk management is important.

The Great Plains is variously defined but the Great Plains Agricultural Council uses a classic definition to guide much of its activities. The Great Plains Proper (GPP) is a 396-county area within the 10-state region that includes the plains portions of the states (Skold 1991a). The GPP is bounded on the west by the Rocky Mountains and extends east to approximately the 98th meridian, where the mixed-grass true prairies give way to the tallgrass prairies farther east (Kraenzel 1955; Kuchler 1964). The southern boundary of the GPP separates the Panhandle of Texas from the remainder of the state.

The agricultural systems and technologies discussed in this chapter refer to the GPP. Whether taken as the whole of the 10-state region or as the GPP, it is an important agricultural region for the United States.

Land Use

The states in which the Great Plains lies include over 705 million acres of land (Table 21.1). More than one-half is in grassland pasture and range, a mix of native (unbroken) rangelands and improved pastures. In 1987, 197 million acres were classed as cropland and 85 million acres were forest land.

Stanley R. Johnson and Aziz Bouzaher (ed.), *Conservation of Great Plains Ecosystems*, 343–364.

Table 21.1. Land and cropland use in the Great Plains states and United States, 1987

	Great Plains States	United States
Land use	thousand acres	
Cropland		
Used for crops[a]	143,761	330,877
Idle	25,618	68,143
Used only for pasture	28,144	64,877
Grassland	361,549	591,083
Forest land	84,649	648,164
Special uses	33,220	278,599
Other land	28,116	283,300
Total land area	705,107	2,265,147
Cropland use		
Cropland harvested	105,548	282,224
Failure	2,107	3,638
Cultivated summer fallow	28,562	35,322
Annual acreage reduction	21,341	43,153
Conservation Reserve Program	5,482	9,871
Irrigated land	20,671	46,386
Irrigated cropland	20,132	38,214

Sources: USDA 1991; U.S. Department of Commerce 1989.

[a]Cropland harvested, crop failure, cultivated summer fallow.

Important amounts of Great Plains cropland are used for cultivated summer fallow. Because of a relatively high incidence of crop failure and the practice of summer fallow, harvested cropland is a smaller proportion of total cropland than is true for other regions in the United States. Further, more than 28 million acres of cropland are used only for pasture, cropland that is annually planted to a small grain, forage sorghum, or a perennial legume-grass mix and is grazed by livestock. More than 20 million acres of land in farms in the Great Plains are irrigated; 97 percent of the irrigated land is cropland. In the 1987 Census of Agriculture, more than 21 million acres of cropland were under the annual cropland set-aside program and 5.5 million acres had been enrolled in the Conservation Reserve Program (CRP).

Since the Soil Bank program of the 1950s, the Great Plains states have had sizable amounts of cropland idled by annual and longer term cropland retirement programs (Figure 21.1) (Heimlich and Kula 1991). The Soil Bank idled over 29 million acres of Great Plains land for extended times, but expiration of the Soil Bank contracts and the attractive commodity prices of the early 1970s brought most of that land back into crop production. Now, however, more than 20 million acres of Great Plains cropland is under CRP contract (USDA 1992).

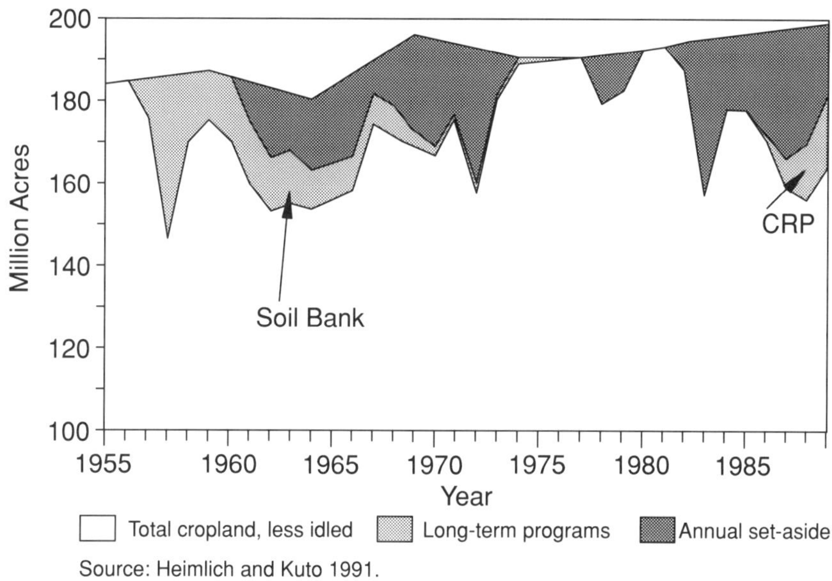

Figure 21.1. Cropland and idled cropland in the Great Plains states, 1954–89

Farm Structure

The 535,000 farms in the Great Plains states average more than 1,000 acres per farm, more than twice the U.S. average. Farm size in the GPP is more than three times the U.S. average (Table 21.2). Based on total value of farm sales, the Great Plains states and the GPP are more livestock oriented than the rest of the United States.

Value of farm marketings per farm is slightly greater for the 10-state region than for the United States; for the GPP sales per farm are 70 percent greater than the U.S. average. However, sales per acre of land are $141 for the United States, $69 for the 10 states, and $76 for the GPP. The GPP leads the 10-state total because of its irrigated land and large-scale beef feedlots.

Even though the Great Plains has substantial amounts of irrigated cropland, value of production per acre is considerably less than the U.S. average. Government programs that offer per acre incentives receive high rates of participation in the Great

Plains. The relatively lower values of output per acre (net as well as gross), together with the climatic uncertainty inherent to the region, make land-based government farm programs attractive and income stabilizing options for many producers.

The Great Plains states and the GPP contribute significantly to U.S. production of several commodities (Table 21.3). Wheat, sorghum, beef cattle, and fed cattle are the products for which the region produces the greatest share of U.S. production.

Table 21.2. Farms and their value of marketings, United States, Great Plains states, and Great Plains Proper, 1987

	United States	Great Plains States	Great Plains Proper
Number of farms (1,000)	2,088	535	214
Land in farms (1,000 acres)	964,471	544,546	310,092
Average farm size (acres)	462	1,020	1,451
Cropland in farms (1,000 acres)	443,318	186,592	123,300
		million dollars	
Market value of products sold	$136,048	$37,742	$23,573
Value of crops sold	$58,931	$11,567	$6,836
Value of livestock/products	$77,177	$26,157	$16,737
		dollars	
Value of sales per farm	$65,165	$70,537	$110,305
Value of sales per acre	$141	$69	$76
Crop sales per cropland acre	$133	$62	$55

Table 21.3. Value of U.S. farm production in Great Plains states and Great Plains Proper, 1987

	Great Plains States	Great Plains Proper
	percent of U.S. total	
Sorghum	87	40
Corn	21	13
Wheat	61	51
Cotton	36	22
All cattle	59	42
Fed cattle	65	50

Agricultural Systems

Previously, researchers (Hines et al. 1991; Sutton 1984) defined 9 and 16 agricultural subregions in the GPP. Their broadly defined physiographic regions reflect identifiable patterns of soils, climate, water resources, and land uses. For purposes of this chapter, however, five agricultural systems are defined by major land use and technological practice. Although each of the five systems is a distinct type of resource use or farming system, considerable variation exists among the systems; most farms possess some combination of two or more.

Each system employs practices that, in one sense, are sustainable in the region. Early settlers attempted and failed at a variety of farming and ranching practices. In general, attempting to survive by adopting management practices that were successful in the eastern United States or Europe did not succeed in the GPP. The systems remaining and the technologies and management practices that are applied are those that are technologically most successful in this fragile and uncertain environment.

The variations existing within each system are associated with the geographic, topographic, and climatic extremes of the region. Elevation and latitude combine to limit crop alternatives, define livestock management practices, and determine the best cultural practices. How one treats these differences determines how many agricultural systems are found within the GPP. The five systems include:

- Range livestock
- Crop fallow
- Groundwater irrigated
- River valley irrigated
- Confined livestock feeding

Range Livestock

Range livestock systems occupy the largest land area in the GPP; more than half of the land in farms is in native grasslands and improved pasture. Beef cattle are the predominant livestock to graze the mixed grasses, forbs, and brush found on these lands. Range scientists would likely identify several grassland ecosystems within the GPP. Based on cluster analysis of Census of Agriculture data, the Hines researchers (1991) defined three range livestock systems geographically dispersed in the Great Plains. In each case, approximately 70 percent of the sales from range livestock farms are from cattle (Table 21.4). Farms tend to be large, especially in New Mexico and northern Plains range areas. From 5 to 26 percent of the land in range livestock farms is cropland; much is used to provide off-grazing-season feed for the livestock herds.

Table 21.4. Range livestock farms in the Great Plains, 1987

	Northern Range Livestock	New Mexico Range Livestock	Texas Range Livestock
Number of counties	69	14	11
Total number of farms	24,063	6,166	6,879
Acres per farm	3,452	3,812	840
Farm sales			
Per farm	$79,053	$64,621	$44,013
Per acre	$23	$17	$52
Farm sales		percent	
Corn	3	1	—
Wheat	11	2	6
Cotton	—	4	6
All cattle	71	66	68
Fed cattle	13	1	19
Farmland			
Cropland	16	5	26
Irrigated land	2	1	—

Source: Hines et al. 1991.

The length of the grazing season varies greatly on these range farms. Farms in the northern Plains may have a three- to four-month season in which winter ranges may have to be supplemented with hay and other harvested feeds. In the southern ranges, nearly year-round grazing is possible most years.

Farms in many parts of the Great Plains may have access to public grazing land. Much of the public grazing land is intermingled, often on an alternate square mile basis, with private land. While grazing of public land is managed according to quantity and season of use by the public land management agencies, it is difficult to separate the management of intermingled public land from the management of private land with which it is mixed.

Crop Fallow

After range livestock in land area, crop fallow systems are next in order. Crop fallow farms vary according to (1) extent of fallow and its efficacy, (2) extent of livestock on farm, and (3) the choice of crops alternated with fallow. Soil scientists would add soil types as another source of variation. In general, as we move east and south in the region, the incidence of fallow diminishes. Cotton is a viable alternative crop in the southern parts of the region; where cotton is grown, very little summer fallowing is practiced. Each of these variations is to a large extent geographically determined, but government policies also affect land use.

Following the cluster analysis by Hines and others (1991), limits of crops define two different clusters of this agricultural system (Table 21.5). Spring wheat is the dominant crop in the northern Plains; winter wheat is the crop of choice in the central and southern Plains. Wheat is the most important cash crop among these clusters of counties, accounting for 20 to 28 percent of the total sales from farms. Both fed and nonfed livestock are important enterprises on these farms, with fed livestock being more important in the central and southern parts of the region than in the north.

Table 21.5. Crop-cattle farms in the Great Plains, 1987

	Winter Wheat and Cattle	Spring Wheat and Cattle
Number of counties	85	84
Total number of farms	57,732	50,447
Acres per farm	848	1,378
Total sales		
Per farm	$69,331	$62,345
Per acre	$82	$45
Farm sales	percent	
Corn	3	4
Wheat	20	28
Cotton	3	—
All cattle	58	36
Fed cattle	19	8
Farmland		
Cropland	57	60
Irrigated land	3	1

Source: Hines et al. 1991.

Salassi (1990) has identified typical farms to characterize agricultural production systems within a given region. Salassi has defined a wheat farm that is representative or typical of wheat farms for the northern, central, and southern Great Plains (Table 21.6). Livestock are important enterprises on representative wheat farms; likewise, wheat and other small grains are also important on representative livestock farms, once they are defined.

Table 21.6. Midsized Great Plains wheat farms with gross annual sales between $100,000 and $250,000, 1986

	Northern Plains	Central Plains	Southern Plains
Land area		acres	
All land	1,882	1,923	2,142
Pasture and hay	706	675	1,184
Cropland		percent	
Wheat	30	31	42
Barley	11	2	—
Corn	7	9	4
Cotton	—	—	10
Silage	1	1	—
Sorghum	—	11	13
Summer fallow	15	26	7
Value of sales			
Crops	$55,852	$69,914	$52,553
Livestock	$49,019	$49,260	$66,554

Source: Salassi 1990.

Summer fallow occurs most often on representative central Plains wheat farms and is of less importance in the northern Plains where the use of spring wheat and other spring crops provides a substantial winter fallow period. Further, snowfall accounts for a greater proportion of total precipitation in the northern Plains; snow management there can be nearly as effective as summer fallow for soil water accumulation (Snyder et al. 1979). Summer fallowing is not as effective on the southern Plains because of higher rates of evapotranspiration caused by the warmer temperatures (Stewart 1990).

The crop fallow system also varies from west to east within a given latitude. Generally, moving east from the Rocky Mountains, annual precipitation increases. As precipitation increases, the soil water accumulation of summer fallowing becomes less important. If annual precipitation is sufficient to produce a crop, the portion of total cropland given to summer fallow decreases. However, land-based government programs encourage summer fallowing because summer-fallowed land can be counted to meet annual set-aside requirements.

Groundwater Irrigated

In addition to range livestock systems and crop fallow systems, the Great Plains also includes a vast irrigated area based on pumping water from underground aquifers. More than 15 million acres are irrigated from groundwater in the Great Plains; most of the groundwater irrigation is from the Ogallala aquifer in the central and southern Plains states (High Plains Associates 1982).

Before groundwater irrigation, which has mostly occurred since World War II, much of the irrigated area was in range livestock or crop fallow systems. Irrigation increased feed grain production, which triggered a marked expansion of grain-fed livestock in the Great Plains.

The physical characteristics of the Ogallala aquifer vary widely over the region; the aquifer itself varies in thickness from a few feet to over 1,000 feet. The recharge rate is slower than withdrawal. Unlike many surface water irrigation systems, groundwater irrigation occurred in response to technical and economic forces. There were no direct government subsides to develop groundwater irrigation.

Pumping depths also vary. Early developments took advantage of relatively shallow lifts. Groundwater laws are the domain of states; states vary greatly in how they have regulated groundwater irrigation development and use. In some states, significant declines have occurred in water tables and in the amount of groundwater pumped for irrigation. Other states, with well spacing and location restrictions, have observed smaller rates of decline.

Most of the irrigated acreage is used for feed grains. Corn dominates the central Plains; grain sorghum becomes more important toward the southern Great Plains. Physical irregularities in the aquifer along with state groundwater regulations result in groundwater irrigation systems being intermingled with dryland agriculture, either cropping or range livestock. Thus, farms with groundwater irrigation are generally mixed enterprise farms producing feed grains (corn and/or grain sorghum), wheat, and beef cattle. Grain feeding of beef cattle for slaughter is often found on farms with irrigated land; important amounts of the feed grains produced on irrigated land are marketed through grain-fed beef (Table 21.7).

Table 21.7. Types of farms and share of sales for farms irrigated by groundwater in the Great Plains, 1987

	Fed Cattle, Northern Plains	Fed Cattle, Southern Plains
	acres	
Land per farm	1,186	1,000
Cropland	723	456
Irrigated land	181	128
Pasture/range	682	416
Farm sales	percent	
Corn	3	5
Wheat	4	4
Cotton	2	—
All cattle	85	66
Fed cattle	70	56

Source: Hines et al. 1991.

Valley Irrigated

Valley irrigated farms are located along the rivers that originate from snowmelt in the Rocky Mountains. Peak seasonal flows from these rivers are diverted, stored, and distributed to farmers under a complex system of rights and allotments. Storage and water distribution facilities are the result of a mix of private and public initiatives.

The valley irrigated farms are the most diversified of any of the farming systems found in the Great Plains, producing a variety of cash crops and feed crops used in fed livestock enterprises. Important crops include corn (grain and silage), barley, dry beans, alfalfa and other improved hays, sugar beets, potatoes, and wheat. Many of the traditional grain-fed livestock systems are associated with these valley irrigated farms.

The availability of irrigation water has enabled these farms to be cropped relatively intensively. Because of their long history and the cropping intensity, valley irrigated farms consume large amounts of fertilizers and chemicals per acre.

These farms are characterized in two ways. The cluster analysis that Hines and others (1991) made of Census of Agriculture data resulted in a fed-cattle and corn farm, and Salassi and others (1991) defined a representative corn farm for Nebraska and the central Plains (Table 21.8). This farming system is dominant along the Platte Valley in northeastern Colorado and across Nebraska. Similar systems exist in the Arkansas River Valley and in the upper Missouri River tributaries in Wyoming and Montana. In the northern Plains, corn is less likely to be the dominant grain crop, however.

Table 21.8. Valley irrigated farms in the Great Plains, 1986 and 1987

	Fed-cattle and Corn[a]	Nebraska Corn[b]	Central Plains Corn[b]
		acres	
Land in farms	770	909	1,212
Cropland	439	316	513
Irrigated land	150	Unknown	Unknown
Other	180	411[c]	456[c]
Farm sales		percent	
Corn	25		
Wheat	3		
Total cattle	55		
Fed cattle	41		
Gross income			
Crop sales		29	31
Livestock sales		47	44
Other		24	25

[a]Hines et al. 1991.
[b]Salassi et al. (midsized farms) 1991.
[c]Includes hay (cropland) and pasture.

The representative corn farms of Salassi et al. (1991) reflect many of the characteristics of valley irrigated farms. Although the two classifications are not directly comparable, similar patterns emerge. Sales of livestock represent approximately one-half of the total gross sales from the farms. Sales of cash crops other than the major grains may be important in local areas, but when averaged across the region, the importance of any one crop is relatively small. For example, sugar beets are grown under contract only in areas near sugar beet processing factories. Soybeans, dry beans, and alfalfa hay for dehydration have similar local markets. Thus, the diversity on these farms is diluted by aggregation.

Confined Livestock Feeding

Livestock, especially beef cattle, play an important role in each of the Great Plains farming systems discussed here. Over the past 30 years the region has become the most important cattle feeding region in the United States. Five states—Texas, Oklahoma, Kansas, Nebraska, and Colorado—marketed 30 percent of the fed cattle in 1955.

By 1989, these five Great Plains states marketed more than 75 percent of the nation's fed cattle (Krause 1991). Most of this growth occurred because of large, specialized beef feedlots. Feedlots have grown at the expense of other cattle feeding regions and in place of a number of farmers feeding cattle as one enterprise on multiple enterprise farms.

There are several reasons why cattle feeding has shifted toward the central and southern Great Plains. The availability of a large and reliable supply of feed grains associated with the expansion of groundwater irrigation (irrigation from the Ogallala aquifer) was a necessary condition to the relocation of cattle feeding. The climate of the region, relatively mild winters and hot but dry summers, offers some advantage to the central and southern Plains relative to other regions for year-round cattle feeding. The Great Plains have for a long time been an important source of feeder cattle. Before expansion of cattle feeding in the central and southern Great Plains, these feeder cattle often moved to feedlots in the Corn Belt. Finally, both technical and pecuniary economies of size enabled large, specialized lots to flourish. Some of the pecuniary economies were the result of the pre-1986 tax code, which facilitated capital acquisition for large-scale operations (Krause 1991).

Cattle feeding, concentrated in the central and southern Great Plains, is further concentrated into a few very large feedlots. One-third of all cattle fed in the five-state area are fed in lots exceeding a 32,000-head (one-time) capacity. More than three-fourths are fed in lots with 8,000 head or more.

These large, specialized cattle feedlots do not exist independently from the other farming systems found in the Plains. They are markets for feeder cattle, forage, and feed grain. They have stimulated the development of slaughter facilities, which may also serve the smaller feeder. Further, many operate as custom feeders, allowing individuals other than the feedlot operators who are interested in feeding cattle to retain ownership or acquire feeder cattle for slaughter.

Protecting the Agricultural Systems

Because of the extremes in elevation, growing season length, precipitation, temperature, and management practices present in the Great Plains, it is difficult to identify problems pervasive to a given agricultural system throughout the region. However, each of the agricultural systems is faced with challenges regarding the sustainability of current practices. Threats may result from

- New understanding of the effects of traditional practices on the sustainability of the resource base
- An invasion of previously unknown or nonthreatening pests

- Changes in practices necessitated by new priorities regarding environmental protection, value of alternative products, and society's attitudes about side effects from existing resource use

Range Livestock

It is increasingly recognized that management of range resources is broader than domestic livestock grazing (Joyce 1989). While public rangeland, often leased to private parties for grazing, has been managed for multiple uses for some time, management of private grazing land also recognizes multiple use objectives. While ranchers continually face problems of range and livestock management in an uncertain environment, other problems have become prominent in recent years.

Maintaining Range Condition
Public land management agencies have experienced reduced budgets for range management, which has curtailed range improvement. Further, multiple use management has reduced the allotments for livestock grazing. As public land grazing is reduced, it places increased grazing pressure on private rangeland (Joyce 1989). Both public and private rangeland could benefit from adopting range improvement practices that can enhance domestic grazing as well as wildlife habitat.

Controlling Rangeland Pests
Undesirable plant species and insect pests can invade and permanently damage rangeland causing serious loss in productivity. For example, leafy spurge is a noxious weed, particularly devastating to northern Plains rangeland; it has now spread to the central Plains states. North Dakota had 200,000 acres of leafy spurge in 1962; 1.1 million acres of North Dakota rangeland were affected in 1990 (Bangsund and Leistritz 1991). In other areas of the Great Plains, rangeland is damaged by invasions of competitive woody plants, snakeweed, knapweed, and yellow star thistle. Grasshoppers also pose a recurring problem throughout the region.

A twofold problem limits rangeland managers' abilities to control these pests. First, the marginal economies of rangeland often make treatment prohibitively expensive. Rangeland grazing is an extensive land use; many acres are often necessary to sustain animals and thousands of acres are required to support a ranch. Profits are small, especially on a per acre basis. Private benefits may not justify the costs of pest control. In addition, one individual may not be able to implement effective treatment. Unless pest management districts require all affected individuals to participate in a treatment, any treatment is ineffective (Davis et al. 1992).

The second limitation to controlling pests on rangeland is the withdrawal of chemicals because of environmental concerns. Available chemicals must be registered only for specific uses. Use of a chemical to control a noxious weed on rangeland may be viewed by a manufacturer as a minor use and not worth the expense to be registered for treating rangeland pests.

Exotic Animal Species

An emerging problem to the range livestock industry is the introduction of exotic species of animals such as nonnative deer and antelope, llamas, and even North American bison. Often introduced by ranchers to enhance their income potential, the species may carry diseases that can spread to domestic livestock. Tuberculosis and brucellosis are two contagious diseases that have been detected in introduced species; the threat to domestic livestock is already present (Price 1992).

Management of Riparian Areas

Grazing animals are known to damage stream banks causing sedimentation and contributing to the pollution of streams (Great Plains Agricultural Council 1992). Because of the scarcity of natural water sources, ranchers may have paid a premium price for a range with natural water available. Regulations on grazing of riparian lands may affect a rancher's operating cost and the value of the rangeland assets.

Crop Fallow

Stewart (1990) identified the important physical factors that make it difficult to achieve sustainable farming in a region like the Great Plains. In natural ecosystems, productivity and sustainability are achieved through a balance between plants extracting nutrients from the soil and simultaneously returning nutrients and organic matter back to the soil. Degradation and conservation occur simultaneously. Because of climatic variability in the Great Plains, it is common to experience several years of below normal precipitation during which soil erosion and organic matter decline exceeds the naturally regenerative process of soil rehabilitation. The history of farming in the Great Plains has been a history of how best to balance soil depletion while harvesting cultivated crops and trying to protect the soil.

For many years agronomists touted summer fallow as a desirable practice, by which soil water could be accumulated and precipitation that falls over two years could be applied to one year's crop. More recently, the practice of summer fallowing has fallen in disfavor, and Peterson and Westfall (1990) cite research revealing that conventional, alternate-year summer fallowing often is not conducive to long-run sustainability.

Summer fallowing is now understood to increase the risk of wind erosion; over time, it reduces the organic matter content of soil. In the northern Plains, summer fallowing has been shown to lead to the creation of saline seeps. Further, summer fallowing is faulted for permitting nutrients to leach below the root zone of crops, and the monoculture associated with the practice has led to grassy weed infestations (Peterson and Westfall 1990).

Recent agronomic recommendations do not eliminate the practice of summer fallow. Rather, more crop residue is retained on the surface during the fallow period. Experiments have also shown that rotations using summer fallow every three to four years are promising (Hickman 1990; Halvorson 1990; Peterson and Westfall 1990). In the northern Great Plains, snow management practices can help as much in increasing soil water as does summer fallowing (Snyder et al. 1979).

Experiments have even claimed that using summer fallow once in three or four years can increase the net returns of farmers. However, farmers may use summer fallow to stabilize income. The extent to which summer fallowing is practiced is also greatly influenced by the nature of the USDA's supply control programs (Skold 1977a,b), as annual set-aside requirements are often met by placing the cropland in summer fallow.

Reducing the amount of summer fallow and even improving residue management when summer fallowing may both increase chemical use for tillage practices (Skold 1991b). If chemical use is further restricted, farmers will need to consider trade-offs between increased soil degradation associated with summer fallow and protecting the environment from increased chemical use.

The monoculture of fallow and small grains has led to some grassy weed infestations. The lack of a tilled crop in rotations with wheat makes it difficult to control these weeds. Thus, cultivated crops are needed in rotation with wheat and fallow to overcome the problem of grassy weeds.

Groundwater Irrigated

Given the nature of the aquifer and its rate of natural recharge, almost any pumping of groundwater from the Ogallala aquifer is not a sustainable activity. Irrigation pumping is mining a slowly replaceable resource. However, several factors led to exploitation of this aquifer (Skold and Young 1987): the presence of an abundant supply of groundwater, the development of irrigation pumping technology, favorable energy and commodity prices, the stability of commodity programs, the meager and uncertain amounts of natural precipitation, and high evapotranspiration rates.

Pumping in excess of recharge has lowered the water table more than 100 feet in some places. However, the Ogallala aquifer is not uniform and there is little lateral water movement. Water table declines have not occurred in many areas and the declines are relatively small in others. Generally, however, groundwater mining (withdrawal in excess of recharge) is present.

Improvements in irrigation practices, changes in irrigation technology, and some shifts in the mix of irrigated crops have tended to offset the effects of higher energy prices and greater pumping depths on pumping costs (Skold and Young 1987). However, the decline in acres irrigated from groundwater has not been as rapid as many analysts projected (Bajwa et al. 1992).

Nevertheless, the rate of exploitation of the Ogallala has to be of concern. The extensive cattle feeding industry of the central and southern Plains is associated with the stable feed grain supply, which comes in large part from its water. But, a valuable resource is being used to produce crops that are, on the margin, not highly valued.

Much of the groundwater irrigated lands are in a virtual monoculture of continuous feed grains, such as corn in the central Plains. Undoubtedly, more chemicals are needed than if alternative crops were grown in sequence with corn. To encourage production of other crops on some of this irrigated land, two changes would be necessary. First, a profitable alternative crop to irrigated corn is needed. Corn responds well to irrigation; available varieties are suited to the Great Plains climate. Needed is an alternative spring crop that can also serve as a feed grain. Most wheat varieties are bred for semiarid dryland conditions or are winter wheat varieties. These varieties do not respond well to irrigation and a fall planted winter crop does not work in sequence with irrigated corn. So, if plant breeders developed a spring-planted small grain to be rotated with corn or grain sorghum, it is likely that this farming system would require fewer chemicals per acre than required by the existing system.

Valley Irrigated

Valley irrigated farms, some of the oldest and most intensively farmed agricultural lands in the Great Plains, are located in broad alluvial valleys of historically meandering rivers. Groundwater, which is part of the underground flow of the river, often is only a few feet below the surface. Continuous cropping for many years with relatively high application rates for animal manure, chemical fertilizers, and pesticides has resulted in chemical leaching into the shallow groundwater. Surface water is also polluted through runoff from irrigated fields being returned to the river for downstream irrigation. Agriculture, and particularly valley irrigated agriculture, is seen as the largest contributor of nonpoint pollution of water (Great Plains Agricultural Council 1992). Nitrate pollution of groundwater has been documented in both Colorado and Nebraska (Schepers et al. 1991; Wylie et al. 1993).

Traditional fertilizer, chemical use, and irrigation practices of valley irrigated farms are not sustainable. Agricultural techniques must be found and implemented that do not result in nonpoint source pollution of surface and groundwater.

A second problem contributing to the nonsustainability of valley irrigated agriculture is the diversion of agricultural irrigation water to urban and industrial uses. Between 1971 and 1986, 144,974 acre-feet of irrigation water in the lower reaches of the Arkansas River in Colorado were transferred from agricultural to nonagricultural uses. In recent years, two-thirds of the applications to transfer water rights in Colorado involved requests to transfer water from agricultural to nonagricultural uses (MacDonnell 1990).

Confined Livestock Feeding

The greatest single threat to sustaining large cattle feeding systems is water quality protection (Great Plains Agricultural Council 1992). Both surface water and groundwater pollution can be associated with confined livestock systems. To meet zero discharge regulations, livestock waste will have to be captured and spread over vast areas of land; land for disposing of livestock waste is not always available to feedlot operators. Some of the size economies present with concentrations of cattle feedlots may dissipate because of the water quality issue, which will depend on the forces of interregional competition. If the water quality standards applied to confined livestock systems are national standards, it is perhaps too early to tell which region(s) will be most adversely affected. Further, the cost of meeting compliance standards may vary considerably among regions of the United States.

In the long run, the concentration of cattle feeding in the central and southern Great Plains depends upon continued irrigation of agriculture. If depletion of the Ogallala and transfers of surface water to urban and industrial uses result in reductions in irrigated agriculture in the Great Plains, cattle feeding is likely to decline as well.

So far, advances in irrigation technologies, pumping efficiencies, and cropping practices have minimized the effects of declining groundwater aquifers on irrigated feed grain production. Moreover, some of the changes seen as desirable in crop fallow systems will increase feed grain production under dryland cropping systems in the Great Plains.

Transferring surface irrigation water from agriculture to urban use is more difficult to mitigate. If these transfers continue and even increase (as some analysts project), reduction in Great Plains feed grain production will be immediately felt by some cattle feeders. The small farmers in the irrigated valleys are the first to be eliminated as their irrigation water is transferred to urban uses.

Needed Policies and Programs

Basis of Commodity Supply Control Programs

Land-based supply control programs have never been very effective in the Great Plains. While the transfer payments associated with program participation have helped stabilize farm income, other problems are present in this approach to supply control.

Annual set-aside and acreage diversion programs are, at best, an inefficient means to reduce the grain output from Great Plains farms. Long-term cropland retirement programs, such as the Soil Bank and the Conservation Reserve Program, are even more limited in achieving supply control (Skold 1989). Long-term land retirement programs do receive high marks for resource protection, since permanent cover greatly reduces soil exposure to wind and water erosion. In addition, long-term programs contribute to improved water quality and enhanced wildlife habitat (Joyce and Skold 1988).

Annual set-aside programs do not possess such desirable environmental side effects. Supply control programs requiring annual acreage diversions have two problems: (1) They add to the amount of summer fallow, exposing additional soil to wind and water erosion, and (2) they limit the adoption of cropping systems that have the potential for reducing the wheat fallow monoculture, thereby increasing the use of other crops in more extended rotations with wheat and fallow.

The potential for replacing land-based supply control programs with quantity- or production-quota-based commodity programs needs to be evaluated. Quantity-based supply control programs would encourage the optimum combination of resource use (land, labor, capital) to be decided by market forces rather than attempting to control product supplies by limiting the use of one input (land) while encouraging the substitution of other inputs, such as chemicals and fertilizers. The use patterns that emerge could be superior to the existing patterns from both an environmental and an economic perspective. Quantity-based commodity programs may also help provide stability to a region with uncertain climate. Since agricultural output from the Great Plains is strongly influenced by climate, farm incomes affected by year-to-year production variability could be smoothed by annual marketing quotas.

Quantity-based commodity programs would also reduce the incentive to bring highly erosive CRP land out of permanent cover and back into cultivated crops. Possibly some of the production base associated with CRP land could be shifted to other cropland. Further, the quality of the land that is cropped would be protected by encouraging cropping systems that better balance nutrient and organic matter depletion.

Rangeland as a Public Good

Great Plains rangeland is a valuable national resource. However, it is not receiving adequate attention because it is not a commodity and therefore does not enjoy the attention that lobbyists pay to commodity groups. Funding for range improvement on public land has not kept pace with needs (Joyce 1989). Also, the Soil Conservation Service (USDA 1987) reports that private rangeland needs range improvement. Degradation of rangeland is often the result of climate variability and rangeland pests; management may or may not be a contributing factor.

Because of the low per acre returns, rangeland protection programs are not always cost-effective. Further, because of space considerations, one individual cannot improve or protect rangeland independent of action by his/her neighbors.

Programs to protect the environment from chemicals can also contribute to the deterioration of rangeland. Furthermore, use of these chemicals to control rangeland pests may not be cost-effective because of low per acre returns.

Policymakers should take a serious look at the value of rangeland as a public good that requires public support for protection. To sustain these resources it may be necessary to (1) identify different sources of revenue to finance improvement practices, such as recognizing the value of protecting rangelands for wildlife habitat or ecosystem diversity, and (2) provide additional public cost-sharing programs to implement range improvements and protect the resource from degradation.

Protection of riparian areas encounters some of the same difficulties as protecting rangeland. Considering designated riparian areas as public goods should also be evaluated.

Future Use of CRP Lands

An immediate and special need is for policies to affect the use of croplands presently enrolled in the CRP program. This policy need is closely related to the basis of commodity supply programs and rangeland improvement. Given the land-based character of current commodity programs, CRP contract holders have a financial incentive to return the land to crop production when the contracts expire. In the 10 Great Plains states, more than 70 percent of the cropland enrolled in CRP is cropland base acres. Base acres—the right to grow commodity program crops—are an important determinant of cropland income potential and land values. CRP contract holders will want to preserve their cropland base.

However, the publicly desired environmental benefits from maintaining some of this CRP land in protective cover will be lost if the land returns to crop production. Soil erosion will increase, wildlife habitat will be lost, and water quality will be reduced. Policies need an economic incentive for CRP contract holders to sustain the public good created by the CRP.

Tailored Water Quality Protection

The region has some serious water quality protection problems. Even though agricultural systems often coexist on the same farm, it would be impossible to write a general water quality protection program that would be effective and meet the needs of each system. Programs to protect and enhance water quality are needed, but the needs are unique among the different agricultural systems of the Great Plains.

Range livestock and crop fallow systems pose little threat to the environment from chemicals and fertilizers. Even groundwater irrigated farms, which use significantly more chemicals and fertilizers per acre, do not generally threaten to pollute the environment. Only when aquifers are relatively shallow is there a problem with groundwater pollution from irrigation and chemical use practices. When aquifers are deep and adequate well-head protections are in place, pollution of groundwater on its irrigated farms is not a major problem.

On the other hand, river valley irrigation systems have the potential to pollute both surface and groundwater. Measures to bring about sustainable fertilizer and chemical use on valley irrigated farms may be redundant on crop fallow farms. Unless changes in practices, incentives, or regulations are specific to a particular system, such changes may only increase costs for systems that do not hold the same environmental threat.

Pesticide Registration

The registration of pesticides limits the use of certain pesticides on minor, but critically needed, uses. Lack of registration for use on pastures and ranges may mean less protection of range resources. The inability to use an insecticide on trees may mean the loss of needed protection for farm windbreaks and shelter belts. Flexibility in pesticide use regulation could mean substantial resource protection.

Educational Programs

Perhaps the greatest program for achieving sustainability for each of the agricultural systems important to the Great Plains is education. Farmers, by and large, are concerned about the environment. Those farmers who remain are the survivors, and they want their heirs to survive. Their practices are a product of what they learn from publicly and privately supported research plus their own experiences. In many cases, incentives and regulations are not necessary to achieve better resource use. Education about irrigation scheduling, low pressure distribution systems, and optimal sizing of power has greatly reduced irrigation pumping costs. Integrated resource management systems have brought the technical and financial aspects of management into a common perspective. The practices are being made more sustainable.

Care must be taken to identify those environmental protection measures that can best be served by education. Other environmental protection measures may need some economic incentives while, for others, regulation is the only choice. Flexibility must be maintained, with respect to the agricultural system in question, the nature of the resource protection measure needed, and ultimately, the use of resources.

References

Bajwa, R. S., W. M. Crosswhite, J. E. Hostetler, and O. W. Wright. 1992. *Agricultural irrigation and water use.* Agricultural Information Bulletin 638. U.S. Department of Agriculture, Economic Research Service.

Bangsund, Dean A., and F. Larry Leistritz. 1991. *Economic impacts of leafy spurge on grazing lands in the northern Great Plains.* Agricultural Economics Report 275-S. North Dakota State University.

Davis, Robert M., Melvin D. Skold, James S. Berry, and William P. Kemp. 1992. The economics threshold for grasshopper control on public rangelands. *Journal of Agricultural and Resource Economics* 17(1):56-65.

Great Plains Agricultural Council. 1992. *Agriculture and water quality in the Great Plains: Status and recommendations.* Great Plains Agricultural Council Publication 140. Water Quality Task Force (Ronald D. Lacewell, Chairman), Fort Collins, Colorado.

Halvorson, Ardell D. 1990. Cropping systems and N fertilization for efficient water use in the central Great Plains. In *Conservation tillage: Proceedings, Great Plains Conservation Tillage Symposium.* Great Plains Agricultural Council Publication 131. Bismark, North Dakota: Great Plains Agricultural Council.

Heimlich, Ralph E., and Olaf E. Kula. 1991. Economics of livestock and crop production on post-CRP land. In *The Conservation Reserve—Yesterday, today, and tomorrow: Symposium proceedings, Washington, D.C., Jan. 14, 1991,* ed. L. A. Joyce, J. E. Mitchell, and M. D. Skold. USDA-FS General Technical Report RM-203. Fort Collins: Rocky Mountain Forest and Range Experiment Station.

Hickman, John S. 1990. Conservation tillage systems in Kansas. In *Conservation tillage: Proceedings, Great Plains Conservation Tillage Symposium.* Great Plains Agricultural Council Publication 131. Bismark, North Dakota: Great Plains Agricultural Council.

High Plains Associates (Camp Dresser and McKee, Inc., Black and Veatch, Arthur D. Little, Inc.). 1982. *Six-state High Plains-Ogallala aquifer regional resources study.* Report to the U.S. Department of Commerce and the High Plains Study Council, Austin, Texas.

Hines, Fred, Judy Sommer, and Mindy Petrulis. 1991. Agriculture in the Great Plains: An assessment of its importance to the region's economy. In *Proceedings, Great Plains Agricultural Council.* Fort Collins, Colorado.

Joyce, Linda A. 1989. *An analysis of the range forage situation in the United States: 1989-2040.* USDA-FS General Technical Report RM-180. Fort Collins: Rocky Mountain Forest and Range Experiment Station.

Joyce, Linda, and Melvin D. Skold. 1988. Implications of changes in the regional ecology of the Great Plains. In *Impacts of the Conservation Reserve Program in the Great Plains: Symposium proceedings, Denver, Colorado, Sept. 16-18, 1987,* ed. J. E. Mitchell. USDA-FS General Technical Report RM-158. Fort Collins: Rocky Mountain Forest and Range Experiment Station.

Kraenzel, Carl F. 1955. *The Great Plains in transition.* Norman: University of Oklahoma Press.

Krause, Kenneth R. 1991. *Cattle feeding, 1962-89 location and feedlot size.* Agricultural Economic Report 642. U.S. Department of Agriculture, Economic Research Service.

Kuchler, August W. 1964. *Potential natural vegetation of the conterminous United States.* New York: American Geographical Society.

MacDonnell, Lawrence J. 1990. *The water transfer process as a management option for meeting changing water demands,* Vols. I and II. Boulder, Colorado: Water Resources Law Center, University of Colorado.

Peterson, G. A., and D. G. Westfall. 1990. Sustainable dryland agroecosystems. In *Conservation tillage: Proceedings, Great Plains Conservation Tillage Symposium.* Great Plains Agricultural Council Publication 131. Bismark, North Dakota: Great Plains Agricultural Council.

Price, Dean (private communication). 1992. Veterinary Services, Animal and Plant Health Inspection Service, U.S. Department of Agriculture, Denver.

Salassi, Michael E. 1990. *Representative U.S. wheat farms.* Statistical Bulletin 786. Washington, D.C.: U.S. Department of Agriculture, Economic Research Service.

Salassi, Michael E., William D. McBride, and Robert A. Pelly. 1991. *Representative U.S. corn farms.* Statistical Bulletin 820. Washington, D.C.: U.S. Department of Agriculture, Economic Research Service.

Schepers, J. S., M. G. Moravek, E. E. Alberts, and K. D. Frank. 1991. Maize production impact on groundwater quality. *Journal of Environmental Quality* 20:12-16.

Skold, Melvin D. 1977a. Natural resource conservation and use implications of price and income policies. In *Farm and Food Policy Symposium: Proceedings.* Great Plains Agricultural Council Publication 84. Fort Collins.

———. *Regional potentials for cropland use shifts.* Natural Resource Economics Working Paper 23. Washington, D.C.: U.S. Department of Agriculture, Economic Research Service.

———. Cropland retirement policies and their effects on land use in the Great Plains. *Journal of Production Agriculture* 2(3):197-201.

———. *Communication to Great Plains Agricultural Council correspondents.* Great Plains Agricultural Council, Fort Collins.

———. Economic issues related to sustainable agricultural systems in the Great Plains. In *Sustainable agriculture for the Great Plains: Symposium proceedings.* USDA-ARS-89. Fort Collins.

Skold, Melvin D., and Robert A. Young. 1987. The role of natural resources in a changing Great Plains economy. In *The rural Great Plains of the future: Symposium proceedings.* Great Plains Agricultural Council Publication 125. Fort Collins.

Snyder, J. R., M. D. Skold, and W. O. Willis. 1979. The economics of snow management: An application of game theory. *Western Journal of Agricultural Economics* 4(2):61-71.

Stewart, B. A. 1990. Limits, bounds, and the future of conservation tillage in the Great Plains. In *Conservation tillage: Proceedings, Great Plains Conservation Tillage Symposium.* Great Plains Agricultural Council Publication 131. Bismark, North Dakota: Great Plains Agricultural Council.

Sutton, John D. 1984. *Land use in the Great Plains.* AGES 840329. Washington, D.C.: U.S. Department of Agriculture, Economic Research Service, Natural Resources Economics Division.

U.S. Department of Commerce. 1989. *1987 Census of Agriculture, Volume 1, Geographic Area Series.* Washington, D.C.: U.S. Department of Commerce, Bureau of the Census.

U.S. Department of Agriculture (USDA). 1991. *Agricultural statistics.* Washington, D.C.: U.S. Department of Agriculture.

U.S. Department of Agriculture, Economic Research Service (USDA/ERS). 1992. *RTD updates: Conservation reserve.* Washington, D.C.: USDA-ERS, Resources and Technology Division.

U.S. Department of Agriculture, Soil Conservation Service (USDA/SCS). 1987. *The second RCA appraisal* (Review draft). Washington, D.C.

Wylie, B. K., M. J. Shaffer, M. K. Brodahl, D. Dubois, and D. G. Wagner. 1993. Regional distributions of NO_3-N leaching. Unpublished manuscript under review.

22. Livestock and the Environment: Emerging Issues for the Great Plains

Allan Butcher
Attorney at Law
Nancy Easterling, Larry Frarey, Kristin Gill, and Heather Jones
Texas Institute for Applied Environmental Research
Tarelton State University

Concentrated animal feeding operations (CAFOs) significantly contribute to the Great Plains economy. In 1987 CAFOs produced a total market value for the region of $16.9 billion: poultry products over $1.4 billion, dairy products over $1.5 billion, swine operations over $4.0 billion, and cattle feedlot operations the greatest contribution—nearly $10.0 billion (U.S. Department of Commerce 1989).

The Great Plains livestock industries have undergone tremendous consolidation in recent years. In 1991, the four largest broiler companies produced 41 percent of total broilers, and the 20 largest companies controlled 79 percent of total production (Amey 1992). Likewise, beef producers will continue to consolidate: "Constraints on environment, stable demand, and the industry's infrastructure will force down the number of cattle producers—eliminating the under 20 head size—and increase the size of the survivors, who will take advantage of the economies of size" (Hurt et al. 1992a).

In contrast, dairy and swine producers are lagging behind in consolidation. However, the national trend in dairies has been toward fewer cows on fewer farms producing more milk. In 1934, 4.5 million farms milked an average 5.4 cows per farm and produced an average 40.3 hundredweight per cow per year. In 1987 approximately 200,000 farms averaged 51 cows each; in 1990 average annual milk production was 140.6 hundredweight per cow per year (TIAER 1992b).

Currently, in the hog industry the number of large operations is growing tremendously. A University of Missouri economist (Marberry 1992) estimates that about 1,100 specialized units (producing more than 10,000 hogs per year) marketed nearly 26 million hogs—more than 25 percent of total hog production.

However, the growth and concentration in the U.S. livestock industries have exacerbated environmental problems. According to a draft report of the U.S. Environmental Protection Agency (EPA), "livestock waste may account for up to 20 percent of water pollution nationally" (USEPA 1993). The report concludes that

Stanley R. Johnson and Aziz Bouzaher (ed.), *Conservation of Great Plains Ecosystems*, 365–390.

> . . . livestock waste poses a significant threat to water-based ecosystems, often causes large economic damages that can take years to reverse, and poses a number of threats to human health. Given the widespread nature of the pollution and the severity of the impacts resulting from poor management of waste, EPA must insure that its NPDES [National Pollutant Discharge Elimination System] program is being implemented adequately and that it should undertake additional measures to assure protection of all media, including groundwater and air resources.

Although the size of livestock operations has increased in the Great Plains, a dichotomous structure still prevails; that is, family farms and large CAFOs coexist in most areas. This fact has important implications for policymakers addressing pollution problems caused by improper manure management by CAFOs. Because of the significant costs involved in managing manure properly, care must be taken to ensure adequate compliance by smaller livestock operations. Moreover, it is evident that "more than economics will influence animal agriculture in the future. The general public's concerns for the environment, animal welfare, and food safety will all play an increasingly influential role in how food is produced in the United States and around the world" (Muirhead 1992).

Production Trends in the Concentrated Feeding Operation Industry

Feedlot Cattle

The total number of feedlot cattle in the Great Plains region and the size of operations are both increasing. In 1980, almost 98 percent of the operations in the Great Plains had herds with fewer than 1,000 head and accounted for 26 percent of the total number of cattle. In the same year, herds with more than 32,000 head accounted for less than 1 percent of the operations in the Great Plains, yet they contained 19 percent of all cattle. By 1991, only 15 percent of the region's cattle were in operations with fewer than 1,000 head while 32 percent of the cattle were in operations with more than 32,000 head (USDA 1992). Reasons for these shifts include economies of size and decreased beef consumption by Americans (Hurt et al. 1992a).

Economies of size play an important part in this shift. Bankers have been skeptical about lending large amounts of capital to smaller operations because of the large financial losses in cattle feeding that occurred in the 1970s and early 1980s. "Total production estimates for feedlots show a $20/cwt [hundredweight] advantage for larger commercial operations (greater than 1,000 head). . . . Thus, the small farmer feedlots are unlikely to make a comeback unless environmental (or other) regulations make smaller scale operations competitive" (Hurt et al. 1992a).

Continued expansion is expected by beef cow herds and feeder cattle operations when the Conservation Reserve Program contracts expire in the mid-1990s. Additional growth could come from export markets: The Japanese already are increasing their demand for U.S. beef, and a number of former Eastern Bloc nations (including Russia) could provide new markets for cattle (Hurt et al. 1992a).

Economics, water resources, and lower population density in the Great Plains are factors in the current shift of the cattle industry from the Corn Belt. Lower population density means less potential for conflict on environmental issues, fewer nonfarm employment opportunities for laborers (Hurt et al. 1992a), and fewer complaints about odors from concentrated animal feeding operations.

Swine

The swine industry is in transition with the trend toward more animals from fewer farms. The nearly 3 million pork producers of 1950 were reduced to only 256,000 in 1992. "Farms have grown in size with about 6 percent of the producers raising 60 percent of the hogs. Nearly 80 percent of the hogs are grown on farms producing 1,000 or more hogs per year" (National Pork Producers Council 1992). According to a University of Missouri survey, more than 25 percent of all hogs marketed come from operations producing 10,000 or more hogs per year (Marberry 1992).

Production costs per animal tend to fall as hog operations become larger. Larger operations report higher production efficiencies and lower per unit costs than smaller hog operations, resulting in greater profits. Operations that have at least 3,000 head of farrow-to-finish production gain about 75 percent of the economies of size of operations raising 10,000 hogs per year (Hurt et al. 1992b).

The University of Missouri study found that contract production clearly is the most dynamic segment of the industry. In 1991, 15 percent of all U.S. hogs were produced on contract. Sixty-five percent of contractors surveyed began raising hogs under contract during the past four years (Marberry 1992).

In addition, the geographic location of pork production is shifting. While the traditional Corn Belt states still represent the majority of production, growth is occurring in nontraditional hog states like Texas, Colorado, and Oklahoma (National Pork Producers Council 1992). New hog operations and processing plants are expected in the Great Plains area because of lower population density, which brings less environmental conflict and lower wage rates (Hurt et al. 1992b).

The North American Free Trade Agreement (NAFTA) may further encourage growth of U.S. operations. Since Mexico produces only a limited amount of grain, hogs may be raised in the United States and transported live to Mexico for processing, so processing plants can take advantage of Mexico's cheaper labor (Hurt et al. 1992b).

Genetic engineering is changing the swine industry as well. The shift to a percent-lean pricing component may cause selection for this trait, while attempting to maintain or enhance the production efficiencies already in the genetic pool. Experiments with hogs from European seedstock suggest that it will be possible to finish hogs at substantially higher weights with low fat content. Also, dramatically fewer sows will be needed to produce our pork supply (Hurt et al. 1992b).

Concentrated swine production over the past two decades, primarily in confined facilities, has made the public aware of odors, the need to protect groundwater quality, and the need to manage manure disposal. An acceptable level of odor (the duration, frequency, and intensity) from confined swine facilities and the lagoons for swine waste has yet to be determined. The size of the operations must be balanced with the ability of the land to utilize the nutrients applied. At present, the loss or decomposition of nitrogen and other nutrients is poorly understood because it varies with different manure handling systems. Increased environmental testing and required management plans are likely in the future (Safley 1992).

Poultry

The poultry industry comprises two separate industries, the broilers and layers. In both the trend is toward larger operations. The poultry industry, like other livestock industries, is driven by technology. "Genetic improvement, labor saving production techniques, feed formulation advances, and processing automation have forced changes. Much of the innovations have resulted in significant economies of size and reduced labor at all levels" (Christensen 1992).

Small chicken enterprises began declining in the late 1950s. Farms with fewer than 400 chickens decreased approximately 62 percent between 1964 and 1969 while large poultry operations rapidly expanded. The trend toward concentrated operations is still the industry standard according to an Egg Industry report (1992) that 168.8 million layers are the product of 55 producer companies that have flocks of one million or more.

Broiler production is expected to continue its current expansion through 2000 (Roenig 1992). In the Great Plains region, the number of layer operations with 100,000 or more layers has increased to 54 percent in 1987 from 25 percent in 1974.

The poultry industry has been expanding to the Midwest, largely fueled by the proximity of feed production, which is the major component of egg production costs. Costs of transporting feed have been increasing due to deregulation of rail freight rates in the mid-1980s and the declining quality of the rail industry's infrastructure. Cooler weather and labor costs not significantly higher than those in the Southeast are additional enticements toward Great Plains production (Clark 1992).

Growth in the market for value-added egg products such as dried and frozen eggs and the separation of whites from the yolks promises new markets for layer operations. Maximizing genetic potential will be the challenge of the future for egg production (Clark 1992).

Environmental regulations have become widespread in the poultry industry. Some states require growers to have a permit from city, county, state, or federal organizations regarding water quality and other environmental protection issues. "This is the new world of poultry production" (Perry 1992).

Dairy

Between 1959 and 1987, the number of dairy operations in the United States decreased from 1.8 million to 200,000. The number of cows also decreased substantially, particularly between 1959 and 1969, declining more than 5.4 million. Cow numbers, however, remained relatively unchanged from 1969 to 1987. Yet the average herd size increased nationwide from 20 cows in 1969 to 50 cows in 1987 (U.S. Department of Commerce 1971, 1989).

The dairy industry has experienced unprecedented moves toward larger operations over the last two decades. Nationwide, the number of large dairies (500 head or more) almost doubled between 1974 and 1987. Several states in the Great Plains region experienced this growth in large operations—Texas (from 30 to 84), New Mexico (from 11 to 35), and Colorado (from 4 to 20) (U.S. Department of Commerce 1977, 1989).

The growth in the size of operations can be attributed in part to economies of size. Cost analysis done by Matulich (1978) showed "substantially lower milk production costs per cow as herd size increased from 375 to 450 cows, and more gradual economies of size as dairies grew to 1200 cow herds."

Declines in the number of farms and cows are due in part to the overwhelming increases in production over the past 50 years. Average annual milk production in 1934 was 40.3 hundredweight per cow per year; in 1990, that average reached 140.6 hundredweight per cow per year (TIAER 1992b).

Opportunities for advancement in the dairy industry will include the use of genetics, as well as the development of ration formulations to increase the protein and reduce the fat content in milk. Rations will also be adjusted to use nutrients efficiently and improve production efficiency to reduce methane losses (Muirhead 1992). The approval of bovine somatotropin (BST) will increase productivity for the dairy industry, and fewer cows will be needed to produce the same amount of milk.

The rapid expansion of the dairy industry has already fueled increasing environmental regulations, specifically about pollution from animal waste disposal sites. EPA Region VI recently revised and adopted a new National Pollutant Discharge Elimination System (NPDES) permit, an extension of the Texas Water Commission's policy to deal with confined feeding operations.

Policy Tools on the Books

Permitting Background

In 1972, the U.S. Congress passed the Federal Water Pollution Control Act Amendments, commonly referred to as the Clean Water Act (CWA) (*Federal environmental laws* 1992).[1] The expressed purpose of the CWA was "to restore and maintain the chemical, physical, and biological integrity of the nation's waters." The CWA would fulfill the mandate by eliminating "the discharge of pollutants into the navigable waters" by 1985. Moreover, Congress set an interim goal of attaining "water quality which provides for the protection and propagation of fish, shellfish, and wildlife and provides for recreation in and on the water" by July 1, 1983.

The CWA prohibits the discharge of any pollutant by any person. The definition of *pollutant* includes both solid waste and agricultural waste discharged into any water. The act defines *point source* as "any discernible, confined and discrete conveyance, including but not limited to any pipe, ditch, channel, tunnel, conduit, well, discrete fissure, container, rolling stock, concentrated animal feeding operation, or vessel or other floating craft, from which pollutants are or may be discharged" (*Federal environmental laws* 1992). Even erosion gullies associated with a structured device that alters the natural flow of water may constitute a point source. (See Sierra Club v. Abston Construction Co., Inc.)

The CWA initially focused on bringing the nation's traditional industrial sector into compliance while providing regulatory predictability (Rogers 1986). One section of the act, the NPDES, established a permitting scheme under which the EPA "may, after opportunity for public hearing, issue a permit for the discharge of any pollutant, or combination of pollutants . . . upon . . . such conditions as the Administrator determines are necessary to carry out the provisions of this chapter" (*Federal environmental laws* 1992).

CAFOs are expressly included under the definition of point source, and therefore must obtain an NPDES permit before discharging any pollutant into navigable waters (Lexis 1993). EPA regulations designate those animal feeding operations that constitute CAFOs according to the number of animal units confined at the operation. However, another section allows the EPA to designate any animal feeding operation as a CAFO

[1]Further details and corroborating cases throughout this section are cited in *Federal environmental laws* (1992) and Lexis (1993).

"upon determining that it is a significant contributor of pollution to the waters of the United States" (Lexis 1993). Moreover, EPA regulations impose a "no discharge" effluent standard on all CAFOs. A CAFO may discharge pollutants to navigable waters, by NPDES permit, only during a "chronic or catastrophic" rainfall event, usually represented by a 25-year, 24-hour storm event (Lexis 1993).

Background of Voluntary Programs

As the federal government's third largest agency, the U.S. Department of Agriculture (USDA) has a wide distribution of employees and offices to provide a direct link among the department, farmers, and ranchers (U.S. GAO 1991). Agricultural agencies have always played a significant role in environmental policy, especially in the last few years. These divisions of the USDA and state counterparts promote environmentally sound agricultural practices through cost-share, technical assistance, educational, and other programs. Advocates of implementing agricultural environmental programs through "traditional program channels" argue that such programs will remain responsive to farmers' needs (Batie et al. 1986). Beck's (1981) study of water treatment management plans under the Clean Water Act showed that soil and water conservation districts are the agencies of choice in addressing agricultural water pollution initiatives.

Regulatory Programs in the Great Plains

In an attempt to gain headway against agricultural nonpoint pollution problems, the EPA has issued a "general" CAFO permit as opposed to the traditional, "individual" NPDES permit. On February 8, 1993, EPA Region VI published its general CAFO permit that applies to the Great Plains states in the region: Texas, Oklahoma, New Mexico. Formerly, the EPA had issued only individual NPDES permits, a practice that severely limited regulatory efforts (Long 1992). "EPA's regional offices and the relevant delegated states have issued roughly 800 permits to CAFOs, although perhaps as many as 10,000 feedlots currently meet the 1,000 animal unit cutoff."

The general permit provides EPA the opportunity to regulate more CAFOs with less bureaucratic effort because the permit automatically covers all dairies that discharge wastewater and meet the minimum size criteria. The EPA would expend incalculably greater effort if it attempted to formulate and enforce individual permit requirements for all of the dairies soon to be covered by the general permit (USEPA 1993).

If vigorously enforced, the general permit could also have a significant impact on runoff from manure application fields used by CAFOs. If wastewater is discharged onto land, the discharge or drainage of the irrigated wastewater is prohibited where the process will result in discharge to water of the United States. Furthermore, the subsection titled "Manure and Pond Solids Handling and Land Application" provides that the storage and land application of manure shall not cause a discharge of significant pollutants to water of the United States or cause a water quality violation in water of the United States. Essentially, the discharge or runoff of waste from the application site is prohibited.

The CWA allows the EPA to delegate NPDES permit authority to the states; thus far, the EPA has delegated that authority to 39 states (*Federal environmental laws* 1992). An informal telephone survey of the applicable regulatory agencies in the Great Plains region indicates a wide range in delegated regulatory efforts. While some states are now strengthening permitting and enforcement procedures, others, such as Missouri, have strong procedures that have remained basically unchanged since 1970.

The minimum criteria for requiring a permit vary widely. Missouri, similar to other states, has species-specific threshold numbers for animals above which a permit is required. Wyoming, however, has very few permitted operations but has many small operators who have fewer animals than the minimum number requiring a permit.

Mere identification of CAFOs is difficult, not to mention inspection and enforcement. Minnesota has a system under which counties inform the state's pollution control agency when a person requests a building permit for a CAFO. The applicant must present a state CAFO permit before the county will issue a building permit. Other states rely on complaints or periodic inspections to identify and locate CAFOs. However, the enforcement and regulatory staffing levels in many states do not allow for proactive investigation. For example, two Great Plains states have only one full-time employee each to handle CAFO permits and enforcement.

Complaints from the public constitute the central enforcement mechanism in many states. Even when a state has an investigatory staff, public complaints usually drive the enforcement system. One state's inspection program makes an unannounced inspection about once every five years. Deficiencies in a complaint-driven system stem from the difficulty of detecting and confirming a violation. By the time an inspector arrives at a site, the operator may have stopped the offending practice.

Many states exercise some type of control over land application of liquid and solid wastes, including prevention of runoff from land application sites, which constitutes a discharge under state permits. Groundwater protection is often a driving force in land application regulations. Instead of statewide regulations, Nebraska relies on local resource districts to promulgate land application regulations on a local level. In all states, the operator who applies waste should consider the soil type, vegetative cover, weather conditions, and rates of application. The most common regulatory method for land application requires the state regulatory agency to write specific regulations into a facility's permit, thus making the regulations enforceable.

Some states employ varying degrees of regulation over land application based on local conditions and uses of the field receiving the waste material. However, because of the lack of enforcement personnel and the difficulty in monitoring the compliance of land application procedures, the actual success of a regulatory program relies on the individual operator.

Some states allow the individual leeway in determining the need for a permit, but an inspector may declare a permit necessary. However, that "self-determination" does not exempt the operator from compliance. If an unpermitted operator is out of compliance with regulations, that operator is susceptible to enforcement like a permitted operator.

Most state regulators attempt to work with a noncomplying operator in an attempt to gain compliance. The level of cooperation exhibited by the operator has a marked effect on the ferocity of enforcement, although the operator may have to pay for actual damages, such as fish killed as a result of noncompliance. One state regulatory official stated that he would rather an operator spend money on coming into compliance than to pay a fine.

Merging Voluntary and Regulatory Strategies

Even with the potential for improvement through existing programs, the distinguishing features of nonpoint and point source pollution suggest that present programs do not provide a perfect solution to address agricultural pollution. Nonpoint pollution is difficult to quantify, monitor, control precisely, and, if addressed through command-and-control regulations, difficult to enforce systematically. Inspecting all sources would require an "army of enforcers" that society is not likely to tolerate (Rogers 1986).

In addition, particularly given the government's ongoing support of agriculture through market intervention activities, subsidization, and educational and technological assistance, financial support will become necessary as the regulatory programs reach further into the agriculture sector. Currently, the brunt of compliance costs have been borne by larger industrial-sized dairies and other CAFOs.

The USDA has a long tradition of befriending the agriculture industry through promoting of price stabilization programs, incentive-based voluntary conservation programs (which may include direct subsidizes), and production research. In an evolving policy environment, the USDA has maintained traditional voluntary program approaches, leaving regulatory enforcement roles to the programs under jurisdiction of the EPA and the state counterparts (U.S. GAO 1990; Rice 1979).

Although voluntary programs are widely recognized as promoting the adoption of best management practices, detractors of the practice remain unconvinced. Davidson (1989) asserts "[t]here is little, if any, precedent in our experience of government to suggest that the problem of erosion and nonpoint pollution can be solved by asking landowners to regulate themselves," making the point that the implementation history of soil conservation practices in agriculture has been largely shaped by the availability of support funding. Of course, the fact that cost-sharing is an important factor in gaining compliance can be argued the other way, namely that without funding support most operations in the industry cannot tolerate the cost of pollution protection.

The state of Washington provides an instructive example of how agencies and approaches work together. Pursuant to the Federal Water Pollution Control Act, section 208, an institutional inventory was conducted in the 1970s in the dairy region north of Seattle. Field assessments confirmed that water quality problems in the river basin were localized, not severe, but large in total amounts of contaminants entering the stream system. Three large dairies in the basin were under NPDES permit. Farmers felt that only a few sites were causing most of the problems and distrusted the state agencies, particularly the Department of Ecology, because of sporadic bouts with zealous enforcement and a perceived lack of farm knowledge. The local conservation districts and other traditional agricultural agencies were not interested in a regulatory role.

However, an "Ag Working committee"—consisting of staff from Soil Conservation Service (SCS), Extension, Agricultural Stabilization and Conservation Service (ASCS), and farmers—began developing a workable institutional approach. The committee recommended a voluntary approach backed up by regulatory mechanisms. The regulatory backup was considered necessary because participants noted the persistence of problem farms that were unlikely to comply with voluntary measures.

The two conservation districts with jurisdiction in the basin, designated as lead agencies, received authority to pursue program objectives: education and technical assistance, financing and funding, incentives, monitoring, and enforcement.

The state changed the applicable regulations to establish that compliance with best management practices (BMPs) would be sufficient to meet water quality standards. By changing the regulations, the state avoided the threat of enforcement against operators in compliance with the established criteria while BMP effectiveness was being evaluated, thus averting an awkward enforcement situation. This overall approach—the involvement of conservation districts followed by potential enforcement from the state regulatory authority—was adopted for the entire State of Washington (Rice 1979).

Recently, Washington fortified its CAFO regulatory program by requiring CAFOs to obtain permits. However, most sites with plans meeting SCS technical specifications will likely be exempt from this requirement. The exemption is part of Washington's delegation under the Federal Water Pollution Control Act and is subject to EPA approval. The new regulatory unit has a small operating staff and does not anticipate significant problems (Velthuzien 1992). Thus, a relatively mature program (in place since the 1970s) has nearly met its objectives through voluntary efforts backed by an enforcement program. The enforcement program is now being implemented and is expected to be successful with a very small staff for a state with more than 1,000 dairies.

Odor Control

The regulation of odor from CAFOs remains a daunting task and probably must be addressed before water quality programs can fully function. Many persons complaining to water quality agencies about CAFO practices have odor complaints as well. One problem with odor control is that large-scale programs to address odor will require technological advances in detection and measurement not now fully developed. Effective odor control may require management practices that will add significantly to the expense of operating concentrated animal feeding operations.

The Missouri Division of Natural Resources actually discontinued enforcement of odor control regulations because of the high cost involved and its inability to develop definitive rules. Recently, the defendant prevailed in a Missouri private nuisance action in which neighbors sought an injunction and money damages against a large hog operation. The court rejected claims that the action was barred by Missouri's right-to-farm law. The most convincing evidence, in the view of the defendant's counsel, was expert and lay testimony that the operation was in compliance with all rules and regulations and was carefully operated, using state-of-the-art technology (Wheeler 1992).

Zoning has been the traditional means of minimizing neighbor conflict caused by odors. To some degree, odors are considered an inevitable consequence of CAFOs. Some states have established agricultural areas, where odor complaints are resolved through an informal administrative process (Iowa Code 1990). Many states, including Texas, have enacted right-to-farm laws designed to protect landowners against nuisance actions if there has been no substantial change in operation for a designated period, typically one year. In practice these laws have not proven effective (Wheeler 1992; Oppedal 1988).

Current Strategies for Agricultural Pollution Abatement

The Clean Water Act amendments of 1987 attempt to facilitate state regulation of nonpoint sources through the use of best management practices. BMPs for CAFOs are intended to control water pollution by containing runoff. However, to date, implementation of BMPs has largely occurred on a voluntary basis. While the CWA provides for federal grants to help states implement regulatory programs mandating the use of BMPs, few states have implemented such programs.

BMPs designed to contain contaminated runoff include the use of lagoons, berms, settling basins, and solid separators for all species. Scraping of corrals is a BMP for dairies that have open confinement areas. While until recently the NPDES program has not required lagoon liners, most states have some stipulations concerning liners.

Land application is a principal BMP for utilizing wastes, making it a significant area of concern. Under the CWA, a distinction is drawn between nonpoint source and point source pollution, but some sources, such as CAFOs and ancillary waste disposal areas, have characteristics of both, depending upon management practices and how we define point source and nonpoint source. A field fertilized with manure from a feedlot may be a nonpoint source, but the waste material is derived from a point source at the lot itself. The terms have different meanings in different contexts and are at best difficult concepts.

Attention is now being given by many states to pollution derived from application fields, the so-called "back side of the dairy." The state of Texas, for example, regulates the entire CAFO without the point source or nonpoint source distinction. Thus, wastewater used for irrigation that runs off the field clearly falls within the purview of the Texas Water Commission.

Problems with Current Strategies

Enforcement of the no-discharge effluent limitation for CAFOs has proven problematic for the EPA and delegated state agencies. Long (1992) estimates that no more than 800 of more than 10,000 CAFOs that should be permitted under the current EPA size criteria are in fact permitted. Further, "initial EPA surveys indicate that NPDES authorities are experiencing implementation problems with interpreting EPA's existing feedlot regulations as well as issuing and enforcing permits for feedlots" (USEPA 1989).

Many pollution abatement programs across the country have equated BMP implementation with success. It has not yet been proven that surface water quality will meet acceptable standards if all up-gradient landowners implement BMPs. Responsible management is an integral part of BMP implementation.

To date, much of the attention focused on application fields has involved edge-of-field strategies designed to prevent contaminated water from leaving the CAFO. Responsible management, which is difficult to document, is a major factor in the success of these strategies. Another regulatory complication is the difficulty in estimating the rate of application after the fact. Also, the success of edge-of-field strategies outside of experimental plots has not yet been established.

The trend toward larger, concentrated operations has led to manure being considered a waste rather than a valuable commodity to enhance feed crop production. Manure and wastewater may be overapplied to cropland in an effort to simply get rid of the waste. Smith (1992) warns that, due to lack of storage capacity, manure may be applied during times of low plant demand. This can lead to surface and groundwater pollution through runoff and leaching. "Manure presents some very real application problems, both in timing and method" (Smith 1992). Indeed, overapplication and improper application have the potential for generating nonpoint source pollution from point sources.

Enforcement programs for point sources are set up to allow semiannual inspections to ascertain compliance with standards. This works well where there are discrete discharge points with consistent effluent. On CAFOs with containment facilities in place, however, discharges rarely occur outside of rainfall events. Because pollution from CAFOs is largely during wet weather, the timing of the inspection is critical to the assessment of compliance.

The accompanying hydrograph (Figure 22.1), characterizing runoff from a rainfall event in a watershed containing a large number of CAFOs, shows that concentrations of nutrients change significantly throughout the period. A single grab sample could give an erroneous picture of the actual loading to the river. Sampling strategies and data for wet weather conditions must be studied thoroughly before meaningful results are obtained.

Enforcement against dischargers of animal waste is potentially awkward administratively. In an internal memo the Texas Water Commission noted its burden of proving a substantial potential for environmental impact, which is difficult without witnesses who have observed actual polluted water discharge. Ideally, a water sample analyzed for pollutants, a measurement or estimate of discharge flow rate, and photographs of the discharge would be available. These data may include the size and storage capacity of basins, acreage of lots and other areas draining into the retention basins, pond liners, soil type, and proximity to aquifers and aquifer recharge areas. In addition, for enforcement purposes the commission requires information on the history of citizens' complaints about environmental practices at the site, inspection history of the facility, and statements from any witnesses to discharge who are willing to testify.

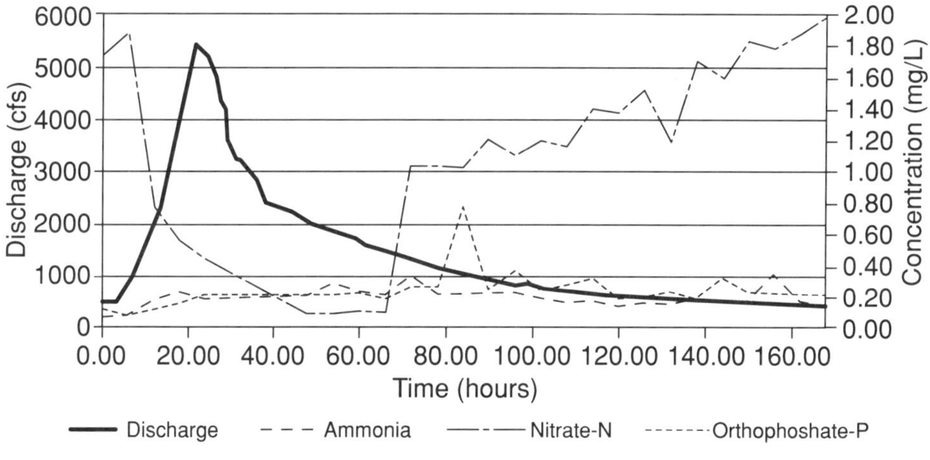

Source: TIAER unpublished data.

Figure 22.1. Hydrographic data

Traditionally, Congress has supported farming operations as they try to come into compliance with national objectives. When EPA and its state counterparts have tried to implement their goals, however, present agency coordination has not been sufficient to ensure that cost-sharing funds and other forms of assistance will be available from USDA. In Texas, focus groups discussing CAFO regulations called for coordination between the regulatory agencies and traditional USDA agencies. A potential problem exists, however, in trying to tie regulatory programs with technical assistance and cost-sharing. Significant effort among government institutions is needed to properly link the regulatory programs of EPA to the assistance programs of USDA.

Costs associated with environmental protection could accelerate the decline of the small producer. Prices for agricultural products will go up in the long term to reflect increased costs. This will not happen, however, in the short term, which more negatively affects small operators. As shown in a Texas study (Davis et al. n.d.), the impact of regulatory programs on a per cow basis is much greater for small than for large operations.

Some CAFO industries (mainly poultry and swine) are separating into growing facilities and production plants. Safley (1992) noted that approximately 30 to 40 percent of the protein imported to farms as swine feed remains in a hog's body, which is sent to the processing plant. The remainder must be disposed of by growers, who cannot internalize this cost and pass it to consumers in the cost of the product. Figure 22.2 shows the movement of protein and the accompanying potential pollutant sources from swine operations.

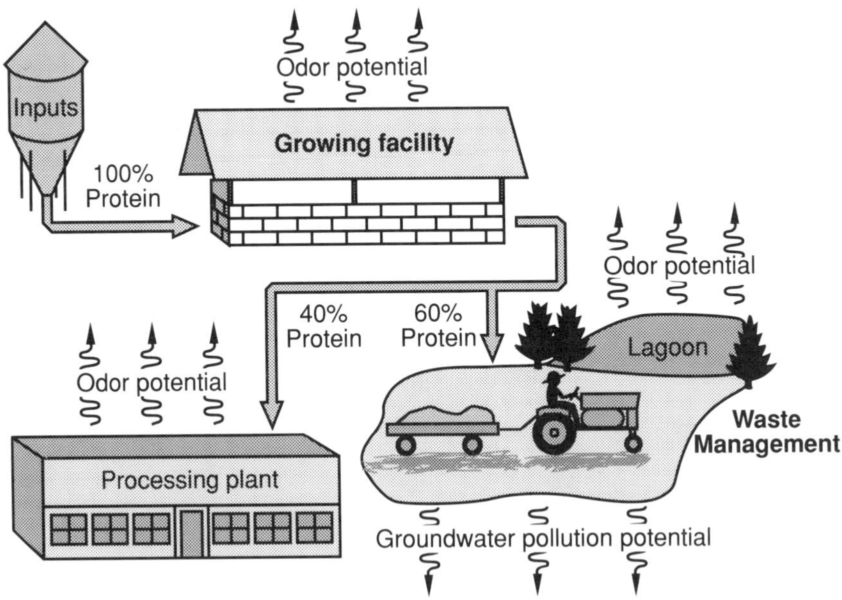

Figure 22.2. Movement of protein and potential pollutant sources

Lagoons have been the primary structural BMP used to comply with the no-discharge effluent limitation. However, odor emanates from both the lagoon and manure application. Factors affecting these new sources of odor include the following: loading rates, proximity of neighbors, wind direction, and time of day and temperature when dewatering. A review of complaints against dairies in Texas revealed that many may be related to the not-in-my-back-yard phenomenon (TIAER 1992b). Because odor complaints are subjective and there are no federal odor regulations, many official water quality complaints begin as odor complaints.

The complaint resolution processes in many parts of the country were established to deal with large industrial and municipal point sources. The conflict resolution process is very expensive, given the need for professional engineers, lawyers, and other legal costs. Additional costs involved in long distance proceedings might be avoided by a local option. Focus group meetings of dairy producers and complaining neighbors in a rural Texas community, for example, resulted in the call from both sides for a local complaint resolution process.

Policy Options to Modify the Status Quo

Appropriate Monitoring and Implementation Scheme

Policy options for the successful abatement of CAFO pollution can most meaningfully be discussed where the criteria for measuring success are clearly identified.

Microwatersheds: Focus of Compliance Efforts

As a result of renewed emphasis on nonpoint source pollution and the inherent difficulty in isolating its diffuse sources, water-quality-based monitoring and assessment programs at the watershed and microwatershed level should be implemented to first *identify* contamination sources and then to *abate* the pollutant loads they produce. This approach represents a significant shift from prevailing regulatory efforts to control pollution from point sources since nonpoint sources must first be identified, and their pollutant contribution isolated, before they can be regulated. Unlike "end-of-the-pipe" pollution sources, the majority of nonpoint sources can deny contributing to the problem until confronted with hard proof of their complicity. The effort to isolate agricultural nonpoint sources will entail significant cost. However, public concern for agricultural nonpoint source pollution is not likely to abate until quantitative and visible improvement in water quality can be linked to implementing BMPs to control runoff from agricultural sources. Thus, the focus of water pollution programs must once again return to an assessment of overall water quality in rivers and streams. This effort can only be accomplished by tracing pollution sources up the watershed on a microwatershed by microwatershed basis.

Measuring Success

Identifying nonpoint sources of pollution within a watershed or microwatershed is but one step in an effective abatement program. Researchers and lay persons alike will ultimately gauge success by improved water quality in receiving streams, rivers, and lakes. Major industries and municipalities have dodged this issue for 20 years by simply adhering to the effluent limitations imposed under the NPDES program. Any comprehensive plan to control CAFO pollution, and nonpoint pollution in general, should be based on water quality standards for major waterways within a watershed that encompass maximum levels for all commonly recognized contaminants. However, this effort must be broad enough to include the more difficult task of establishing wet weather standards—particularly because nonpoint source pollution is precipitation based. Once water quality standards for major waterways within a watershed are established, maximum permissible loading rates for individual stream segments feeding those waterways can be established, as based on the assimilative capacity of stable water bodies into which the stream segments flow. At that point, after nonpoint sources of pollution within the microwatershed have been identified, efforts can be undertaken to implement BMPs that will result in sufficient nonpoint source abatement to achieve the designated water quality standard. CAFO producers should insist on quantitative water quality standards, for without them no predictable level of pollutant load reduction or BMPs exists.

Most CAFOs will achieve pollutant load reductions by implementing practices to control runoff after land application of manure. In Erath County, Texas, the EPA is currently sponsoring a multidisciplinary project titled, "Livestock and the Environment: A National Pilot Project." At present, project researchers are determining appropriate water quality indicators within the North Bosque River and its tributaries. These efforts will result in the establishment of wet weather water quality standards for the North Bosque River and major stream segments in the watershed. Researchers will monitor these indicators before and after the implementation of selected BMPs to control manure runoff after land application.

In most watersheds, CAFOs make up a relatively small portion of the potential sources of water pollution. Private septic tanks, municipal runoff, municipal waste treatment facilities, and agricultural runoff from sources other than CAFOs all contribute significantly to the degradation of receiving water. All of these sources must be included in any calculation of the maximum allowable pollutant loading rate for any particular microwatershed.

Loading rates for stream segments must be based on the assimilative capacity of the water bodies into which the segments flow. The wet weather, storm-driven nature of nonpoint source pollution results in the runoff of contaminated sediment and soluble nutrients and their transport to stable water bodies, such as lakes or reservoirs, where the pollutants settle out. In most cases, the assimilative capacities of reservoirs, lakes, and estuaries will determine permissible upstream loading rates. Obviously, the development of overall nutrient loading rates for an entire watershed will be time consuming and expensive. Unlike the establishment of effluent limitations for traditional point sources, this process does not permit facile monitoring and immediate recognition of pollution abatement. In contrast to point source control efforts, the successful control of nonpoint source pollution will be measured with less precision and over more extended time. Some observers believe that at least a portion of this cost and monitoring effort must eventually be borne by contributing polluters within a microwatershed (Foran et al. 1991; Mallory 1992).

Focusing on the Problem

To date, federal regulatory efforts concerning CAFOs have focused on contaminated runoff from confinement areas. Implicit in that focus is the assumption that pollution from manure application fields is unlikely if sufficient land exists for the disposal of manure solids and liquids. However, that notion is underpinned by a second assumption: that farmers who apply manure to the land treat the manure as a crop fertilizer and apply it at appropriate agronomic rates. Discussion about field carrying capacity, cows per acre, and producers owning enough land to properly dispose of manure may be increasingly less meaningful in light of current production practices and management priorities. Increasing numbers of CAFO operators purchase commodities for their feeding operations and grow less feed. These producers are not row crop farmers; their expertise lies in livestock and poultry husbandry.

Another assumption that has hindered the successful regulation of pollution from CAFOs is that BMP implementation is synonymous with successful nonpoint source pollution control. To the contrary, edge-of-field BMPs to impede nutrient runoff provide no guarantee of reducing contaminated runoff as such. Undoubtedly, specific BMPs are crucial for controlling nonpoint source runoff, and modeling efforts based on field calibration can help predict the best combination of BMPs for any given application. However, BMPs are only as good as the management expertise associated with their implementation and long-term use. Data compiled by the Texas Institute for Applied Environmental Research (TIAER) from the North Bosque River watershed indicate that dairy managers are more likely to implement appropriate structural BMPs than managerial BMPs (TIAER 1992a). This fact may in part be explained by the greater visibility of structural BMPs to regulators. The agricultural extension service obviously has an important role to play here in educating operators on managerial techniques.

Current regulations to control nutrient runoff after land application of manure must necessarily rely heavily on producers' voluntary compliance with designated BMPs. Compliance surveys by TIAER (1992a) of permitted dairies in Erath County, Texas, applying manure to application fields indicate that determination of the rate of manure application to a particular field is extremely difficult once application is completed.

Thus, any attempt by regulatory agencies to ensure proper land application through site inspection would require prohibitive allocations of agency resources.

Experience teaches that education and demonstration programs are insufficient to induce voluntary compliance with existing water pollution regulations by most producers (Magette et al. 1989). The substantial expense entailed in constructing a comprehensive waste management system is simply too great for many producers to confront without the threat of penalties for noncompliance. For example, EPA recently estimated that the cost of compliance with the new Region VI general permit for CAFOs will total about $100 per cow (USEPA 1993). Moreover, CAFO producers are often unable to recoup the cost of compliance through the market by raising prices. This fact, as well, militates against voluntary compliance with environmental regulations.

Regulatory programs through EPA and their state counterparts will drive pollution abatement associated with CAFOs; however, successful BMP programs will likely occur through programs focused on microwatersheds by local agencies. Local districts and local landowners will play a large role in sorting out sources and loads of pollution and comparisons with local monitoring results. Regulation, ultimately, may be reserved for "poor performers." The wet weather nature of most CAFO pollution will largely negate the usefulness of traditional command-and-control inspection programs of regulatory agencies. Poor performers may eventually come under regulatory and permitting programs specifying technologies that guarantee pollution prevention. State CAFO pollution programs may look much like current "cooperative federalism" at federal and state levels where EPA establishes programs and provides for delegation to the state.

Two Fundamental Alternatives

Solutions for the current CAFO pollution problem caused by land application of manure can be addressed under the following dichotomy: (1) policies promoting better on-farm manure management, and (2) policies promoting off-farm use of manure. A combination of voluntary, regulatory, and market-based programs promoting both alternatives represents the greatest chance for success. The Texas pilot project mentioned earlier includes policy analysis designed to drive other research efforts. Currently, the project's working group is examining potentially viable alternatives to the present practice of applying manure to crop and pasture land.

BMPs to Control Runoff

A wide range of BMPs, both structural and managerial, can alleviate the problems of contaminated rainwater and nutrient runoff after land application of manure. For example, freestall dairy housing is a capital-intensive BMP that shows promise in both increasing milk production and controlling waste runoff. Because corrals are covered in a freestall facility, the amount of rain runoff from feedlot areas is dramatically reduced, freeing the producer from problems associated with rainfall. Freestall operations often incorporate other technological advantages as well. Concrete lanes within a freestall facility may be flushed into a central sump where liquid waste is then separated from solid waste by a mechanical separator. The wastewater is then recirculated as flush water while the separated solids can be used as bedding material, composted, or applied to the land as manure solids. On-farm composting of solid manure prior to land application can reduce some nitrogen runoff because composting converts the nitrogen in manure to a more stable, organic form.

Other less costly structural BMPs show promise in reducing nutrient runoff from land application of manure. Vegetative filter strips between application fields and adjacent waterways are a relatively simple and effective BMP (Sweeten et al. n.d.). Filter strips may prove particularly effective when used in conjunction with such managerial BMPs as selecting application fields with little slope, applying manure only at agronomic rates, utilizing cropping systems for optimum nutrient uptake, and reducing the nutrient content in animal rations. The pilot project is testing the effectiveness of vegetative filter strips, moderate slope for application fields, and selected manure application rates in Erath County, Texas.

Policies to Promote the Use of BMPs

Considering the elaborate state and federal regulatory apparatus now in place to control water pollution from CAFOs, it is safe to assume that regulation will not be displaced by a wholesale shift to market-based charges (negative incentives) to induce CAFO operators to adopt BMPs. However, within the current regulations, space exists for indirect charges and novel regulatory approaches for improvement.

Direct Financial Assistance

Federal and state financial assistance must continue at present levels or increase to aid small operators in installing the structural BMPs necessary to comply with the no-discharge effluent limitation for CAFOs. Direct financial assistance is particularly important for small CAFOs because of their position as price takers in the marketplace. At present, the USDA directs much of this assistance through the ASCS. However, the EPA indicates that additional funding for BMPs may be available through state revolving loan programs (*Inside EPA* 1992). Both the EPA and Texas A&M University predict that the cost of compliance per animal is higher for smaller CAFOs than for similarly situated larger operations (Davis et al. n.d.). If this holds true, then special assistance for smaller operations may be necessary to avoid serious hardship for the smaller, "family farm" operations. Moreover, smaller producers are often at a disadvantage with larger operations because of smaller cash flows and less ability to obtain commercial financing due to fewer assets.

To avoid placing the financial onus of environmental compliance solely on CAFOs, Congress could approve a federal tax on CAFO products at the wholesale level, the proceeds from which could be rebated to producers certifying that they have implemented a minimum level of pollution control technologies and management practices. As a result, regulated markets could be manipulated to aid CAFOs to recoup environmental compliance costs that they would otherwise be forced to absorb.

Tax Breaks as Incentives

Tax breaks may also provide positive incentives to CAFO operators to implement BMPs. Investment tax credits and accelerated write-offs for infrastructural investment are options at the state and federal levels. Local property tax exemptions may work as well. In Illinois, both the state EPA and the Extension Service certify facilities as being in sufficient compliance to qualify for property tax exemptions. In Minnesota, CAFOs constructing waste storage structures with at least a 120-day storage capacity qualify for local property tax exemptions (Wilson 1990).

Regulatory Alternatives

One important regulatory issue is whether a federal water quality standard is necessary or possible to avoid the "tragedy of the commons" where CAFOs might migrate to states with more lenient environmental regulation. EPA developed federal effluent limitations for traditional point sources to address this very problem. While the NPDES no-discharge effluent limitation for CAFOs arguably serves this purpose, spotty enforcement, insufficient staff in the EPA, and the difficulty inherent in regulating runoff all support a federal wet weather water quality standard to augment the no-discharge limitation. In practice, the standard could prove quite beneficial to the states by establishing an initial baseline for intermittent streams by which to measure the reduction in runoff after BMP implementation. However, the difficulty in establishing a uniform standard for the myriad intermittent waters nationwide may prove too great.

The development of a two-tier permitting system may represent another policy option. Under this alternative, CAFOs employing a predesignated level of BMPs and technologies guaranteeing minimum pollution would pay far less for a waste discharge permit than CAFOs failing to employ that level of technology. A minimum size could be included to exempt very small producers from the permit process, similar to that currently employed by both EPA and various states. Once implemented, the two-tier system would require less field inspection, thereby reducing the current strain on agency resources. For example, in the dairy industry, the implementation of freestall technology with guttered roofs coupled with the management practice of hauling solid manure off-farm for processing would go far toward eliminating the potential for wet weather pollution by dairies, with a corresponding cost reduction in the regulatory sector. However, any combination of BMPs that includes the land application of manure is inherently more difficult and costly to regulate. Once manure is applied to the land, the rate of application is difficult to detect. However, because of its lower potential for nitrate contamination, composted manure might represent an acceptable BMP under the low-cost permit.

Two-tier wholesale pricing of products produced by CAFOs represents an indirect, market-based incentive toward sufficient BMPs. For example, USDA could monitor a system under which dairy producers who fail to implement sufficient BMPs would receive less for their milk than producers who have implemented those BMPs. However, the price differential would have to be sufficient to spur the majority of producers to invest in BMPs; otherwise, noncomplying CAFO operations could ignore the no-discharge effluent limitation indefinitely.

Any change based simply on the size of a CAFO or the amount of total manure produced must be avoided. Sufficient data demonstrate efficiencies of size and scale in most CAFO sectors. Thus, to penalize an operation based on size alone flies in the face of clearly prevailing economic trends.

Trading of NPDES and similar state permits is an often-discussed yet seldom used policy option that is simultaneously part regulation and part incentive. The theory behind permit trading is that potential waste dischargers in a particular watershed or stream segment should be allowed to buy and sell the right to discharge among themselves so long as the overall water quality in the receiving water does not fall below a predesignated level (Hahn 1989). However, trading has not proven practical in most cases, particularly between point and nonpoint sources. First, a minimum water quality standard must be established for any area where trading might occur. Second, in order for a nonpoint source to be able to trade allowable discharge with a point source, the nonpoint source must first determine the pollutant load it represents. However, difficulty in determining that load is one of the defining characteristics of nonpoint sources. Yet nonpoint sources will often predominate in rural watersheds where many CAFOs likely exist. Ultimately, the problems entailed in establishing a water discharge trading system in most cases have proven to be a greater burden than simply complying with the prevailing effluent limitation.

Off-farm Manure Management as a BMP

While the BMPs discussed here show promise in reducing nutrient runoff from manure application fields, policymakers could eliminate much of the problem at its source by promoting alternatives to land application as a disposal mechanism. Obviously, those producers who conscientiously apply manure to crops as fertilizer at agronomic rates should continue to do so. This practice reduces the purchase of commercial fertilizers, which pose nutrient runoff problems of their own. However, those producers who apply manure to land simply as a convenient means of disposal should be given alternatives.

A variety of off-farm uses for manure produced by CAFOs exist. While some of the alternatives appear more immediately feasible than others, none should be eliminated from consideration simply because additional research may be necessary to bring that option to fruition. On-farm manure management, because of trends toward consolidation, may be an ever-growing problem; all alternative uses should be explored until proven economically impractical.

Perhaps the simplest alternative to land disposal of manure by CAFOs is not really an off-farm alternative in its purest sense. Manure produced by CAFOs can be marketed directly to crop producers; this is currently under way in Lancaster County, Pennsylvania. Every six months a local agency produces two lists: one identifying livestock producers wishing to sell manure, and a second identifying crop producers in need of manure as fertilizer. As long as the cost of hauling the manure does not exceed the cost of alternative commercial fertilizer, this arrangement offers a promising alternative to CAFOs and crop producers alike.

A second alternative to on-farm manure application is off-farm composting at a central facility. If marketed properly, this alternative might produce sufficient revenue to cover transportation costs and return some profit to CAFO producers. Centralized composting requires considerable organization and financing for facility construction. The City of Stephenville, Texas (in Erath County), is examining the feasibility of a central composting facility as part of a larger municipal composting facility. Considerable interest has been expressed by local dairy producers who would welcome an alternative to land disposal of manure.

Technology is available to burn solid manure in a large power plant to produce electricity. The Mesquite Lake Cattle Manure Power Plant in El Centro, California, burned 1,000 tons of manure per day in February 1992, providing electricity to more than 20,000 people (Gehringer 1992). A similar facility was planned for central Texas during the mid-1980s, but fell victim to the insolvency of project financiers.

Manure can be used as a component in animal feed ration or as bedding material in animal stalls, but greater effort needs to be expended in increasing its viability.

Finally, manure can be run through a municipal or private waste treatment plant. This alternative entails considerable capital investment to upgrade municipal facilities or to construct private facilities. However, under certain conditions that investment may prove feasible. For example, if large dairy production units were constructed on a large tract of land, the initial investment for a treatment facility might be cost effective.

Because few alternatives to land application of manure presently exist, government policies need to promote the creation of off-farm alternatives as well as to promote their use by CAFO operators.

To promote the creation of off-farm manure management facilities, local governments could provide long-term property tax exemptions and utility-related incentives to encourage commercial composting facilities. Many local governments currently take this approach to lure new businesses to their communities. Further, state and federal governments could provide tax breaks to promote such facilities. Finally, cost-share financing may be available to construct alternative manure processing facilities. The EPA has indicated that it may extend its State Revolving Fund loans to support these efforts as well.

Policies to promote the increased use of municipal waste treatment facilities for manure disposal are more difficult to envision, although the Clinton administration may provide some funding in this area. However, a large privately financed CAFO park could incorporate a treatment facility from the park's inception. Apart from federal grants for the construction of larger treatment facilities in areas producing large quantities of manure, few incentives come to mind.

Policymakers must promote the use of alternative manure disposal by producers. As mentioned, a two-tier price structure for milk might be created under which those dairy producers not disposing of manure on land would receive a higher price for their milk. Because national milk markets are highly regulated by the federal government, this type of intervention appears feasible. A deterrent might work as well, where a farmer disposing of manure by land application would be required to pay an annual tax for such disposal. This tax is similar to the higher-priced discharge permit discussed earlier. The current lack of composting facilities or other alternatives to land application in most areas makes these incentives and disincentives quite speculative. Reviewing the costs of off-farm waste disposal in areas where such alternatives presently exist would help determine the level of incentives necessary to ensure using alternative waste disposal facilities once established. Ultimately, a farmer could be required to use alternative waste disposal where available.

Odor Regulation

The federal Clean Air Act does not regulate CAFO odors; odor regulation has traditionally been the exclusive province of the states. However, many of the technologies and management practices currently required to control water pollution by CAFOs (such as the creation of wastewater lagoons and the land application of manure) aggravate dairy odors. Thus, research efficiency and thoroughness dictate a simultaneous consideration of both problems.

Nuisance odor is a concern with large CAFOs. However, odor regulation is a difficult task because of the subjective human factor in judging odor offensiveness and intensity. Moreover, few reliable, quantifiable indicators of nuisance odor currently exist. Odor frequency, intensity, duration, and objectionableness—the FIDO factors—must all be considered in determining whether a bona fide nuisance exists (Watts and Sweeten 1993). Proposed solutions to nuisance odor range from sophisticated indoor filtration and manure treatment systems to land-use planning specifying minimum distances from CAFOs to the nearest adjacent landowner.

Zoning for a large CAFO park falls under the rubric of land-use planning. Dairy CAFO parks are currently under consideration for Erath County, Texas, and Hawaii. Establishing a CAFO park would alleviate much of the present regulatory uncertainty prevailing in many areas of the country concerning both water pollution and nuisance odor. Producers would benefit from the expedited permit process and protection from nuisance lawsuits. On the other hand, non-CAFO rural landowners could rest easier knowing that producers were locating in an area with specific technologies and an adequate buffer zone. At present, in Erath County, both sides claim to be victimized by inequitable land prices. Dairies argue that adjacent landowners request exorbitant purchase prices when dairies attempt to buy their neighbors out to avoid nuisance lawsuits. Adjacent landowners claim that the value of their land decreases whenever a dairy locates nearby.

Institutional Arrangements

Public and private institutions provide four services that are essential for the long-term prosperity of the CAFO industry: (1) educational and technical assistance, (2) funding and financing, (3) incentives, and (4) regulation (Rice 1979). CAFOs can benefit greatly when the agencies providing these services work together efficiently.

Traditionally, state and federal agencies aligned with the USDA have concentrated on providing education and technical assistance, and some positive incentives, while EPA and its state counterparts have filled the regulatory role. However, a strict division along these lines with little interaction between the two ultimately works to the detriment of the agencies' constituents and the regulated community. For example, extension services can provide continuing education to CAFO operators to keep them informed of changes in regulations and sources of assistance to help satisfy those regulations. However, in the event a water quality violation by a CAFO is uncovered by a state or federal regulatory agency, the CAFO needs sufficient time to turn to the SCS and ASCS for technical and financial assistance before enforcement proceedings begin. This is particularly true where a regulatory agency embarks on a broad enforcement initiative. Prior to taking any action, the regulatory agency should coordinate its efforts with the SCS, ASCS, and state agencies representing local soil and water conservation officials to be sure assistance will be immediately available. Linkages between the regulatory agencies and assistance agencies should be close enough to permit continuous monitoring by all concerned, with sufficient flexibility to extend the period initially designated for compliance under appropriate circumstances. Thus, the key to optimal agency performance on behalf of its constituents is a clear protocol outlining both the role of all agencies involved and the time frame for each step in the process, coupled with open and continuous communication among all agencies concerned.

Under certain circumstances, additional agencies may be included in the preexisting protocol to increase overall efficiency. For example, in the dairy industry, health department officials charged with inspecting milk production facilities at dairies might be trained to conduct inspections of waste control operations during the same visit. Combining the two functions obviously reduces agency expenditures while possibly providing more frequent inspections of waste control operations.

References

Amey, David. 1992. Nation's broiler industry—Industry concentration. *Broiler Industry* (Dec.):26A-26D.

Batie, S. S., L. A. Shabman, and R. Kramer. 1986. U.S. agriculture and natural resource policy: Past and future. In *The future of North American grange,* ed. Ford Runge. Ames: Iowa State University Press.

Beck, Robert E. 1981. Agricultural water pollution control law. In *Agricultural law,* ed. John H. Davidson. Colorado Springs, Colorado: Shepard's/McGraw-Hill.

Christensen, Lee A. 1992. Regional shifts; are they real? *Watt Poultry Yearbook,* USA edition, pp. 44-55.

Clark, Edward. 1992. Poultry production returns to the Midwestern states. *Feedstuffs* 64(49):12-14.

Davidson, John H. 1989. Commentary: Using special water districts to control nonpoint sources of water pollution. *Chicago-Kent Law Review* 65:503, 513-14.

Davis, E., R. Schwart, and R. Lacewell. No date. Economic impacts of proposed NPDES general permit on Texas livestock industry. Texas A&M University, College Station.

Egg Industry. 1992. Too many layers, too many eggs and too little money . . . *Egg Industry* (Nov./Dec.): 11-14.

Foran, J., P. Butler, L. Cleckner, and W. Buckley. 1991. Regulating nonpoint source pollution in surface waters: A proposal. *Water Resources Bulletin* 27(3):479, 481.

Gehringer, M. 1992. Manure to burn. *The Dairyman.* Feb.

Hahn, R. 1989. Economic prescriptions for environmental problems: How the patient followed the doctor's orders. *Journal of Economic Perspectives* 3:95-96.

Hurt, Chris, Kenneth A. Foster, John E. Kadlec, and George F. Patrick. 1992a. Continuing transition. *Feedstuffs* 64(38):1-17.

———. 1992b. Industry evolution. *Feedstuffs* 64(35):18-23.

Inside EPA. 1992. Congress eyes sweeping expansion of state wastewater loan fund. Dec. 11. Iowa Code. 1990. S176B. 1-13.

Lexis. 1993. Computer database.

Long, Catherine. 1992. Livestock waste pollution: A nationwide problem (draft). U.S. Environmental Protection Agency, Office of Policy, Planning, and Evaluation.

Magette, W., A. Shirmohammadi, B. Lessley, and R. Weismiller. 1989. Managing groundwater quality in relation to agricultural activities. *Water Resources Bulletin* 25(3):599, 604.

Mallory, N. 1992. Ideas for the livestock compact (draft). National Center for Food and Agricultural Policy, Washington, D.C.

Marberry, Steve. 1992. Pork production shifts to larger farms. *Feedstuffs* 64(49):1-5.

Matulich, Scott C. 1978. Efficiencies in large-scale dairying: Incentives for future structural change. *American Journal of Agricultural Economics* 60(4):642-647.

Muirhead, Sarah. 1992. Advancements in animal agriculture to be influenced by outside concerns. *Feedstuffs* 64(48):1-13.

National Pork Producers Council. 1992. *Pork facts* (fact sheet). National Pork Producers Council, Des Moines, Iowa.

Oppedal, A. 1988. Wynot versus Hans. *Hog Farm Management.* Nov., pp. 24-29.

Perry, Charles. 1992. Regs, regs, and more regs. *Watt Poultry Yearbook.* USA edition.

Rice, J. M. 1979. Management and financing of agricultural BMPs. In *Best management practice for agriculture and silviculture, proceedings of the 1978 Cornell Agricultural Waste Management Conference.* Ann Arbor, Michigan: Ann Arbor Science Publishers.

Roenig, Bill. 1992. Some other product has to go. *Watt Poultry Yearbook.* USA edition.

Rogers, W. 1986. *Environmental law* 2:4.9, 4.10.

Safley, L. M., Jr. 1992. Evaluation of swine waste management: Facilities and methods. In *National livestock, poultry and aquaculture waste management: Proceedings of the national workshop,* ed. John Blake, James Donald, and William Magette. St. Joseph, Michigan: American Society of Agricultural Engineers.

Sierra Club v. Abston Construction Co., Inc. 620 F.2d 41, 44 (5th Circuit, 1980).

Smith, J. Julian. 1992. Our responsibility in animal waste management. In *National livestock, poultry and aquaculture waste management: Proceedings of the national workshop,* eds. John Blake, James Donald, and William Magette. St. Joseph, Mich.: American Society of Agricultural Engineers.

Sweeten, John M., M. McFarland, B. L. Harris, G. Westmoreland, C. Baird, and L. Manning. No date. *Animal waste application.* Publication L-5044. Texas Soil and Water Conservation Board and the Texas Agriculture Extension Service.

Texas Institute for Applied Environmental Research (TIAER). 1992a. *Final report on Section 319 non-point source management program for North Bosque watershed.* TIAER, Tarleton State University, Stephenville, Texas.

———. 1992b. *Livestock and the environment: Rethinking environmental policy, institutions, and compliance strategies.* TIAER, Tarleton State University, Stephenville, Texas.

U.S. Department of Agriculture, Crop Reporting Board, Soil Research Service (USDA). 1992. Cattle on feed by size and state.

U.S. Department of Commerce. 1971, 1977, and 1989. *U.S. Census of Agriculture—United States summary and state data,* vol. 1(51), (Issued Dec. 1971, Dec. 1977, and July 1989).

U.S. Environmental Protection Agency, Office of Water (USEPA). 1989. *Report to Congress: Activities and programs implemented under Section 319 of the Clean Water Act* (Aug.).

U.S. Environmental Protection Agency (USEPA). 1993. Responsiveness summary for Region VI general CAFO permit. *Federal Register* 58:7610.

U.S. General Accounting Office (U.S. GAO). 1990. *Agriculture: USDA needs to better focus its water quality responsibilities.* GAO/RCED-91-168.

———. 1991. *U.S. Department of Agriculture: Revitalizing structure, systems, and strategies.* GAO/RCED-91-168.

Velthuzien, A. M. 1992. Washington dairies brace for tough new laws. *The Dairyman* (Feb.):23.

Watts, P., and J. Sweeten. 1993. Towards a new regulatory definition of odor nuisance. Submitted to the *Journal of the Air and Waste Management Association.*

Federal environmental laws. 1992. St. Paul, Minnesota: West Publishing Co.

Wilson, L. 1990. *Livestock waste control programs of ten midwest and western states.* Iowa Department of Natural Resources, Des Moines, Iowa.

Wheeler, J. Patrick. 1992. Livestock odor and nuisance actions versus "right-to-farm" laws: Report by Defendant Farmer's Attorney. *North Dakota Law Review* 68:459.

Environmental Management Initiatives

23. Regulatory and Nonregulatory Approaches to Ecosystem Conservation: Institutions and Funding Mechanisms

Brendan Doyle, Peter Kuch, and Robert Wolcott
U.S. Environmental Protection Agency

As resource managers in the Great Plains develop a strategy to address ecosystem conservation issues, they may employ both regulatory and nonregulatory approaches to achieve their goals. The U.S. Environmental Protection Agency's (EPA) experience in other geographic initiatives (such as the Chesapeake Bay and Great Lakes programs) suggests that coordinated initiatives on the part of federal and state regulatory and resource management agencies, as well as cooperative efforts with allied public and private organizations, special districts, and private landowners can conserve and protect ecosystems and yield substantial improvements in environmental quality. These geographic initiatives have focused attention and resources on ecological resources that have attracted broad-based public support.

At the outset, Great Plains program managers may want to define ecosystem management units such as the ecoregions described by EPA's Ecological Monitoring and Assessment Program (see Figure 23.1) in EPA regions VI, VII, and VIII, Canadian provinces, and 13 states. Each ecoregion has its own unique characteristics and interrelated functions, fauna, airsheds, watersheds, and landscapes. By coordinating and targeting environmental management efforts within each ecoregion, program participants can ensure the continued functioning and productivity of each region's ecosystem, sustain economic uses of ecoregion resources, and marshal program support within a manageable area. One initiative could involve participants focusing on species that are at risk of becoming endangered or threatened both within and among the Great Plains ecoregions. Collaborative efforts might include integrating EPA's water quality regulatory programs—monitoring water quality, implementing water quality standards—with U.S. Fish and Wildlife's species conservation planning and U.S. Army Corps of Engineers' wetlands programs. In addition, collaboration might extend to state water resources agencies, irrigation districts, and private landowners roles in managing and conserving aquatic species habitats on a watershed-by-watershed basis. Similar efforts to implement EPA air quality regulations across airsheds and waste management programs within watersheds and airsheds could also be coordinated with other parties' interests in ecoregion conservation.

Stanley R. Johnson and Aziz Bouzaher (ed.), *Conservation of Great Plains Ecosystems*, 393–402.

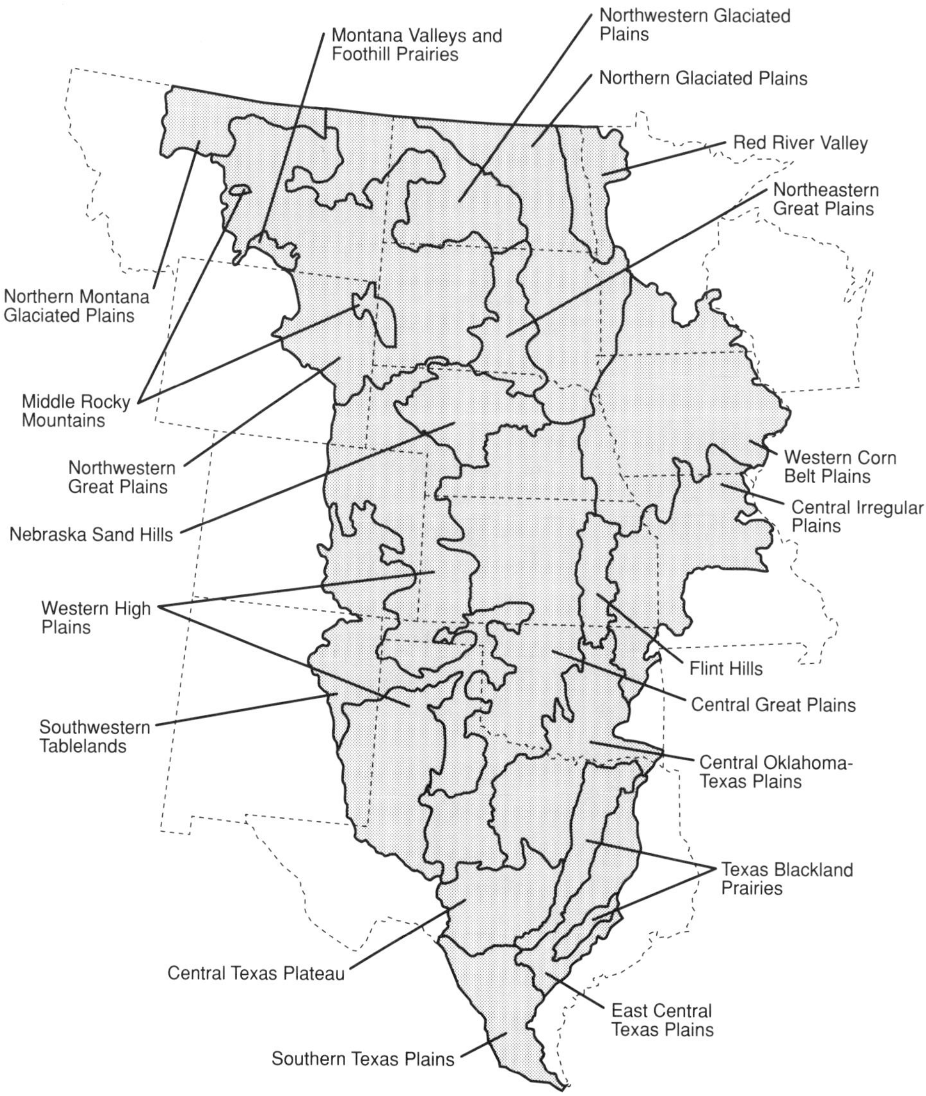

Source: Omernik 1987.

Figure 23.1. Great Plains ecoregions

In addition to coordinating implementation of traditional "command and control" regulatory programs, resource managers could apply economic incentives or seek market-based approaches to meeting ecoregion conservation goals. If, for example, critical habitat areas or ecoregion functions were affected by irrigation return flows, program participants could create a market "to buy and sell discharge rights" by providing participating irrigators with the flexibility to find the most cost-effective ways to meet applicable regulatory limits on the total amount of phosphate or nitrogen to be discharged to a nearby water-quality-impaired lake or river. Participating landowners, organized in a manner similar to an irrigation district, would each examine the costs to reduce their discharges and work with other district members on allocating amounts of discharge up to the *total* level of pollutants allowed into the lake or river. Landowner A, for example, may find that it is more cost-effective to pay the cost of Landowner B's additional reductions than it is to meet the same discharge limit and pay the full cost alone. A market of discharge "credits" could be created, whereby participants buy and sell or trade the rights to discharge varying amounts of polluted irrigation water up to an established limit. Irrigators in the San Joaquin Valley of California are experimenting with these market-based approaches (Young and Congdon 1992). Because irrigated agriculture extends throughout the western Great Plains ecoregions, the same concept could be applied in areas where irrigated agricultural practices produce pollutant loadings to critical groundwater resources or aquifers.

Purposes for Creating Great Plains Funds

Much baseline information describes the unique characteristics and resources of the Great Plains. Once this information has been integrated for analysis, program participants will be able to focus attention on those ecosystems or ecoregions most at risk, and in particular, which economic activities such as agriculture, urban development, mining, manufacturing, and the like are posing the greatest risks to sustaining ecosystem functions. For instance, many in the region will be interested in trying to identify critical habitats of species that may be becoming threatened or endangered and in finding ways to target resources and management practices on attaining well-defined habitat conservation goals. Thus, interstate and regional collaborative efforts will need to be coordinated and directed in order to be effective. The availability and quality of data to describe baseline conditions and the effects of alternative management practices on ecoregion functions will be critical to the success of Great Plains conservation initiatives.

New Institutional Arrangements

Great Plains states (throughout EPA Regions VI, VII, and VIII and the central Canadian provinces) may wish to explore creating a Great Plains Ecosystem Conservation Council with members appointed by each state's or province's governor to define program goals and guide collaborative efforts. Such a council could act as a Board of Directors or trustees for a Great Plains Trust or Ecosystem Conservation Fund to provide funding and support. The council, in turn, could have several types of advisory committees and coordinate goal-setting, planning and budgeting, and program management with other program participants. With input from individual landowners, business, agriculture, and industry, as well as public and private agencies on advisory committees, the council could provide the political and economic framework for fashioning an efficient plan for realizing scientifically based (sustainable development) objectives for the region.

In addition to coordinating regulatory and nonregulatory (economic incentive) approaches to ecosystem conservation efforts, Great Plains program participants may also want to develop new organizations and funding mechanisms to support program projects and activities. Coordinating geographically targeted programs will be complex, and the direct and indirect, or "transaction," costs of obtaining ecosystem or ecoregion improvements (data collection, data analysis, field demonstration projects, public-private ventures, and public education programs related to ecosystem conservation) may be substantial. However, recent experience in EPA's Great Lakes and Chesapeake Bay programs suggests that they are manageable. Drawing on the Great Lakes and Chesapeake Bay program examples and having a Great Plains Fund administered by members who represent the public and private stakeholders in Great Plains ecosystems will provide a foundation for building program support.

Assuming that creating such a council is desirable, and that the council succeeds in defining goals and policies for achieving them, a Great Plains Fund could underwrite several projects:

- Developing strategic planning for conserving Great Plains ecosystems
- Developing programs for public participation, education, and awareness
- Buying and trading of water rights to support in-stream uses, particularly where aquatic species' habitats are at risk
- Buying and trading of air rights to meet regional air quality standards and airshed improvement goals
- Developing model farm, model community, or model state projects or programs
- Purchasing cropping or ranching rights
- Sharing costs of management practices
- Acquiring easements or fee-simple title to lands and wetlands needed to link fragmented pieces of habitats and ecoregions at risk, such as buying out long-term, critical Conservation Reserve Program leases

- Leveraging investments in needed research and data collection and analysis
- Paying the transaction costs involved in swapping land that is currently under public ownership for private lands that are needed to implement Great Plains conservation policies
- Providing offsets or supplements to lost local government and landowner revenues in areas where property tax exemptions or other locally provided conservation activities occur

The Great Lakes and Chesapeake Bay Programs

The Chesapeake Bay and Great Lakes programs have involved federal agencies, state agencies, local governments, private and nonprofit organizations, and have attracted broad public support and generated corresponding levels of funding. Their success demonstrates that innovative funding mechanisms can be established to underwrite ecosystem conservation costs. Innovative mechanisms include pooling federal, state, or local government resources; joint-venture project funding among industry, agriculture, nonprofit organizations, and private foundations; capitalizing revolving loan and grant funding programs; and implementing user fees at the local and state level. Previous experience in these and other geographic initiatives suggests that Great Plains program participants may want to create a series of funds to pool available funding at the federal, state, or local government levels. Creating Great Plains funds would provide mechanisms for program participants to dedicate existing and new financial resources to carry out program activities and provide the funding vehicle(s) needed to address Great Plains airshed or watershed conservation needs. The eight Great Lakes states[1] and Ontario have used general funds to capitalize a Great Lakes Protection Fund with a $100-million endowment. State contributions are apportioned by the volume of Great Lakes water used by each state or province. Interest earnings on the endowment fund are expended on Great Lakes initiatives such as developing a mathematical model to describe the chemical mass balance of toxic substances in the lakes, developing water quality standards tailored to protect Great Lakes aquatic ecosystems, and restoring threatened and endangered fish species' aquatic habitats. Five percent of annual fund interest earnings are returned to each participating state or province to underwrite their participation in the program. Great Lakes Protection Fund managers solicit proposals from project sponsors. The governing board (with representatives from each participating state or province) makes funding decisions.

Businesses, nonprofit organizations, and private foundations in the Great Lakes region are also funding innovative collaborative projects targeted at ecosystem and ecoregion conservation, watershed improvement projects, and pollution prevention initiatives. The

[1]Minnesota, Wisconsin, Illinois, Indiana, Michigan, Ohio, Pennsylvania, and New York

Great Lakes Council of Governments, the Environmental Defense Fund, and the printing industry, for example, have joined efforts in the "Great Print Project" to explore creative options for reducing air and water emissions in areas where the region's printing industry facilities are located.

In the Chesapeake Bay program, 11 federal agencies[2] and 8 state agencies pool their funding (nearly $400 million per year) to coordinate individual agency programs across those parts of Pennsylvania, Maryland, the District of Columbia, and Virginia that drain water to the Chesapeake Bay. Coordinated programs include monitoring of water and air quality, applied research, and the development of shared databases on Chesapeake Bay environmental conditions and trends. The federal agencies, the District of Columbia, and the three state partners in the Chesapeake Bay program have formed a steering committee (chaired by EPA) to plan and budget funding of federal resources and to avoid duplication of management programs. This consortium of agencies is currently developing revenue raiser proposals such as doubling Chesapeake Bay bridge tolls, taxing water effluent discharges, and other fees and charges to link the *polluter pays* or *beneficiary pays* economic principles to the bay's program financing.

Maryland and Virginia contribute proceeds from the sale of motor vehicle vanity plates (nearly $22 million to date), which convey "Save the Bay" or other wildlife conservation messages, to a dedicated Chesapeake Bay Trust Fund. The fund is managed by representatives of each governor's office who, in turn, sponsor innovative community and regional projects and programs, such as wetlands restoration and environmental education.

The Chesapeake Bay Commission, with representation from each state legislature, coordinates legislative issues that need to be addressed, such as banning TBT boat-bottom paint. The Chesapeake Bay program also has three advisory committees: a scientific and university research committee, a citizen's advisory committee, and a committee composed of representatives from county and local governments. More than 50 nongovernmental organizations, ranging from research to education to resource management, are involved in "Save the Bay" programs.

State, local, and regional governments alone cannot conserve ecosystems. Often nonprofit organizations, public interest groups, and others play a key role. Since 1977, the Chesapeake Bay Foundation (a nonprofit organization with 86,000 members) has raised funds from private sector sources and individuals for Chesapeake Bay environmental education programs such as operating a "skipjack" sailboat for bay touring, ad campaigns such as "Save the Bay" labeling on all Coca-Cola™ products produced by

[2]U.S. Environmental Protection Agency, National Oceanic and Atmospheric Administration, U.S. Forest Service, U.S. Park Service, Agricultural Soil Conservation Service, U.S. Geological Survey, U.S. Army Corps of Engineers, U.S. Navy, U.S. Department of Defense, U.S. Coast Guard, and the Soil Conservation Service (350 full and part-time employees with a budget of nearly $250 million annually).

bay area bottlers (partial proceeds of soda can recycling sales go the Foundation), lobbying, and litigation.

Application of Other Funding Mechanisms to the Creation of Great Plains Funds

Funding of the Great Lakes and Chesapeake Bay programs provides important models to use in implementing Great Plains programs. In the case of the Great Plains, money from existing programs, as well as new money derived from a variety of sources, could be pooled into a series of funds, the organization of which depends upon the wishes of program participants. A portion of existing federal agency funds could be set aside for these new programs. Such funds could include payments from the Conservation Reserve Program; expenditures of the U.S. Land and Water Conservation Fund; state program grants from the EPA such as air and water quality and solid waste programs; program funding provided by the National Park Service; funds of the U.S. Fish and Wildlife Service, U.S. Army Corps of Engineers, and Bureau of Land Management; disaster-relief funding of the Federal Emergency Management Agency from the 1993 floods; and funding from the U.S. Department of Transportation.

Likewise, state agency funding related to ecosystem conservation programs and education efforts could be pooled among participating state agencies. Great Lakes and Chesapeake program managers indicate that their participating states prefer to pool their state funds separately from federal funds. This approach gives states the opportunities to use those funds to meet federal matching requirements (e.g., awarding of EPA public water supply program grants is conditioned upon states providing 25 cents of each dollar being granted) and to find ways to support and administer state health and environmental protection programs not funded by federal programs. Regional and local funds, such as land trusts and other nongovernmental organizations' revenues (raised by watershed improvement districts, irrigation districts, highway improvement districts, and state, regional or local conservation districts) are also being created to fund the development and management of waterfowl habitats, groundwater recharge areas, well-head protection areas, watersheds capable of providing safe drinking water, endangered species' habitat restorations, waste-site remediation, and other conservation activities.

Opportunities for New Partnerships

Creation of the different funds suggested here can also provide innovative support for intergovernmental partnerships. If, for instance, a municipality offered lower real estate taxes, tax rebates, or payments for easements on private land, local government could be reimbursed for lost revenue through state funding. Habitat restoration projects that are administered by state agencies could become eligible for federal nonpoint source clean water funds, if the project can be shown to effectively reduce nonpoint source contributions to rivers not currently meeting water quality standards.

The city of Boulder, Colorado, situated at the western edge of the Great Plains region, has completed an innovative stream restoration project in conjunction with meeting advanced wastewater treatment requirements at their facility. The city, a local chapter of Trout Unlimited, regional nonprofit organizations, and university staff all collaborated in abating an ammonia discharge problem and in restoring two miles of stream beds and fisheries at less than 30 percent of the cost of installing traditional control technology. Several innovative federal-state-local fund transfers (such as retargeting U.S. Forest Service payments to county and local governments toward ecoregion conservation projects) are being discussed in the timber management plans now under consideration in the Northwest.

Highway trust funds, state park and lands acquisition funds, special district levies, and the development of a Great Plains ecotourism industry could also provide innovative revenue for funding Great Plains conservation activities. Federal transportation grants, for example, are being used to develop a heritage trail system through several states highlighting the economic and natural history of various regions. For example, the Southwestern Pennsylvania Heritage Route—a joint venture by the Southwestern Pennsylvania Heritage Preservation Commission and the U.S. Department of the Interior—commemorates the development of the mining, iron casting, and railroad industries. In the Great Plains, program participants could organize an ecoregion trail highlighting the unique natural systems and economic activities in an ecoregion, perhaps organized around early pioneers' migration routes such as the Oregon Trail. The ecosystems of the Great Plains grasslands offer a wide range of ecotourism opportunities.

Opportunities for Applying Economic Principles

Opportunities to develop new sources of revenues for any of the Great Plains funds are also opportunities to apply economic principles such as polluter pays, whereby the party creating an adverse ecosystem or ecoregional impact pays the cost of mitigating the effects it has on ecosystem functions. Another economic principle has the beneficiaries of an ecosystem or ecoregional improvement pay; in short, the recipients of the benefits of ecosystem conservation foot the bill.

Defining the costs of global, national, regional, and local benefits is an important aspect of developing Great Plains ecosystem conservation programs. By examining those economic activities that affect ecosystem functions or ecoregions targeted for conservation improvements, Great Plains participants have the opportunity to assign the social costs of ecosystem impacts to parties responsible for creating them (to the extent they can be attributed to anthropogenic causes). Creatively applying user fees, licenses, and tolls are examples of the funding mechanisms that could be employed (USEPA 1991, 1992).

Assuming that past and current effects on Great Plains ecosystems can be tied to identifiable economic activities, then these activities could establish funds to mitigate their effects, thereby having the polluter pay. For example, if mining activities affect the aquatic ecosystems in an area of the Great Plains region and regulations require past or current mining operations to prevent or mitigate these effects, then a tax or surcharge on mined products collected and dedicated for ecosystem improvements could be used to fund the mitigation of such effects in critical areas. Several states have already established mine reclamation funds for these purposes.

To illustrate how a beneficiary pays principle may be applied, assume that the expansion of residential and commercial land uses encroach on the terrestrial habitats of critically endangered species. State or local governments with jurisdiction over these habitat areas could tax real property transfers throughout the state or in a designated region to fund habitat conservation activities. In this way, those who affect habitat areas pay to conserve them and receive the direct benefits of their being conserved. Several states have dedicated portions of the sales of fishing and hunting licenses to resource management; special license fees for recreational vehicles may also provide revenues. EPA is promoting the development of watershed management districts where the beneficiaries of water quality improvements share the costs of creating them. Similarly, compacts and cost-sharing among landowners and resource managers in other ecologically defined subareas of ecoregions, such as airsheds or aquifer recharge areas, may yield conservation benefits.

In short, fund revenues may also be generated through carefully designed economic incentives, user fees, taxes, price adjustments, or the like, across the full range of activities that affect the sustained functioning of the Great Plains ecosystems targeted for conservation.

Separate funds could be developed for pooling public and private resources in accordance with the bylaws and accounting and auditing systems adopted by Great Plains program participants. Such funds will only be successful if all participants realize the benefits of paying into the fund and, in turn, fund managers are accountable for the resources they allocate to Great Plains conservation activities.

Moving toward Sustainable Development

In the long term, Great Plains program participants may want to find ways to shift away from economic activities that permanently impair the functioning of Great Plains ecosystems and ecoregions toward an economy that produces sustainable development dependent on renewable resources such as grasslands, wetlands, forests, and other biomass. Collaborative efforts among all economic sectors and local, state, and federal governments throughout the Great Plains could underwrite the costs of researching and developing sustainable economic development. Taxes, levies, and surcharges on ecosystem-impairing or resource-depleting activities (mining, sod farming, oil and gas production, continuous cropping) could also provide long-term incentives to develop renewable and sustainable energy and economic resources.

Acknowledgments

The authors thank Ed Stickle, EPA Chesapeake Bay Program, and Chris Grundler, EPA Great Lakes Program, for their interviews.

References

Omernik, J. M. 1987. Ecoregions of the conterminous United States. *Annals of the Association of American Geographers* 77(1):118-125.

U.S. Environmental Protection Agency (USEPA). 1991. *Economic incentives: Options for environmental protection.* Washington, D.C.: Office of Policy, Planning, and Evaluation.

———. 1992. *The United States experience with economic incentives to control environmental pollution.* Washington, D.C.: Office of Policy, Planning, and Evaluation.

Young, Terry F., and C. H. Congdon. 1992. Using tradable discharge permits to regulate agricultural nonpoint source pollution: A case study of the grasslands subbasin in California's San Joaquin Valley. Environmental Defense Fund (draft).

24. We Get What We Organize For: Linking Central Bureaucracies to Local Organizations

David M. Freeman
Colorado State University

The vast web of ecosystems that constitutes the Great Plains region is complex, dynamic, and theory-defying. It is approached with disconnected slices of knowledge, pieces of human experience representing little historical depth, tentative hypotheses, best available estimates, and the desire to get things done at the least possible cost and greatest dispatch. It is a given that errors will be made in the formulation and implementation of policies that are intended to enhance ecosystem sustainability. The question is: how will policymakers organize to minimize and correct these errors in the most timely way so that whatever damage done to humans and other living things does not fundamentally threaten the sustainability of Great Plains society and the biotic pyramids upon which our social order depends? This chapter briefly addresses the twin problems of sources of policy error and organizational response to error in policymaking and implementation of policy for sustainable ecosystem management.

Why Policy Planning for Ecosystem Management Generates Error

Error is built into the structure of human decision making without regard to political or social ideology, party affiliation, economic interest, or nobility of value commitments. The reasons are fundamental.

First of all, citizens, politicians, and planners—while all claiming to be operating in rational ways—function with very different and changing mixes of rationalities. There is simply more than one way to be rational (Diesing 1973, 1982), and most people are rational in different, and incompatible, ways at different times even in the same policy planning process. Technical rationality means finding the best means to pursue a given objective; economic rationality involves making trade-offs among conflicting objectives that can be priced in marketplaces; political rationality has to do with forging and sustaining winning coalitions; and legal rationality centers on reasoning from precedent. Policymaking for sustainable ecosystem management is a cacophony of voices speaking in languages of incompatible rationalities.

Stanley R. Johnson and Aziz Bouzaher (ed.), *Conservation of Great Plains Ecosystems*, 403-409.

Second, there are among us no socially neutral individuals or organizations—in the public or private sectors—interested in pursuing a given vision of sustainable ecosystem management for the welfare of society as a whole. In assessing a proposed course of policy action, individuals and organizations project to themselves anticipated gains and losses. But self-interested analysts seek to highlight positive spillovers or externalities, which make friends and allies, while ignoring as much as possible the negative impacts that will harm people and other living things downwind, downstream, or in the future. Negative externalities make enemies and, to the extent that they are acknowledged, they impose costs upon the decision makers because it takes money and time to cope with them (Freeman 1992). Given limited budgets and pressing deadlines, it is simply not rational in any sense of the word to highlight the negative externalities associated with what one proposes to do. There is, therefore, a strong tendency of decision makers to plunge ahead while calling all possible attention to the positive spillovers and simultaneously refusing to look at the real costs.

Third, decision makers commit serious policy error simply due to the nature of scientific knowledge. Science gains knowledge by defining discrete subparts of larger wholes, holding confounding factors constant while manipulating those pieces that are of interest, and drawing conclusions that X is related to Y in some predictable fashion. Scientific knowledge is extraordinarily valuable, but it is gained by abstracting slices of reality from large wholes, and manipulating those slices under controlled conditions. There are at least two fundamental problems here that will not go away:

1. The very factors that have been held aside during scientific studies are very much active in the real world, and they affect the variables in logical scientific arguments in unexpected and unanticipated ways. What is true in the lab when X and Y were studied does not predict what will happen in the real world when factors P, Q, and R are interacting with X and Y.

2. Many social and ecological phenomena operate on probability, not deterministically. Even if society is prepared to conduct the expensive kinds of studies that bring into play many more factors, and study their interrelationships with greater attention to nuance over longer time periods, policy analysts cannot combine factors easily because they cannot predict even modest chains of probabilistic events (Freeman 1974). If analysts know a good deal about particular sets of probabilistic phenomena that hang together to the point that they can predict to a level of .9 within such a set (not bad in the range of subjects involved in ecological management), analysts lose their predictive capacity quickly when they start putting their separate pieces of knowledge together. For example, if A and B are related at the level of .9, as are each of another three sets of phenomena that are known to be interacting with A and B, analysts construct a probabilistic chain of only four links, which is very short in the world of ecosystem management. Yet, the four-piece chain [$.9 \times .9 \times .9 \times .9$] yields an outcome of only .66—roughly a two-thirds chance of being able to predict the outcome. Not too many people will enjoy crossing streets if they believe they have only a .66 chance

of making it without serious trouble. The same chain, consisting of .8 probabilities, yields only a .41 chance of being right about the outcome, while one with .6 values (much more realistic in the real world of the several ecological sciences) generates a predictable outcome only to the level of .13—not exactly confidence-enhancing.

In sum, rationality is multiple and conflicting; there are no socially, politically, and ecologically neutral decision makers; and science abstracts slices out of the socio-ecological systems but never fully comprehends the whole of them. Furthermore, policy analysts run into serious problems when they try to put together multiple slices of probabilistic knowledge. Error, therefore, is built into the very structure of human attempts to plan and manage. The issue is not how to eliminate error (an impossible task) but how to organize policymakers to be as adaptive and error-correcting as possible.

Coping with Error—On the Importance of Local Organizations

Many scholars have examined the problems of complexity, error, planning, and action (see Friedmann 1987; Lindblom and Cohen 1979; Lindblom 1990). For purposes here, one key theme is addressed: managing ecological policy across several levels of human organization.

Socio-ecological policy takes on life in the interaction of several levels of human organization. Policy ideas, emerging out of the rough and tumble of political life, mean little until encoded in law by a legitimate political authority. But policy encoded in law is little more than ink stains in a book until a central bureaucracy is established that will be responsible for administering it to the appropriate political unit. Central bureaucracies, however, are staffed by civil servants who are located in capital or other major cities are remote from the ecological systems for which they are responsible. This means that state bureaucracies must necessarily manage in collaboration with a coalition of local public, quasi-public, and private organizations that deal with resources in site-specific socio-ecological patches.

The thesis is that properties of local organizations—their linkages with state and federal bureaucracies, their mediation between government bureaucratic agendas and those of the local community—have everything to do with whether or not there will be rapid and appropriate adaptation in error correction and, thereby, whether or not there will be productive and sustainable use of resources in ecosystem management.

The term *local organization* refers to myriad public, quasi-public, and private local organizations that have their own budget and responsibilities for natural resource management in particular patches of the landscape, which include water conservation districts, irrigation districts, mutual irrigation companies, associations of livestock grazers, soil conservation districts, wildlife and recreation associations, and production and marketing cooperatives. It is beyond the scope of this chapter to discuss what makes for effective local organizational design and management. The critical factors for organizing human beings for ecological sustainability have been identified elsewhere (Bromley 1992; Freeman 1989; Ostrom 1990), and there is much agreement about strategic attributes.

The problem of organization and error can now be simply stated: public policy for enhancing the sustainability of human communities must, by necessity, be general and aimed at central tendencies across broad national and regional areas. Yet, policy impacts are always local, unique, and site-specific. The tension between knowledge of central tendencies and local site-specific intelligence is unavoidable. The measure of central tendency that permits a general overview from afar becomes inadequate when coping with particular local circumstances. Central bureaucracies can set general directions, specify standards to be fulfilled, and make funds available to assist with efforts needed for sustainability. But no central bureaucracy can gather and comprehend sufficient high-resolution knowledge of all the particular socio-ecological patches, employ sufficient personnel to manage all the patches, and develop management schemes that would permit effective rapid adaptation of general policy to the needs of the countless dynamic and variable conditions in all local areas.

An example is in order. The U.S. Forest Service, in conjunction with the U.S. Fish and Wildlife Service, has taken the position that the Big Bend reach of the Platte River in western Nebraska must be served by enhanced flow from Colorado; in a dry year the increase may have to be as much as 500,000 acre feet of water (*Water News* 1994). The objective of securing the additional water would be to sustain populations of fish and migratory waterfowl—several of which represent threatened and endangered species. The preservation of riparian habitat critical to these populations is certainly a laudable objective. However, these central bureaucracies have no way to secure or deliver the flow without negotiating arrangements with a web of local organizations consisting of, among others, the Northern Colorado Water Conservancy District, several cities, and a number of mutual irrigation companies that pull water from the Poudre and Platte Rivers. These organizations, which stand between individual water users and the central agencies, possess legitimate water rights under state appropriation doctrines and have a history of important investment in water development. Furthermore, these local organizations are critically important repositories of local information essential to making actual deliveries to given points in the system. Rather than attempt to establish workable linkages with these local water entities, in 1991 the Forest Service—largely blind to the dynamics of the local organizational web—simply threatened to deny special use permits that are required under federal law for the dams, headgates, pipelines, and canals that were built years ago on federal land high in the watershed. Such special use permits were routinely approved by the federal agencies, but with environmental sustainability for western Nebraska in mind, the central agencies are now prepared to deny permits in order to compel the Colorado water releases that agency decision makers find to be in the region's best interest. The threat is real—local water organizations had better release flows as demanded by the federal agencies or face losing the right to operate their water collection and conveyance facilities.

The problem has three dimensions: (1) the central bureaucracies (Forest Service and Fish and Wildlife Service) have no real way of moving water to the desired location except by working through the intricate web of organizational relationships that has developed around the rivers in question; (2) the central bureaucracies do not have enough local river knowledge to understand how the river can be managed so as to actually get the water to the Big Bend segment of the Platte; and (3) the threat to local organizations creates hostility on the part of those who are essential to any reasonable river management plan. This kind of confrontational stance is not conducive to the kind of information sharing and option appraisal that is required if solutions are to be found. Even if the federal agencies were to win the confrontation on their terms, there is serious doubt that water sacrificed upstream would reach the targeted habitat in the absence of effective linkages and working relationships with local water organizations along the rivers. Clearly, the solution lies in the direction of building linkages with the local organizations, finding ways to purchase water shares within the framework of state water law and given the willingness of local organizations to free up some of the available supply, and developing exchanges such that water can be delivered to the endangered habitats with minimum shrinkage to all parties.

This overall problem of linking general policy of central bureaucracies to individual situations via intermediary organizations has been confronted and resolved in the private sectors of virtually all industrial and post-industrial societies whatever their ideology. Nowhere is it imaginable that factory managers would start and stop their central assembly lines according to the demand dictates of individual or particular groups of customers. The need of central factory management for smooth and predictable production is meshed with individual customer demand via the functioning of intermediate wholesale and retail organizations that match general production policy to the site-specific needs of individual situations. Yet, in the public sector, central agencies all too often make and attempt to implement general policy without having constructed effective linkages to the webs of local organizations that are already in place or must be constructed. This has resulted in: (1) maladaptive policy in specific niches; (2) real or threatened costly legal confrontations between central and local organizations; (3) expensive and befuddled discussion of local matters in central arenas where harassed agency managers do not have the time and resources to address voluminous site-specific problems—which exasperates locals; and (4) expensive and befuddled discussion of central policy intentions in local arenas where harassed citizens do not have the time and resources to compel central dictates to be fitted to their particular circumstances—which drives local people to desperation or cynical resignation.

In all of this, public discourse becomes corrosive, the sense of civic commitment weakens, public exchanges become all too often quiet, cynical deal-making contests, and the civic process becomes laced with destructive conflict among contending local groups that lack effective organizational channels for mediating their interests with each other and with the political-administrative centers.

Conclusion—We Must Make a Mesh of Things

To avoid this destruction, organizations must overcome at least two kinds of arrogance. First, central agencies must overcome the kind of arrogance that holds that their knowledge and interest is inherently superior and must, therefore, ride roughshod over local knowledge and interest. Central system knowledge and general policy commitment to ecosystem sustainability are critically important, but they must be blended with, and adapted to, site-specific realities; local organizations provide the social space within which central system knowledge and policy can be adapted to local environmental necessities. Central agencies must make a concerted effort to establish linkages with local organizations and see them as potential allies in the struggle for sustainability, not as adversaries.

Second, local organizations and interests must overcome the kind of arrogance that holds that they should be permitted to continue generating past patterns of negative externality downwind, downstream, upon the soil, other living things, and coming generations. While it is true that all policy impacts are local, and must be so adjusted, it is equally true that the self-seeking behavior generating ecological problems in the first instance comes from local people abusing specific habitats in ways that may be individually rational in the short run but unacceptably undercut the biota to which we are all tethered. Accountability is owed to much more than the shareholders or the economic bottom line of profit and loss because we all know that today's market prices do not reflect the true costs of what we are doing to other living things and succeeding generations. Local organizations, properly designed and linked to central government agencies, can create incentives to do in concert what cannot be done individually, and assert discipline that creates confidence on the part of other locals—and central decision makers—that if one individual refrains from wrongful abuse (and thereby sacrifices short-term advantage) that individual can count upon the organization to ensure that others will do likewise and thereby contribute to the common good. The sacrifice will not go in vain but will be combined with others so that sustainability will have been achieved through mutual effort and shared cost. Only effective local organization can ensure that the "tragedy of the commons" will be avoided (Hardin 1968; Freeman 1989; Ostrom 1990).

In sum, then, the business of making and implementing policy for advancing sustainable practices in the Great Plains region resolves into meeting an organizational challenge. Good science is necessary, but it is not sufficient. The kind of local organizations we recognize and create, the nature of their linkages to central bureaucracies, the manner in which they make social and political space within which the agendas of the center and local niche can be blended will have everything to do with the productivity and sustainability of our investments in ecosystem improvement. In the final analysis we will get what we organize for.

References

Bromley, Daniel W. (ed.). 1992. *Making the commons work: Theory, practice, and policy.* San Francisco: Institute for Contemporary Studies Press.

Diesing, Paul. 1973. *Reason in society.* Westport, Connecticut: Greenwood Press.

Diesing, Paul. 1982. *Science and ideology in the policy sciences.* Chicago: Aldine Publishing.

Freeman, David M. 1974. *Technology and society: Issues in assessment, conflict, and choice.* Chicago: Rand McNally.

Freeman, David M. 1989. *Local organization for social development: Concepts and cases of irrigation organization.* Boulder: Westview Press.

Freeman, David M. 1992. *Choice against choice: Constructing a policy assessing sociology for social development.* Niwot: University Press of Colorado.

Friedman, John. 1987. *Planning in the public domain: From knowledge to actio*n. Princeton: Princeton University Press.

Hardin, Garrett. 1968. The tragedy of the commons. *Science* 162:1243-48.

Lindblom, Charles E., and David K. Cohen. 1979. *Usable knowledge.* New Haven: Yale University Press.

Lindblom, Charles E. 1990. *Inquiry and change: The troubled attempt to understand and shape society.* New Haven: Yale University Press.

Ostrom, Elinor. 1990. *Governing the commons: The evolution of institutions for collective action.* New York: Cambridge University Press.

Water News. 1994. 11:2. Northern Colorado Water Conservancy District, Loveland, Colorado.

25. Conclusions and Recommendations

Aziz Bouzaher and S. R. Johnson
Iowa State University

What actions of private citizens, nongovernmental organizations, government, and other institutions are implied by the concepts and information provided by the symposium and this book? Scientists from all disciplines represented at the symposium concurred that the trends in use and the status of the major resource complexes in the Great Plains point conclusively to problems of severe ecological risk. Using ecosystems concepts in comprehensive, participatory, and community-based management initiatives is essential for developing policies and institutions to deal effectively with these risks. Effective approaches to managing the Great Plains landscape must reflect complex ecosystem concepts. The consensus was that science and practical experience, along with participatory and environmental management initiatives, will lead to sounder conventions and concepts.

The participants also concluded that it is unwise to wait for full development of ecosystems concepts, valuation, and other elements of management approaches for the Great Plains initiative. A more pragmatic approach should be directed toward ecosystems and regions according to categories of value and perceived risk. An effective way to view and organize the regions and ecosystems within the Great Plains is as a set of overlays. For example, small ecosystems may exist in streams that are part of larger watersheds that are in turn a part of river basins or flyways for migrating species. The effective approach to policy and management and the necessary scientific work to support better informed intervention is to define common ecosystems, to classify them by risk and value, and then to explore their interconnectedness.

Using this risk-value framework, priorities can be determined for science and policy intervention. Of course, it is natural that priorities will be directed toward resources and ecosystems of high value and at high risk. Integrated environmental management and related science initiatives in the Great Plains requires the use of simple and clear principles for coordinating private and public activity. The framework also encourages action to promote resource conservation and improve opportunities to develop economies and communities that are more compatible with good husbandry of Great Plains natural resources.

The chapters in this book also reflect organized group discussions during the symposium that helped to identify concepts and recommendations for how to best

Stanley R. Johnson and Aziz Bouzaher (ed.), *Conservation of Great Plains Ecosystems*, 411-414.

manage, protect, and conserve the high-risk, high-value ecosystems and regions of the Great Plains. A number of consensus or near consensus recommendations and ideas for coordinated action emerged. It is best, however, to interpret these "recommendations" not as a consensus of the participants but as a "harvest" from their ideas. These recommendations are summarized here and are linked to the discussion of the foregoing chapters.

Policy Recommendations

Participants acknowledged that many existing policies are narrowly focused and, therefore, not consistent with sustainable development of the Great Plains in an ecosystems context. These policies should be reviewed, emphasizing coordination and cooperation, incentive compatibility versus regulation, broader consideration of the resources affected or addressed, and new information on risk and valuation developed within an ecosystems context. An ecosystems management approach, essential to a viable future for the Great Plains, involves innovative cooperation among government authorities, broader applications of ecosystems concepts, and new policy to enhance opportunities for current and subsequent generations and cultures.

Specific policy recommendations are:

- Formulate innovative policies to address biodiversity and test them on a pilot basis using the risk status and value approach;
- Assess agricultural policy changes likely in the 1995 Farm Bill and encourage active participation in the associated debate by Great Plains representatives and stakeholders;
- Undertake cross-state management of ecosystems, especially for migratory waterfowl;
- Assess economic development initiatives for their ecological implications and impacts;
- Reexamine federal land use, including trade-offs for alternative uses and full costing of external environmental costs (such as grazing fees);
- Reexamine water rights compacts between states on a more comprehensive basis;
- Examine U.S. Army Corps of Engineers projects relative to integrated management of recreation, economic, ecosystem, and water management uses;
- Undertake river basin initiatives;
- Develop and implement nutrient management laws;
- Pursue privatization of environmental sources by certification and standards laws;
- Test and implement landscape and land management in the form of use incentives and restrictions; and
- Introduce a variety of demonstration programs, including those for institutions.

Research and Development Recommendations

The research and development recommendations are motivated by the rich set of scientific issues that emerged from the symposium. Many recommendations focus on how to transfer scientific developments to practical applications in both the public and private sectors. If the sustainability of the Great Plains is to receive improved scientific support, a concerted effort should be devoted to defining and cataloging ecosystems, measuring biodiversity, developing matrixes for risk and valuation, managing information, merging experimental and statistical data with data developed by monitoring ecosystems, and creating new models of individual and collective action that are efficient and sensitive to the status of ecosystems. New forms of scientific collaboration are needed. This collaboration will effectively use available resources to solidify and extend the conceptual and technical foundations for the Great Plains Initiative.

Specific recommendations for research and development are:

- Systematically assemble databases for the Great Plains describing resources, ecosystems, and economic conditions;
- Develop and implement a broadly accepted system for classifying and defining ecosystems as they relate to management initiatives;
- Complete research metrics measuring risk and value of ecosystems;
- Design pilot biological methods to use in a proactive approach to biodiversity and endangered species;
- Evaluate and compare existing policies that foster preservation of ecosystems to identify innovations and best practices;
- Develop and test performance indicators for assessing progress of the Great Plains Initiative;
- Develop and test ecosystem monitoring systems that use biological and geophysical process models and advanced statistical methods;
- Undertake pilot management projects for selected ecosystems using cross-disciplinary teams; and
- Study human resources and culture in the Great Plains with the objective of upgrading management systems with values and traditions of associated communities.

Institutional Recommendations

The institutional recommendations support broadened public and private participation in the decisions that will determine the future of the Great Plains. Cooperation is needed among federal, state, and local authorities and newly formed groups for more effective management of ecosystems and natural resources. Private initiatives that use market incentives and education are areas of opportunity. At the same time, we must recognize the importance of market failure and the possible role of government and nongovernment collective action as approaches to the diverse environmental, cultural, and economic problems and opportunities in the Great Plains.

Specific recommendations for institutions are:

- Establish a Great Plains Trust Fund to support innovative and exploratory ecosystem projects;
- Develop alternative institutions to administer governance and rights redistribution that address the application of complaints-driven regulations for nonpoint source environmental problems;
- Stimulate and support nongovernmental organizations with interests in the Great Plains and its associated ecosystems;
- Conserve the culture and tradition of the human population through better institutions;
- Investigate and introduce alternative forms of governance by focusing on key ecosystems;
- Establish a Great Plains Initiative Council that includes the highest level state and federal leaders;
- Develop coordinating mechanisms to facilitate both research and development and policy formulation;
- Create constituency groups to bring together stakeholders and interest groups for practical ecosystems to develop ideas for policy design and implementation; and
- Introduce institutional systems for developing and publishing baseline information on the condition of ecosystems.

This book is itself a resource to the residents of the Great Plains on issues of environmental protection and more basic science, and for those in the public and private sectors responsible for developing policies and institutions that can contribute to conservation of ecosystems and a higher quality of life for residents and other users. Advances in the policy sciences and in disciplinary research are essential for improving the management strategies for the Great Plains. Science, by its nature, is compartmentalized. Thus, the challenge for the Great Plains is to use new concepts of integrated environmental management to advance the knowledge of life systems and their responses to various forms of intervention. Perspectives from other disciplines and those responsible for ecosystem, community, and economic management can enrich and deepen the research and build a stronger foundation for organizing human activity in the Great Plains. Our development of the Great Plains contributes to the residents' quality of life. We must recognize that current residents are responsible to those who will occupy this vast and rich area and use these unique resources in future generations.

List of Contributors

Donald A. Becker
Omaha District, U.S. Army Corps of Engineers, 215 N. 17th Street, Omaha, NE 68102

James R. Brandle
Associate Professor, Department of Forestry, Fisheries, and Wildlife, University of Nebraska—Lincoln, 101 Plant Industry Building, Lincoln, NE 68583-0814

Gerald F. Bratton
Technology Transfer Leader, U.S. Forest Service, Rocky Mountain Forest and Range Experiment Station, Center for Semiarid Agroforestry, East Campus, University of Nebraska—Lincoln, Lincoln, NE 68583-0822

Allan Butcher
Attorney at Law, P.O. Box 1642, Stephenville, TX 76401

Bruce Cannon
Chief, Legislation and Strategic Planning Division, Office of Policy Development, Federal Highway Administration, 400 Seventh Street SW, No. 3318, Washington, DC 20590

C. V. Cole
Agricultural Research Service, U.S. Department of Agriculture, Natural Resources Ecology Lab, Colorado State University, Fort Collins, CO 80523

Thomas D. Crocker
Professor, Department of Economics, and Director, School of Environmental and Natural Resources, University of Wyoming, Laramie, WY 82071

Brent J. Danielson
Assistant Professor, Department of Animal Ecology, Iowa State University, 3 Science II, Ames, IA 50011

A. J. Dornbusch Jr
Director, Midwest National Technical Center, Soil Conservation Service, U.S. Department of Agriculture, 100 Centennial Mall, N-152, Lincoln, NE 68508-3866

Brendan Doyle
Office of Policy Analysis, U.S. Environmental Protection Agency, 401 M Street SW, Washington, DC 20460

Mark Drabenstott
Vice President and Economist, Federal Reserve Bank of Kansas City, 925 Grand Avenue, Kansas City, MO 64198-001

Nancy Easterling
Special Projects Coordinator, Texas Institute for Applied Environmental Research, Tarleton State University, Box T258, Stephenville, TX 76402

Dan Flores
Hammond Chair of American Western History, University of Montana, Missoula, MT 59812-1022

Larry Frarey
Policy Analyst, Texas Institute for Applied Environmental Research, Tarleton State University, Box T258, Stephenville, TX 76402

David M. Freeman
Professor, Department of Sociology, Colorado State University, Clark Boulevard, Fort Collins, CO 80523

Kristin Gill
Research Assistant, Texas Institute for Applied Environmental Research, Tarleton State University, Box T258, Stephenville, TX 76402

Richard D. Gorton
Chief, Environmental Analysis Branch, Omaha District, U.S. Army Corps of Engineers, 215 N. 17th Street, Omaha, NE 68102

Jay R. Harman
Professor of Geography, Michigan State University, East Lansing, MI 48825

John A. Harrington Jr.
Associate Professor, Department of Geography, Kansas State University, Dickens Hall, Manhattan, KS 66506-0801

Dexter Hinckley
Senior Ecologist, Climate Change Division, Office of Policy Analysis, U.S. Environmental Protection Agency, 401 M Street SW, Washington, DC 20460

Thomas L. Huntzinger
Project Chief, National Water Quality Assessment, U.S. Geological Survey, 4821 Quail Crest Place, Lawrence, KS 66049-3839

Kenneth S. Johnson
Associate Director, Oklahoma Geological Survey, 100 E Boyd Street, Room N-131, Norman, OK 73019

Heather Jones
Research Assistant, Texas Institute for Applied Environmental Research, Tarleton State University, Box T258, Stephenville, TX 76402

J. L. Kearney
Engineering Geologist, Midwest National Technical Center, Soil Conservation Service, U.S. Department of Agriculture, 100 Centennial Mall, N-152, Lincoln, NE 68508-3866

Erwin E. Klaas
National Biological Service, Iowa Cooperative Fish and Wildlife Research Unit, Department of Animal Ecology, Iowa State University, 09 Science II, Ames, IA 50011

Fritz L. Knopf
Leader, Vertebrate Ecology, National Biological Service, 4512 McMurray Avenue, Fort Collins, CO 80525-3400

M. M. Kothmann
Professor, Department of Rangeland Ecology and Management, Texas A&M University, 225 Animal Industries Building, College Station, TX 77843-2126

Sagar V. Krupa
Professor, Department of Plant Pathology, University of Minnesota, 495 Borlaug Hall, 1991 Buford Circle, St. Paul, MN 55108-6030

Peter Kuch
Office of Policy Analysis, U.S. Environmental Protection Agency, 401 M Street SW, Washington, DC 20460

Allen H. Legge
President, Biosphere Solutions, 1601 11th Avenue Northwest, Calgary, Alberta, Canada T2N 1H1

Charles J. Mankin
Director, Oklahoma Geological Survey, 100 E Boyd Street, Room N-131, Norman, OK 73019

G. A. Peterson
Professor, Department of Soil and Crop Sciences, Colorado State University, Fort Collins, CO 80523

Richard Rathge
Professor, Departments of Sociology and Agricultural Economics, and Director, North Dakota Data Center, PO Box 5636, North Dakota State University, Fargo, ND 58105-5636

William A. Reiners
Professor, Department of Botany, University of Wyoming, PO Box 3165, Aven Nelson Building, Laramie, WY 82071-3165

Fred B. Samson
U.S. Forest Service, PO Box 7669, Missoula, MT 59807

Peter R. Schaefer
Associate Professor, Department of Horticulture, Forestry, Landscape, and Parks, South Dakota State University, Box 2207C, Brookings, SD 57007

Jason F. Shogren
Associate Professor, Department of Economics, Iowa State University, 478D Heady Hall, Ames, IA 50011

Melvin D. Skold
Professor, Department of Agricultural and Resource Economics, Colorado State University, and Executive Director, Great Plains Agricultural Council, Fort Collins, CO 80523-0002

Robert D. Sopuck
Executive Director, Sustainable Development Coordination Unit, Province of Manitoba, 305-155 Carlton Street, Winnipeg, Manitoba, Canada R3C 3H8

Margaret G. Thomas
Senior Resource Planner, Midwest Research Institute, 425 Volker Boulevard, Kansas City, MO 64110-2299

B. M. Vining
Water Quality Specialist, Midwest National Technical Center, Soil Conservation Service, U.S. Department of Agriculture, 100 Centennial Mall, N-152, Lincoln, NE 68508-3866

Robert Wolcott
Office of Policy Analysis, U.S. Environmental Protection Agency, 401 M Street SW, Washington, DC 20460

Subject Index

F

Great Plains (continued)

K

L

M

N

O

(Oklahoma continued)

P

Q

R

S

T

(Texas continued)

U

V

W

Ecology, Economy & Environment

1. C. Folke and T. Kåberger (eds.): *Linking the Natural Environment and the Economy: Essays from the Eco-Eco Group.* 1991 ISBN 0-7923-1227-9
2. U. Svedin and B. Hägerhäll Aniansson (eds.): *Society and the Environment: A Swedisch Research Perspective.* 1992 ISBN 0-7923-1796-3
3. M. Ruth: *Integrating Economics, Ecology and Thermodynamics.* 1993 ISBN 0-7923-2377-7
4. C.A. Perrings, K.-G. Mäler, C. Folke, C.S. Holling and B.-O. Jansson (eds.): *Biodiversity Conservation.* Problems and Policies. 1995 ISBN 0-7923-3616-X
5. S.R. Johnson and A. Bouzaher (eds.): *Conservation of Great Plains Ecosystems: Current Science, Future Options.* 1995 ISBN 0-7923-3747-6

KLUWER ACADEMIC PUBLISHERS – DORDRECHT / BOSTON / LONDON